Andreas Vogel
Abu K. M. Sarwar
Rudolf Gorenflo
Ognyan I. Kounchev (Eds.)

**Theory and Practice of
Geophysical Data Inversion**

Theory and Practice of
Applied Geophysics

Edited by Andreas Vogel

Volume 1
Andreas Vogel (Ed.)
Model Optimization in Exploration Geophysics

Volume 2
Andreas Vogel (Ed.)
Model Optimization in Exploration Geophysics 2

Volume 3
Andreas Vogel, Rudolf Gorenflo,
Bernd Kummer, Charles O. Ofoegbu (Eds.)
Inverse Modeling in Exploration Geophysics

Volume 4
Andreas Vogel, Charles O. Ofoegbu,
Rudolf Gorenflo, Bjorn Ursin (Eds.)
Geophysical Data Inversion
Methods and Applications

Volume 5
Andreas Vogel, Abu K. M. Sarwar,
Rudolf Gorenflo, Ognyan I. Kounchev (Eds.)
Theory and Practice of
Geophysical Data Inversion

Andreas Vogel
Abu K. M. Sarwar
Rudolf Gorenflo
Ognyan I. Kounchev (Eds.)

Theory and Practice of Geophysical Data Inversion

Proceedings of the 8th International Mathematical Geophysics Seminar on Model Optimization in Exploration Geophysics 1990

Die Deutsche Bibliothek – CIP-Einheitsaufnahme

Theory and practice of geophysical data inversion:
proceedings of the 8th International Mathematical Geophysics
Seminar on Model Optimization in Exploration Geophysics 1990 /
Andreas Vogel . . . (ed.). – Braunschweig; Wiesbaden:
Vieweg, 1992
  (Theory and practice of applied geophysics; Vol. 5)
  ISBN 978-3-528-06454-9   ISBN 978-3-322-89417-5 (eBook)
  DOI 10.1007/978-3-322-89417-5
NE: Vogel, Andreas [Hrsg.]; International Mathematical
  Geophysics Seminar on Model Optimization in Exploration
  Geophysics < 08, 1990, Berlin, West >; GT

Softcover reprint of the hardcover 1st edition 1992

Vieweg is a subsidiary company of the Bertelsmann Publishing Group International.

Produced by Lengericher Handelsdruckerei, Lengerich

Foreword

The contributions to this volume cover a wide spectrum of recent developments in geophysical data inversion, including basic mathematics and general theory, numerical methods, as well as computer implementation of algorithms.

Most of the papers are motivated by problems arising from geophysical research and applications both on a global scale and with respect to local geophysical surveys, underlining the increasing importance of geophysical exploration methods in various fields, such as structural geology, prospecting for mineral and energy resources, hydro-geology, geotechnology, environmental protection and archaeology.

The first section of the book deals with basic mathematics and general theory underlying geophysical data inversion. Papers presented here are concerned with stabilization algorithms to solve ill-posed inverse problems, sensitivity of kernel function estimations to random data errors and reduction of errors in inverse modelling of response functions by linear constraints, numerical procedures for approximating the solution to boundary value problems, accuracy and stability of inverse ill-posed problems constituted by problems of moments, and fast Fourier transforms for solving potential field problems.

The second section contains papers on gravity and magnetics, dealing with the solvability of the inverse gravimetric problem for sources represented by point masses and other elementary, solution of the inverse problem in cases of nonuniformly distributed data as obtained by palaeomagnetic studies, satellite observations, and surface projections of buried archaeological targets by inverse filtering of geomagnetic data.

Section three contains two papers on the inversion of electrical and electromagnetic data, concerned with the improvement of the resolution of model parameters by joint inversion of magnetotelluric and transient electromagnetic data as well as apparent complex resistivity data from dipoles energized over a wide range of transmitter frequencies.

Section four comprises a number of contributions on several aspects of seismic data inversion, including papers on the improvements of the efficiency in the reconstruction of acoustical impedance profiles by inversion of reflection seismograms, imaging of the subsurface velocity structures by inversion of seismic crosshole data, solution of the inverse seismic problem in cases of undersampled and noisy data, treatment of ill-posedness by regularization methods in cases of seismic modelling and inversion by integral equation methods, recursive inversion procedures for determination of seismic reflection coefficients, an optimization approach to the solution for the geometrical and mechanical parameters of earthquake sources, and spectral analysis for determination of the characteristics of seismic wave radiation from large explosions.

The paper of Section five is on geothermics, being concerned with the determination of subsurface temperature fields by inversion of geothermal data, thus allowing the recon-struction of past climates.

The monograph concludes with two papers on constraining conditions for joint inversion of gravity and seismic data.

The editors express their gratitude to George Wagenhauser BSc MA, editorial consultant, for his diligence in preparing the manuscripts for publication. Thanks are also extended to Dipl.-Ing. Dipl.-Geophys. Mansour Ohadi for his efforts and contribution in the seminar's organization. The editors are aware that the continuing engagement of Mrs. Elka von Hoyningen-Huene-Vogel contributed greatly to the organization and consequent success of the seminar.

Andreas Vogel
Abu K. M. Sarwar
Rudolf Gorenflo
Ognyan I. Kounchev   Berlin, Fall 1991

Contents

1. Basic Mathematics and General Theory

Effective Numerical Methods for Solving Implicit Ill-Posed Inverse Problems

U. Trautenhahn, D. Schweigert

Technische Universität Chemnitz, Sektion Mathematik, 9010 Chemnitz, Germany

Abstract. The identification of functional parameters in differential equations from noisy data arises in a number of contexts including reservoir engineering, seismology and environmental protection. The numerical treatment of such in general implicit ill-posed inverse problems requires special identification techniques. The Tikhonov's regularization method is known as one of the stabilizing algorithms to solve these ill-posed problems. In this paper, we investigate this method from the numerical point of view. For solving the regularized non-linear minimization problems the Gauss-Newton method is analyzed. We show how to use an implicit differentiation to allow the computation of the Gauss-Newton correction by solving only two problems which are sparse, in general. We apply our results to the numerical identification of the transmissivity coefficient in a two-dimensional elliptic problem.

1   Introduction

In this paper, we consider some numerical methods connected with the identification of unknown coefficients (which are in general functions) in distributed systems from noisy data. Distributed systems are governed by differential equations, in general, which may be described by an operator equation of the form

$$F (q , u) = b \qquad\qquad (1.1)$$

where F maps the couple (q,u) from the product space $Q \times U$ into the space of the right hand side of equation (1.1). This is of course formal and has to be made precise in each particular case.

In the direct problem, for given $b \in B$ and given coefficient $q \in Q_{ad} \subset Q$ the solution $u \in U$ is to be determined. Here, Q_{ad} denotes the set of physically admissible parameters. In this paper we assume that for every $q \in Q_{ad}$ there exists a unique solution $u \in U$ to (1.1) that continuously depends on q and denote this solution by $u = G(q)$. Here, G denotes the solution operator of the direct problem. In (1.1), $F(q,.)$ can be linear or nonlinear and can be stationary or an evolution operator where corresponding initial- and boundary conditions should be also incorporated in the equation (1.1).

In the inverse problem, equation (1.1) and some information on the state u of the form

$$z = C (u) + \delta \qquad (1.2)$$

are given where the continuous observation operator C of the observation equation (1.2) maps the state $u \in U$ into the observation space Z, $\delta \in Z$ denotes the noise, z is the given observation result and $q \in Q_{ad}$ has to be determined. Hence, inverse problems of this type consists in finding $q \in Q_{ad}$ from the equation

$$z = A(q) + \delta \quad , \quad A(q) := C(G(q)) \ . \qquad (1.3)$$

A number of applications lead to problems (1.3) which are ill-posed in the sense of Hadamard, i.e. problems where at least one of the following three properties

  (i)   existence of a solution

  (ii) uniqueness of a solution

  (iii) continuous dependence of the solution on the data

cannot be guaranteed.

2   Linear ill-posed problems

Linear ill-posed inverse problems, in which the operator A of the equation (1.3) is linear have received much attention in the last vew years and their theory seems to be fairly complete. A typical example of this type is an integral equation of the first kind with a compact operator A. The reason why linear equations $Aq = z$ with a compact operator A are ill-posed is given in the following Theorem (cf. e.g. Baumeister 1987):

Theorem 1. Let A: $\mathbb{Q} \longrightarrow \mathbb{Z}$ be compact and injective. Then A^{-1}: $A(\mathbb{Q}) \longrightarrow \mathbb{Q}$ is bounded if and only if $\mathbb{Q}$ is finite dimensional.

The numerical solution of ill-posed problems requires the use of special regularization methods. For linear problems, different regularization methods have been proposed, e.g.:

(1) Tikhonov's regularization (Tikhonov et al. 1979, Friedrich et al. 1979, Groetsch 1984, Morozov 1984, Hofmann 1986, Baumeister 1987, Louis 1989, Wahba 1990)

(2) Iterative regularization (Hoffmann et al. 1984, Vainikko et al. 1986)

(3) Regularization by projection (Natterer 1977, Baumeister 1987)

Here we will focus our attention to the Tikhonov's regularization method for ill-posed inverse problems (1.3) where A: $\mathbb{Q} \longrightarrow \mathbb{Z}$ denotes a linear continuous operator between Hilbert spaces $\mathbb{Q}$ and $\mathbb{Z}$ with dim($\mathbb{Q}$) = ∞ . In this method one seeks an approximation q_α^δ to equation (1.3) by solving the problem

$$\inf_{q \in \mathbb{Q}} J_\alpha(q); \quad J_\alpha(q) = \| Aq - z \|_{\mathbb{Z}}^2 + \alpha\| q - \bar{q} \|_{\mathbb{Q}}^2 , \qquad (2.1)$$

where the regularization parameter $\alpha > 0$ has to be chosen appropriately ($\bar{q} \in \mathbb{Q}$ denotes a suitable approximation of the unknown element q^*). It is well known that (2.1) admits a unique solution q_α^δ for every $z \in \mathbb{Z}$ which is the solution of the Euler equation

$$(A^*A + \alpha I) q_\alpha^\delta = A^*z + \alpha\bar{q} \qquad (2.2)$$

where A^*: $\mathbb{Z} \longrightarrow \mathbb{Q}$ denotes the adjoint of A and I the identity in $\mathbb{Q}$. The Tikhonov's regularization method allows us to determine approximate solutions q_α^δ to (1.3) which stably depend on the data z . Let us suppose that the equation (1.3) with $\delta=0$ admits a unique solution q^* , then following properties are valid (cf. e.g. Groetsch 1984):

(P1) Let $\dfrac{\|\delta\|_{\mathbb{Z}}^2}{\alpha} \to 0$ $(\alpha \to 0)$ then $\|q_\alpha^\delta - q^*\|_{\mathbb{Q}} \to 0$ $(\delta \to 0)$

(P2) If there exists an element $v \in \mathbb{Q}$ satisfying $\bar{q}-q^*=(A^*A)^\gamma v$, $\gamma \in (0,1]$, and $\alpha = \delta^{2/(1+2\gamma)}$, then

$$\| q_\alpha^\delta - q^* \|_Q = O\left[\delta^{2\gamma/(2\gamma+1)}\right] \quad \text{for } \delta \to 0 \quad (2.3)$$

(P3) If α is chosen from Morozov's discrepancy principle, i.e. α solves the equation

$$h(\alpha) = \| Aq_\alpha^\delta - z \|_Z^2 - \| \delta \|_Z^2 = 0 \quad (2.4)$$

(note that (2.4) is uniquely solvable provided $\|A\bar{q} - z\|_Z > \|\delta\|_Z$ holds), then $\|q_\alpha^\delta - q^*\|_Q \to 0$ ($\delta \to 0$).

(P4) If there exists an element $v \in Q$ satisfying $\bar{q} - q^* = (A^*A)^\gamma v$, $\gamma \in (0,1]$ and α is chosen from Morozov's discrepancy principle, then

$$\| q_\alpha^\delta - q^* \|_Q = \begin{cases} O(\sqrt{\delta}) & \gamma \in [\frac{1}{2},1] \\ O\left[\delta^{2\gamma/(2\gamma+1)}\right] & \gamma \in (0,\frac{1}{2}) \end{cases} \quad (2.5)$$

Remark 1. Instead of solving (2.1) one could minimize

$$\| Aq - z \|_Z^2 + \alpha \| q - \bar{q} \|_R^2$$

subject to $q \in R$ where $R \subset Q$ is a densely imbedded space. Then analogous results with respect to convergence and convergence rates in the stronger norm $\|.\|_R$ can be proved.

Remark 2. We note that the convergence rate given in (2.5) is not optimal in case $\gamma \in (\frac{1}{2},1]$. Recently an a posteriori parameter choice strategy has been proposed that leads to optimal convergence rates (cf. Neubauer 1988): If there exists an element $v \in Q$ satisfying $\bar{q} - q^* = (A^*A)^\gamma v$, $\gamma \in (0,1]$ and α is chosen from the equation

$$h(\alpha) = \alpha^3 \| (AA^* + \alpha I)^{-3/2} z \|_Z^2 - \|\delta\|_Z^2 \quad (2.6)$$

((2.6) is uniquely solvable provided $\|Aq-z\|_Z > \|\delta\|_Z$ holds), then (2.3) is valid.

In the case $\|\delta\|_Z$ is unknown Wahba's cross-validation criterion (cf. Golub et al. 1979) which chooses α by minimizing the function

$$V(\alpha) = \frac{(Aq_\alpha^\delta - z)^T (Aq_\alpha^\delta - z)}{\text{tr } (I - A(A^*A + \alpha I)^{-1} A^*)}$$

can be used for choosing a good regularization parameter in

the discretized case $A: \mathbb{R}^m \longrightarrow \mathbb{R}^k$.

The numerical computation of regularized solutions q_α^δ requires in general a discretization of problem (2.2) to obtain a corresponding finite dimensional linear equation

$$(A_h^* A_h + \alpha I)q_{\alpha,h}^\delta = A_h^* z_h + \alpha \bar{q}_h \tag{2.7}$$

with $A_h \in \mathbb{L}(\mathbb{R}^m, \mathbb{R}^k)$, $q_{\alpha,h}^\delta \in \mathbb{R}^m$, $\bar{q}_h \in \mathbb{R}^m$, $z_h \in \mathbb{R}^k$. In many problems, however, the linear dependence between vectors to be determined and the measured data is only known in an implicit manner such that the matrix A_h cannot be given explicitely. Let us suppose in the following of this chapter that the operator equation (1.1) is given in the form

$$Lu + Mq = b, \quad L \in \mathbb{L}(U, \mathbb{B}), \quad M \in \mathbb{L}(Q, \mathbb{B}) \tag{2.8}$$

where $L^{-1} \in \mathbb{L}(\mathbb{B}, U)$ exists and let $C \in \mathbb{L}(U, Z)$. Furthermore let us suppose that the discretization of (2.8) leads to the finite dimensional equation

$$L_h u_h + M_h q_h = b_h, \quad L_h \in \mathbb{L}(\mathbb{R}^n, \mathbb{R}^n), \quad M_h \in \mathbb{L}(\mathbb{R}^m, \mathbb{R}^n)$$

where L_h is regular and that the discretization of the observation equation (1.2) leads to

$$z_h = C_h u_h + \delta_h, \quad C_h \in \mathbb{L}(\mathbb{R}^n, \mathbb{R}^k).$$

Then, the numerical computation of regularized solutions would require
(i)   the generation of the matrix A_h according to
$$A_h := -C_h L_h^{-1} M_h$$
(ii) the numerical solution of the dense linear system

$$(A_h^* A_h + \alpha I)q_{\alpha,h}^\delta = A_h^*(z_h - C_h L_h^{-1} b_h) + \alpha \bar{q}_h \tag{2.9}$$

For growing dimension numbers m, n and k, the amount of work in computing $q_{\alpha,h}^\delta$ is growing rapidly and there arises the question if it is possible to avoide the steps (i), (ii) in order to find $q_{\alpha,h}^\delta$. The answer is given in the following Therem 2 and the following Corollary 1 and Corollary 2.

Theorem 2. Let $\alpha > 0$ and L_h regular, then the solution $q_{\alpha,h}^\delta$ of problem (2.9) is given by $q_{\alpha,h}^\delta := \bar{q}_h - M_h^* w_h$ where w_h is the unique solution of the symmetric system

$$
\begin{pmatrix} C_h^* C_h & \alpha\, L_h^* \\ \alpha\, L_h & -\alpha\, M_h M_h^* \end{pmatrix} \begin{pmatrix} u_h \\ w_h \end{pmatrix} = \begin{pmatrix} C_h^* z_h \\ \alpha(b_h - M_h \bar{q}_h) \end{pmatrix}
$$

Corollary 1. Let $\alpha > 0$, L_h regular and $C_h = I_n$, then the solution $q_{\alpha,h}^\delta$ of problem (2.9) is given by $q_{\alpha,h}^\delta = \bar{q} - M_h^* w_h$ where w_h is the solution of the symmetric positive definite system

$$
(M_h M_h^* + \alpha L_h L_h^*) w_h = M_h \bar{q}_h + L_h z_h - b_h
$$

Corollary 2. Let $\alpha > 0$, L_h regular and $M_h = I_n$, then the solution $q_{\alpha,h}^\delta$ of problem (2.9) is given by $q_{\alpha,h}^\delta = b_h - L_h u_h$, where u_h is the unique solution of the symmetric positive definite system

$$
(C_h^* C_h + \alpha L_h^* L_h) u_h = C_h^* z_h + \alpha L_h^* (b_h - \bar{q}_h)
$$

Remark. The proofs of Theorem 2, Corollary 1 and Corollary 2 use the fact that $q_{\alpha,h}^\delta$ is the solution of the equality constrained problem

$$
\left. \| C_h u_h - z_h \|_{\mathbb{R}^k}^2 + \alpha \| q_h - \bar{q}_h \|_{\mathbb{R}^m}^2 \xrightarrow{q_h,\ u_h} \min \right\}
$$

subject to $\quad L_h u_h + M_h q_h = b_h$

hence $(q_h,\ u_h,\ v_h)$ is the solution of the Kuhn-Tucker conditions

$$
\begin{pmatrix} \alpha I & 0 & M_h^* \\ 0 & C_h^* C_h & L_h^* \\ M_h & L_h & 0 \end{pmatrix} \begin{pmatrix} q_h \\ u_h \\ v_h \end{pmatrix} = \begin{pmatrix} \alpha \bar{q}_h \\ C_h^* z_h \\ b_h \end{pmatrix}
$$

and v_h is a vector of Lagrange multipliers.

In the infinite dimensional case Corollary 1 is not valid since $z \notin D(L)$ in general. In order to understand the difficulties in the infinite dimensional case let us discuss the case $C = I$ and $M = -I$ (i.e. $U = Q = Z = B$) and let $\bar{q}=0$ and $b=0$. We suppose that $L^*: Z \longrightarrow Z$ is positive definite, i.e.

$$\| L^*z \|_Z \geqslant Y \| z \|_Z \qquad (2.10)$$

with $Y>0$. We introduce the set $V = \left\{ v: v=(L^*)^{-1}z , \quad z \in Z \right\}$ and define the scalar product on V by

$$(v_1 , v_2)_V = (L^*v_1 , L^*v_2)_Z \qquad (2.11)$$

to obtain a Hilbert space V . Then we have the relations $V \subset Z \subset V^*$, where $V^* = \left\{ w: w = Lz , \quad z \in Z \right\}$ is the dual space of V . Next we define the bilinear functional

$$a(v_1,v_2) = (v_1, v_2)_Z + \alpha (L^*v_1, L^*v_2)_Z \qquad \forall v_1,v_2 \in V \qquad (2.12)$$

and the linear functional

$$b_z(v) = (z , L^*v)_Z \qquad \forall v \in V , z \in Z \qquad (2.13)$$

and regard the variational problem

$$\text{Find} \quad q \in V : \qquad a(q,v) = b_z(v) \qquad \forall v \in V \qquad (2.14)$$

Then there holds:

Theorem 3. Problem (2.14) is uniquely solvable .

Proof. From (2.11) and (2.12) we have $a(v,v) \geqslant \alpha \| v \|_V^2$ for all $v \in V$, hence a is coercive. Moreover, from (2.10) and (2.11) it follows

$$| a(v_1,v_2) | \leqslant \| v_1 \|_Z \| v_2 \|_Z + \alpha \| L^*v_1 \|_Z \| L^*v_2 \|_Z$$

$$\leqslant (\frac{1}{Y^2} + \alpha) \| v_1 \|_V \| v_2 \|_V$$

for all $v_1,v_2 \in V$ and from (2.13), (2.11) we obtain for all $v \in V$ the inequality

$$| b_z(v) | \leqslant \| z \|_Z \| L^*v \|_Z = \| z \|_Z \| v \|_V .$$

The result follows from the Lax-Milgram Lemma. ∎

Theorem 4. The solution of the problem

$$
\left.\begin{array}{c}
\displaystyle\inf_{q\in Z} J_\alpha(q); \quad J_\alpha(q) = \| u - z \|_Z^2 + \alpha \| q \|_Z^2 \\[2mm]
\text{subject to} \quad Lu - q = 0
\end{array}\right\} \qquad (2.15)
$$

is equivalent to the variational problem (2.14).

Proof. From Theorem 3 it follows that the uniquely determined solution $q\in Z$ of problem (2.14) satisfies the variational equation

$$
(q,v)_Z + \alpha(L^*q,L^*v)_Z = (z,L^*v)_Z \qquad \forall v \in V . \qquad (2.16)
$$

We substitute $L^*v = w$ and find that (2.16) is equivalent to the varitional equation

$$
(L^{-1}q,w)_Z + \alpha(L^*q,w)_Z = (z,w)_Z \qquad \forall w \in Z ,
$$

If we denote by $\langle \cdot , \cdot \rangle_{V^*V}$ the duality between V and V^*, then we obtain the equivalent equation

$$
\langle Lw,(L^*)^{-1}L^{-1}q \rangle_{V^*V} + \alpha\langle Lw,q \rangle_{V^*V} = \langle Lw,(L^*)^{-1}z \rangle_{V^*V} \qquad \forall Lw \in V^*
$$

hence

$$
[\ (L^*)^{-1}L^{-1} + \alpha I \ ] q = (L^*)^{-1}z \qquad \text{in } V . \qquad (2.17)
$$

Since (2.17) is the Euler equation of the quadratic functional $\| L^{-1}q - z \|_Z^2 + \alpha \| q \|_Z^2$ we obtain the expected result. ■

Remark. If $z \in D(L)$, i.e. $Lz \in Z$, then problem (2.15) is equivalent to the solution of the problem
$(I + \alpha LL^*)q_\alpha^\delta = Lz$ which minimizes the quadratic functional $\| \alpha L^*q - z \|_Z^2 + \alpha \| q \|_Z^2$ over $q \in V$ and posesses the property $LL^*q_\alpha^\delta \in Z$.

In order to illustrate the advatages of the variational problem (2.14) over the conventional solution of the Euler equation (2.17) we regard following simple example

$$
\left.\begin{array}{c}
\displaystyle\inf_{q\in L^2(0,1)} \int_0^1 \left\{ [u(x)-z(x)]^2 + \alpha q^2(x) \right\} dx \\[3mm]
\text{subject to} \quad u_{xx} = q , \quad u(0) = 0 , \quad u(1) = 0
\end{array}\right\} \qquad (2.18)
$$

where $z(x) \in \mathbb{L}^2(0,1)$ is given. If we define the operator $L: \mathbb{L}^2(0,1) \longrightarrow \mathbb{L}^2(0,1)$ by

$$Lu = u_{xx} \, , \qquad D(L) = \left\{ u \in \mathbb{H}^2(0,1) : u(0) = u(1) = 0 \right\} \, ,$$

then we have $L = L^*$ and Theorem 4 shows that (2.18) has a unique solution which belongs to $\mathbb{H}^2(0,1)$, i.e. the regularized solution q_α^δ of (2.18) is smoother than the functions in the class over which the minimization is carried out. This solution can effectively be found by solving the variational problem (2.14) where $V = \left\{ v \in \mathbb{H}^2(0,1) : v(0) = v(1) = 0 \right\}$ and

$$a(q,v) = \int_o^1 qv \; dx \quad + \quad \alpha \int_o^1 q_{xx} v_{xx} \; dx$$

$$b_z(v) \quad = \int_o^1 z v_{xx} \; dx \; .$$

If $z \in D(L)$, then the solution of (2.18) is equivalent to the solution of the 4th order problem

$$q + \alpha q_{xxxx} = z_{xx} \, , \qquad x \in (0,1)$$
$$q(0) = q(1) = q_{xx}(0) = q_{xx}(1) = 0$$

and belongs to the space $\mathbb{H}^4(0,1)$.

Now let us extend the results of Theorem 3 and Theorem 4 to the case $M \neq -I$. Again we have to distinguish between the cases $z \in D(L)$ and $z \notin D(L)$. In the case $z \in D(L)$ we suppose $M\bar{q} \in \mathbb{B}$ and

(i) $\quad \| L^* b \|_Z \geq \gamma \; \| b \|_{\mathbb{B}} \qquad \forall b \in \mathbb{B} \quad \text{with} \quad \gamma > 0$ $\qquad\qquad$ (2.19)

(ii) $\quad \| M^* b \|_{\mathbb{Q}} \leq c \; \| b \|_{\mathbb{B}} \qquad \forall b \in \mathbb{B} \quad \text{with} \quad c < \infty$ $\qquad\qquad$ (2.20)

Then there holds:

Theorem 5. The regularized solution q_α^δ of the problem

$$\left. \begin{array}{c} \displaystyle \inf_{q \in \mathbb{Q}} \left\{ \| u - z \|_Z^2 + \alpha \, \| q - \bar{q} \|_{\mathbb{Q}}^2 \right\} \\[2mm] \text{subject to} \quad Lu + Mq = b \end{array} \right\} \qquad (2.21)$$

is given by $q_\alpha^\delta = \bar{q} + M^* w$ where w is the unique solution of the linear operator equation

$$(MM^* + \alpha LL^*)w = b - M\bar{q} - Lz \qquad (2.22)$$

Moreover, the correction $q_\alpha^\delta - \bar{q}$ belongs to the range of the mapping $M^*(L^*)^{-1}$.

Proof. Obviously, q_α^δ satisfies the Euler equation

$$\left[M^*(L^*)^{-1}L^{-1}M + \alpha I \right] \left[q_\alpha^\delta - \bar{q} \right] = M^*(L^*)^{-1}\left\{ L^{-1}(b-M\bar{q})-z \right\}$$

which is equivalent to the representation

$$q_\alpha^\delta - \bar{q} = M^*(L^*)^{-1} \left[L^{-1}MM^*(L^*)^{-1}+\alpha I \right]^{-1} \left\{ L^{-1}(b-M\bar{q})-z \right\} \qquad (2.23)$$

hence, $q_\alpha^\delta - \bar{q} \in R(M^*(L^*)^{-1})$.

In order to prove (2.22) we use $z \in D(L)$, i.e. $Lz \in B$ and find in this case

$$q_\alpha^\delta - \bar{q} = M^*(L^*)^{-1} \left[L^{-1}MM^*(L^*)^{-1}+ \alpha I \right]^{-1} L^{-1} \left\{ b-M\bar{q}-Lz \right\}$$

$$= \left[M^*(LL^*)^{-1}M + \alpha I \right]^{-1} M^*(LL^*)^{-1} \left\{ b-M\bar{q}-Lz \right\}$$

$$= M^* \left[MM^*+ \alpha LL^* \right]^{-1} \left[b-M\bar{q}-Lz \right] \qquad \blacksquare$$

In the more interesting case where $z \notin D(L)$, which is the case in practical applications, (2.22) cannot be used. In this case one has to switch over to a generalized solution concept. We suppose that (2.19) and (2.20) are valid and introduce the set $V = \left\{ v: v=(L^*)^{-1}z , z \in Z \right\}$. We define the scalar product in V by

$$(v_1 , v_2)_V = (L^*v_1 , L^*v_2)_Z$$

to obtain a Hilbert space V . Then we have the relations $V \subset B \subset V^*$ where V^* denotes the dual of V . Next we define the bilinear functional

$$a(v_1,v_2) = (M^*v_1, M^*v_2)_Q + \alpha(L^*v_1, L^*v_2)_Z \quad \forall v_1,v_2 \in V$$

and the linear functional

$$b_z(v) = (b-M\bar{q},v)_B - (z,L^*v)_Z \qquad \forall v \in V , z \in Z$$

and regard the variational problem

$$\text{Find} \quad w \in V : \quad a(w,v) = b_z(v) \quad \forall v \in V \qquad (2.24)$$

Then, analogously to the proofs of Theorem 3 and Theorem 4 one shows following results in the case $z \notin D(L)$:

Theorem 6. Under the assumptions (2.19) and (2.20) the regularized solution q_α^δ of the problem (2.21) is given by $q_\alpha^\delta = \bar{q} + M^* w$ where w is the uniquely determined solution of (2.24). This solution is equivalent to (2.23).

3　Nonlinear ill—posed problems

Nonlinear ill-posed inverse problems, in which the operator $A : D(A) \subset Q \longrightarrow Z$ of equation (1.3) is a nonlinear operator between Hilbert spaces Q and Z, have been studied in different papers. Typical examples of such problems are nonlinear integral equations of the first kind or parameter identification problems in differential equations. The numerical solution of nonlinear ill-posed problems requires the use of special regularization methods such as Tikhonov's regularization methods (Morozov 1984, Kravaris et al. 1985, Engl et al. 1989, Neubauer 1989, Seidmann et al. 1989), Iterative regularization methods (Hoffmann et al. 1984, Vainikko et al. 1986) or Quasi inversion methods (Alessandrini 1986, Kohn et al. 1988). One of the most popular methods is Tikhonov's regularization in which analogously to (2.1) the nonlinear equation (1.3) is approximated by a solution of the minimization problem

$$\inf_{q \in D(A)} J_\alpha(q) ; \quad J_\alpha(q) = \|A(q) - z\|_Z^2 + \alpha \|q - \bar{q}\|_Q^2 \qquad (3.1)$$

Under the assumtion that A is weakly closed, i.e. for any sequence $\{q_n\} \subset D(A)$, weak convergence of q_n to q in Q and weak convergence of $A(q_n)$ to z in Z imply $q \in D(A)$ and $A(q) = z$, it can be shown that problem (3.1) admits a solution. Since the functional $J_\alpha(q)$ is not convex, uniqueness is not to be expected in general. Aspects of stability, convergence and convergence rates (as $\alpha \to 0$), analogously to the properties (P1)-(P4) of Chapter 2, have been studied e.g. in Engl et al. 1989 and Neubauer 1989.

The numerical computation of regularized solutions q_α^δ of problem (3.1) requires Frechet derivatives $A'(q)$, in general. If the nonlinear operator A is only known in an implicit manner and $A'(q)$ cannot be given explicitely, then $A'(q)$ could be approximated by finite difference approximations which requires to solve the associated state equation (1.1) many times. One way to avoide the cumbersome and time consuming evaluation of approximated derivatives $A'(q)$ consists in applying gradient-like methods for minimizing $J_\alpha(q)$. A powerful result which comes from control theory is that the gradient for the regularization functional $J_\alpha(q)$ can be computed by solving the state equation (1.1) together with one associated adjoint equation. Such derivations have been investigated for instance by Lions 1971, Chavent 1980 and Kravaris et al. 1985.

Of course, the convergence of gradient-like methods may be very slow, hence we ask the question how to compute the Gauss-Newton correction of the functional $J_\alpha(q)$ efficiently in order to improve the convergence properties and consequently the overall costs in solving (3.1). Applying the ideas given in Chapter 2 our strategy avoides the cumbersome and time consuming evaluation of $A'(q)$ and requires only the solution of two problems, one direct problem (1.1) and one special linear normal equation problem of the type (2.24).

Given a guess q for q_α^δ, then in the Gauss-Newton method (cf. Schwetlick 1979, Schaback 1985) a new approximation q_{new} to the solution of (3.1) is given by $q_{new} := q + \gamma \Delta q$ where γ is the steplength parameter to be chosen appropriately (cf. Schwetlick 1979) and Δq is the Gauss-Newton correction which is the solution of the quadratic programming problem

$$\inf_{\Delta q \in Q} J_\alpha^1(\Delta q) \; ; \quad J_\alpha^1(\Delta q) = \| A'(q)\Delta q - (z - Aq) \|_Z^2$$

$$+ \alpha \| \Delta q - (\bar{q} - q) \|_Q^2 \quad . \quad (3.2)$$

Problem (3.2) has the same structure as (2.1), hence, Δq is the solution of the following linear operator equation

$$(A'^* A' + \alpha I) \Delta q = A'^*(z - A(q)) + \alpha(\bar{q} - q) \quad (3.3)$$

Since Δq is the solution of the equality constrained problem

$$\|C_u' \Delta u - (z - A(q))\|_Z^2 + \alpha \|\Delta q - (\bar{q} - q)\|_Q^2 \xrightarrow{\Delta u, \Delta q} \min$$

$$\text{subject to} \quad F_u' \Delta u + F_q' \Delta q = 0 \tag{3.4}$$

with $C_u' = C_u'(G(q))$, $F_u' = F_u'(q, G(q))$, $F_q' = F_q'(q, G(q))$, the numerical solution of (3.3) can now be done along the lines of Chapter 2.

Finally we regard the model problem

$$\inf_{q \in Q_{ad}} \int_o^1 \left\{ [u(x) - z(x)]^2 + \alpha [q(x) - \bar{q}(x)]^2 \right\} dx \tag{3.5}$$

$$\text{subject to} \quad -u_{xx} + qu = f \quad , \quad u(0) = u_o \ , \ u(1) = u_1 \tag{3.6}$$

where $z \in L^2(0,1)$, $\bar{q} \in L^2(0,1)$ and $f \in L^2(0,1)$ are given. We chose $Q_{ad} = \left\{ q \in L^2(0,1) \colon q \geqslant 0 \ \text{a.e.} \ x \in (0,1) \right\}$ such that for given q and f (3.6) admits a unique solution $u \in H^2(0,1)$. Given a guess $q \in Q_{ad}$ for q_α^δ , then the Gauss–Newton correction of problem (3.5), (3.6) is the solution of the problem

$$\inf_{\Delta q \in L^2(0,1)} \int_o^1 \left\{ [\delta u - (z - u)]^2 + \alpha [\Delta q - (\bar{q} - q)]^2 \right\} dx \tag{3.7}$$

where u is the solution of (3.6) and δu is the solution of

$$- \delta u_{xx} + q \delta u = -u \Delta q \qquad x \in (0,1)$$

$$\delta u(0) = \delta u(1) = 0$$

Now we define the operator $L \colon L^2(0,1) \longrightarrow L^2(0,1)$ by $L\delta u = -\delta u_{xx} + q\delta u$, $D(L) = \left\{ \delta u \in H^2(0,1) \colon \delta u(0) = \delta u(1) = 0 \right\}$, and the operator $M \colon L^2(0,1) \longrightarrow L^2(0,1)$ by $Mw = uw$. Then we have $L = L^*$ and $M = M^*$ and the application of Theorem 6 shows that the Gauss–Newton correction Δq of problem (3.5), (3.6) (which is the solution of (3.7)) is given by $\Delta q := \bar{q} - q + uw$ where u is the solution of (3.6) and w is the solution of the variational problem (2.24) with

$$V = \left\{ v \in H^2(0,1) \colon v(0) = v(1) = 0 \right\}$$

$$a(w,v) = \int_o^1 u^2 wv \ dx \ + \ \alpha \int_o^1 (-w_{xx} + qw)(-v_{xx} + qv) \ dx$$

$$b_z(v) = \int_o^1 u(q - \bar{q})v \ dx \ + \ \int_o^1 (u - z)(-v_{xx} + qv) \ dx$$

4   Application

In this chapter we investigate the inverse problem of identi-
fying the transmissivity coefficient $q(x)$ in the following
second order elliptic problem

$$\left. \begin{array}{rcll} - \text{div} (q(x) \nabla u(x)) & = & f(x) & x \in \Omega \subset \mathbb{R}^2 \\ u(x) & = & g(x) & x \in \Gamma \end{array} \right\} \qquad (4.1)$$

where Ω is a bounded domain with smooth boundary Γ and f
and g are given sufficiently smooth functions. We suppose
that noisy data $z(x) = u(x) + \delta(x)$, $z \in L^2(\Omega)$, are given
where $\delta(x) \in L^2(\Omega)$ denotes the noise.
Problems of this type are important in different applications.
Equation (4.1), for example, describes the steady flow in a
confined nonuniform aquifer Ω , where q denotes the spatially varying transmissivity coefficient of the aquifer, $u = u(x)$
the piezometric head and $f = f(x)$ the source term. In different papers identifiability questions, questions of the non-
continuous dependence of q on measurements z and questions
concerning the application of regularization methods to obtain
regularized solutions are studied (cf. Chavent 1980, Richter
1981, Falk 1983, Hoffmann et al. 1984, Alessandrini 1986,
Chicone et al. 1987, Colonius et al. 1989). To illustrate our
theory we consider the problem

$$\left. \begin{array}{l} \inf_{q \in \mathbb{Q}_{ad}} J_\alpha(q); \ J_\alpha(q) = \|u(x;q)-z\|^2_{L^2(\Omega)} + \alpha \ \|q-\bar{q}\|^2_{L^2(\Omega)} \\[2mm] \qquad \text{where} \ \ u(x;q) \ \ \text{solves (4.1)} \end{array} \right\} \qquad (4.2)$$

where $\mathbb{Q}_{ad} = \left\{ q \in H^1(\Omega): 0 < q_{min} \leqslant q(x) \leqslant q_{max} < \infty \right\}$
so that problem (4.1) admits a unique solution $u \in H^2(\Omega)$ deno-
ted by u(x;q) and ask how to compute the Gauss-Newton correction Δq of the functional (4.2) which is the solution of the
following linearized problem (compare (3.4))

$$\inf J^1_\alpha(\Delta q); \ \ J^1_\alpha(\Delta q) = \|\delta u(x;\Delta q)-(z-u(x;q))\|^2_{L^2(\Omega)}$$

$$+ \alpha \|\Delta q-(\bar{q}-q)\|^2_{L^2(\Omega)} \qquad (4.3)$$

where u(x;q)   is the solution of (4.1) and $\delta u(x;\Delta q)$ is the
solution of

$$- \text{div} (q \, \nabla \delta u) = \text{div} (\Delta q \, \nabla u) \quad \text{in} \quad \Omega \quad \Big\}$$
$$u = 0 \quad \text{on} \quad \Gamma$$

Applying the ideas given at the end of Chapter 2 we can prove the following result.

Theorem 7. Let $\alpha > 0$ and $q, \bar{q} \in Q_{ad}$. Then there exists a unique element $\Delta q \in \mathbb{H}^1(\Omega)$ which minimizes the functional (4.3) over $\mathbb{H}^1(\Omega)$. This element Δq is given by

$$\Delta q = \bar{q} - q + \sum_{i=1}^{2} \frac{\partial u}{\partial x_i} \frac{\partial w}{\partial x_i}$$

where $u = u(x;q) \in \mathbb{H}^2(\Omega)$ is the classical solution of (4.1) and $w = w(x;q) \in \mathbb{H}^2(\Omega)$ is the solution of the variational problem (2.24) where $V = \mathbb{H}_0^2(\Omega)$ and

$$a(w,v) = \alpha \int_{\Omega} \text{div}(q \, \nabla w) \, \text{div}(q \, \nabla v) \, dx + \int_{\Omega} (\nabla u \circ \nabla w)(\nabla u \circ \nabla v) \, dx$$

$$b_z(v) = \int_{\Omega} (q - \bar{q})(\nabla u \circ \nabla v) \, dx + \int_{\Omega} (z - u) \, \text{div}(q \, \nabla v) \, dx .$$

For a complete proof of Theorem 7 and other results in this area we refer the paper of Tautenhahn et al. 1990; for a detailed discussion of numerical experiments we refer to Schweigert 1990. We note that similar results have been obtained for other applications such as identification problems in two point boundary value problems (cf. Goebel et al. 1988, Tautenhahn 1989, Friedrich et al. 1989) or identification problems in parabolic equations (cf. Tautenhahn 1988).

References

Alessandrini,G., 1986: An identification problem for an elliptic equation in two variables. Ann. Mat. Pura Appl. 145, 265-296.
Baumeister,J., 1987: Stable Solution of Inverse Problems. Vieweg Verlagsgesellschaft mbH, Braunschweig.
Chavent,G., 1980: Identification of distributed parameter systems: About the output least square method, its implementation and identifiability. In: Proc. 5th IFAC Symposium on Identification and System Parameter Estimation, Vol.I (Isermann,R. ed.), 25-97, Pergamon Press, New York.
Chicone,C. and J.Gerlach, 1987: A note on the identifiability of distributed parameters in elliptic equations. Siam J. Math. Anal. 18, 1378-1384.

Colonius,F. and K.Kunisch, 1989: Output Least Squares Stability in Elliptic Systems. Appl. Math. Optim. 19, 33-63.

Engl,H., K.Kunisch and A.Neubauer, 1989: Convergence rates for Tikhonov regularisation of non-linear ill-posed problems. Inverse Problems 5, 523-540.

Falk,R.S., 1983: Error estimates for the numerical identification of a variable coefficient. Math. Comp. 40, 537-546.

Friedrich,V. and U.Tautenhahn, 1989: Regularized Parameter Identification in Elliptic Boundary Value Problems. Z. Anal. Anw. 8, 3-11.

Goebel,M. and U.Tautenhahn, 1988: Parameter Identification in Two Point Boundary Value Problems by Regularization. Preprint 63 TU Karl-Marx-Stadt, 1-15.

Groetsch,C.W.: The Theory of Tikhonov Regularization for Fredholm Equations of the First Kind. Pitman, Boston.

Golub,G.H., M.Heat and G.Wahba, 1979: Generalized cross-validation as a method for choosing a good ridge parameter. Technometrics 21, 215-223.

Hoffmann,K.H. and J.Sprekles, 1984: On the identification of the coefficients of elliptic problems by asymptotic regularization. Num. Funct. Anal. Optim. 7, 157-177.

Hofmann,B., 1983: Regularization for Applied Inverse and Ill-Posed Problems. Teubner, Leipzig.

Hsiao,G.C. and J.Sprekles, 1988: A Stability Result for Distributed Parameter Identification in Bilinear Systems. Math. Meth. Appl. Sci. 10, 447-456.

Ito,K. and K.Kunisch, 1990: The Augmented Lagrangian Method for Parameter Estimation in Elliptic Systems. Siam J. Control Optim. 28, 113-136.

Kohn,R.V. and B.D.Lowe, 1988: A Variational Method for Parameter Identification. Math. Modelling and Num. Anal. 22, 119-158.

Kravaris,C. and H.J.Seinfeld, 1985: Identification of parameters in distributed systems by regularization. Siam J. Control Optim. 23, 217-241.

Kunisch,K. and L.W.White, 1986: Regularity properties in parameter estimation of diffusion coefficients in elliptic boundary-value problems. Appl. Anal. 21, 71-88.

Lions,J.L., 1971: Optimal Control of Systems Governed by Partial Differential Equations. Springer-Verlag, Berlin.

Louis,A.K., 1988: Inverse und schlecht gestellte Probleme. Teubner, Stuttgart.

Morozov,V.A., 1984: Methods for solving incorrectly posed problems. Springer-Verlag, New York.

Natterer,F., 1977: Regularisierung schlecht gestellter Probleme durch Projektionsverfahren. Num. Math. 28, 511-522.

Neubauer,A., 1989: Tikhonov regularisation for non-linear ill-posed problems: optimal convergence rates and finite-dimensional approximation. Inverse Problems 5, 541-557.

Richter,G.R., 1981: Numerical Identification of a Spatially Varying Diffusion Coefficient. Math. Comp. 36, 375-386.

Schaback,R., 1985: Convergence Analysis of the General Gauss-Newton Algorithm. Num. Math. 46, 281-309.

Schweigert,D., 1990: Numerische Identifikation 2D-elliptischer Aufgaben. Preprint TU Chemnitz, (in preparation).

Schwetlick,H., 1979: Numerische Lösung nichtlinearer Gleichungen. VEB Deutscher Verlag der Wiss., Berlin.

Seidmann,T.I. and C.R.Vogel, 1989: Well posedness and convergence of some regularization methods for nonlinear ill posed problems. Inverse Problems 5, 227-238.

Tautenhahn,U., 1988: Parameter identification in parabolic equations. In: Numerical Treatment of Differential Equations 104 (Strehmel, K. ed.), Teubner, Leipzig.

Tautenhahn,U., 1989: A Fast Iterative Method for Solving Regu-
   larized Parameter Identification Problems in Elliptic Boun-
   dary Value Problems. Computing 43, 47-58.
Tautenhahn,U., 1990: Numerical solution of implicit ill-posed
   inverse problems by regularization. Wiss. Z. d. TU Karl-Marx-
   Stadt 32, 24-28.
Tautenhahn,U. and D.Schweigert, (submitted): Numerical Identi-
   fication of Elliptic Systems by Regularization. Optimization.
Tikhonov,A.N. and V.Y.Arsenin, 1979: Solution of Ill-Posed
   Problems (in Russian). Nauka, Moscow.
Vainikko,G.M. and A.Y.Veretennikov, 1986: Iteration procedures
   in ill-posed problems. Nauka, Moscow.
Wahba,G., 1989: Regularization and Cross Validation Methods
   for Nonlinear Implicit, Ill-Posed Inverse Problems. Techn.
   Report 852 Univ. Wisconsin, 1-13.
Wahba,G., 1990: Spline Models for Observational Data. Siam,
   Philadelphia.

On the Data Error Influence in Determining a Specific Response Function

B. Hofmann[1], R. Hausding[2]

[1] Sektion Mathematik und Naturwissenschaften, Technische Hochschule Zittau, Th.-Körner-Allee 16, 8800 Zittau, Germany
[2] Sektion Mathematik, Technische Universität Chemnitz, 9010 Chemnitz, Germany

Abstract

The computational solution of integral equations of the first
kind may be very sensitive to random data errors, particularly
when the kernel function is also estimated. For an example, we
refer to a convolution integral equation for which the kernel
is obtained as derivative of a quotient of observable functions. Equations of this kind arise from response function
models in the context of analysing the volume-pressure dependence for a gas-occupied aquifer. We deduce that smoothing
techniques are required and investigate the utility of linear
constraints for the reduction of errors in the solution. Some
phenomena of error propagation are demonstrated by theoretical
and experimental arguments.

1   Introduction

Ill-posed problems are generally difficult to solve on a
computer, since a small random noise on the given data lead to
significant errors in the solution. The classical case of an
ill-posed problem is the Fredholm integral equation of the
first kind. It is well-known that the solution of such an equation requires the use of regularization or smoothing methods.

On the other hand, several authors have recognized that the problem is not nearly as serious for Volterra equations of the first kind as it is for the Fredholm case (see e.g. Linz, 1985, p. 162). This interpretation, however, may fail when the right-hand side and the kernel of the Volterra equation both are to be estimated from noisy data.

For an example, we refer to a convolution integral equation arising in some models of reservoir mechanics. We are going to consider a triple of positive functions
$$(p,v,x) \in C \; [0,T] \times C \; [0,T] \times C^1 \; [0,T]$$
and a pair of auxiliary functions
$$(\Omega,b) \in C \; [0,T] \times C \; [0,T]$$
satisfying

$$\int_0^t x(t-\tau) \; d\Omega(\tau) = b(t) \qquad (0 \leq t \leq T) \qquad (1.1)$$

where

$$\Omega(t) := \frac{v(t)}{p(t)} \; , \quad b(t) := p(t) - p(0) \; (0 \leq t \leq T) \; (1.2)$$

An interdependence between p, v and x of the form (1.1)-(1.2) is already used by Coats et al., 1964, for the computation of aquifer influence functions. In this context, the functions p and v represent the pressure of gas in a gas-occupied aquifer and the volume which this gas would have under atmospheric pressure. The quotient $\Omega(t)$ is proportional to the porevolume of this gas for the actual pressure p(t). All functions under consideration depend on the time t. On the other hand, x can be considered as a monotone and convex response function. This function x(t) gives an idea of the pressure-reaction at time t to a volume-change at time 0. In this sense, it is a memory function.

In view of the small amount of required geophysical data, the method of influence functions is in spite of its rough (one-dimensional) modelling character still of interest for some classes of underground reservoirs. In any case, the associated mathematical problems are worth analysing.

For a time interval $0 \leq t \leq T$, let us consider the domains

$$\mathcal{D}_v := \left\{ v \in C\ [0,T]:\ v(0) = v_0,\ v(t) \geq 0 \atop (\ 0 \leq t \leq T\ ) \right\}$$

$$\mathcal{D}_p := \left\{ p \in C\ [0,T]:\ p(0) = p_0,\ p(t) > 0 \atop (\ 0 \leq t \leq T\ ) \right\}$$

and

$$\mathcal{D}_x := \left\{ x \in C^1\ [0,T]:\ x(0) > 0,\ x(t) \geq 0,\ (\ 0 \leq t \leq T),\ \atop x'(t) \leq 0,\ x'(t_1) \leq x'(t_2),\ (\ 0 \leq t_1 < t_2 \leq T\ ) \right\}$$

Throughout this paper, v_0 and p_0 are assumed to be known exactly. Note that for all $x \in \mathcal{D}_x$, $\Omega \in C\ [0,T]$

$$\int_0^t x(t-\tau)\ d\Omega(\tau) = \int_0^t x'(t-\tau)\ \Omega(\tau)\ d\tau + x(0)\Omega(t) - x(t)\Omega(0)$$

(1.3)

attains finite values such that the left-hand side of (1.1) is a continuous function on $0 \leq t \leq T$.

For a triple $(\ v,p,x\ ) \in \mathcal{D}_v \times \mathcal{D}_p \times \mathcal{D}_x$ satisfying (1.1)-(1.2) there are three problems of practical interest:

(PA) Find p when v and x are given !

(PB) Find v when p and x are given !

(PC) Find x when p and v are given !

These problems are considered from an analytic point of view in the paper by Hofmann, 1988.

The problem (PA) is a direct problem, whereas (PB) and (PC) are inverse problems. For the solution of (PA) one can show existence, uniqueness and continuous dependence upon the data. Assertions of this kind are derived from the implicit-function theorem applied to equation (1.1) which is a nonlinear inte-gral equation with respect to p when (1.2) is taken into ac-

count. In this context, the particular maximum principle
(Lemma 1) for the considered Volterra-Stieltjes equation plays
an essential role.

_Lemma 1: For a triple (v,p,x) ∈ $\mathcal{D}_v \times \mathcal{D}_p \times \mathcal{D}_x$ satisfying
(1.1)-(1.2) let $0 < v_{min} \leq v(t) \leq v_{max} < \infty$ ($0 \leq t \leq T$).
Then we have_

$$0 < p_{min} := \frac{v_{min}}{v_{max}} p_0 \leq p(t) \leq \frac{v_{max}}{v_{min}} p_0 =: p_{max} < \infty$$

$$(0 \leq t \leq T).$$

□

Although problem (PB) is of inverse nature, uniqueness of solution and continuous dependence upon the data may be shown.
In view of formula (1.3) the equation (1.1) subject to (1.2)
is a Volterra integral equation of the second kind with respect to **v**.

However, an integral equation of the first kind with respect to **x** possessing the obvious ill-posedness properties
appears whenever we are dealt with problem (PC).

Lemma 2: The integral operator $\mathcal{A}: C^1 [0,T] \longrightarrow C [0,T]$,

$$[\mathcal{A}x](t) := \int_0^t x(t-\tau) \, d\Omega(\tau) \qquad (0 \leq t \leq T) \qquad (1.4)$$

_is completely continuous for all $\Omega \in C [0,T]$.
Consequently, $\mathcal{A}$ is not surjective. This operator is injective
iff there is no real number $\varepsilon > 0$ such that $\Omega(t) \equiv \Omega(0)$
($0 \leq t \leq \varepsilon$). In the injective case, the inverse operator $\mathcal{A}^{-1}$
is unbounded._

□

The injectivity condition of Lemma 2 is a consequence of
Tichmarsh's theorem (see e.g. Berg, 1974, p. 138) concerning the null-space of a convolution operator.

Corollary 1: For given $p \in \mathcal{D}_p$ and $v \in \mathcal{D}_v$ let $x \in \mathcal{D}_x$ solve problem (PC). Then this solution is uniquely determined if and only if there is no real number $\varepsilon > 0$ such that $v(t) \equiv v_0$ ($0 \leq t \leq \varepsilon$).

□

Owing to Lemma 2 small perturbations on the image $\mathcal{A}x$ may lead to large changes of x. For the given problem, however, the structure of $\mathcal{D}_x$ involving monotonicity and convexity conditions may help to overcome the ill-posedness of the integral equation by these shape constraints (cf. also Gorenflo, 1979; Rutman, 1988). This regularization approach based on qualitative information about the solution to be determined is sometimes called *'descriptive regularization'*. A stabilization effect may be thus achieved by embedding the solutions to a compact set of L_p. Namely, due to Helly's theorem the subset of uniformly bounded monotone or convex functions on $0 \leq t \leq T$ is compact in L_p [0,T].

We are going to study discrete versions of (PC) in the following sections where the experimental data errors are taken into account.

For the behaviour of solutions to ill-posed problems obtained by descriptive regularization we also refer to Tikhonov et al., 1983 and Hofmann, 1986.

2   The discretized problem

We consider an equally spaced grid $t_i := ih$
(i = 0,1,...,n) with step-length $h := \dfrac{T}{n}$ and midpoints
$\bar{t}_i := t_i - \dfrac{h}{2}$ (i = 1,2,...,n).
Setting $v_i := v(\bar{t}_i)$, $p_i := p(t_i)$, $b_i := b(t_i) = p_i - p_0$ and
$\Omega_i := \dfrac{v_i}{p_i}$ (i = 1,2,...,n) the midpoint rule provides an

approximation

$$\sum_{i=1}^{j} x_{j+1-i} (\Omega_i - \Omega_{i-1}) = b_j \quad (j = 1,2,\ldots,n) \quad (2.1)$$

for equation (1.1). The solution vector $x = (x_1, x_2, \ldots, x_n)^T$ represents an approximation of the exact midpoint values

$\bar{x} = (x(\bar{t}_1), x(\bar{t}_2), \ldots, x(\bar{t}_n))^T = (\bar{x}_1, \bar{x}_2, \ldots, \bar{x}_n)^T$. Moreover, we have

$$D_v := \{ v \in \mathbb{R}^n : v_i \geq 0 \ (i = 1,2,\ldots,n) \},$$
$$D_p := \{ p \in \mathbb{R}^n : p_i > 0 \ (i = 1,2,\ldots,n) \}$$

and

$$D_x := \left\{ x \in \mathbb{R}^n : Gx \geq 0, \ G = \begin{bmatrix} G_1 \\ G_2 \end{bmatrix} \right\}, \quad (2.2)$$

where

$$G_1 = \begin{bmatrix} 1 & -1 & & & \\ & 1 & -1 & & \mathbf{0} \\ & & \ddots & \ddots & \\ & & & \ddots & \ddots \\ \mathbf{0} & & & & 1 & -1 \\ & & & & & 1 \end{bmatrix}, \quad G_2 = \begin{bmatrix} 1 & -2 & 1 & & & \\ & 1 & -2 & 1 & & \mathbf{0} \\ & & \ddots & \ddots & \ddots & \\ \mathbf{0} & & & \ddots & \ddots & \ddots \\ & & & & 1 & -2 & 1 \end{bmatrix}$$

as discrete analogues of domains $\mathcal{D}_v$, $\mathcal{D}_p$ and $\mathcal{D}_x$, respectively. The discretized form of (PC) is for (2.1):

(PD)   Find $x \in \mathbb{R}^n$ when $p \in D_p \subset \mathbb{R}^n$ and $v \in D_v \subset \mathbb{R}^n$ are given !

Lemma 3: The problem (PD) is uniquely solvable whenever $\Omega_1 \neq \Omega_0$. Then we have

$$x_1 = \frac{b_1}{(\Omega_1 - \Omega_0)}, \quad x_{j+1} = \frac{b_{j+1} - \sum_{i=1}^{j} (\Omega_{j+2-i} - \Omega_{j+1-i}) x_i}{\Omega_1 - \Omega_0}$$

$$(j = 1,2,\ldots,n-1) \quad (2.3)$$

□

Note that the uniqueness condition $\Omega_1 \neq \Omega_0$ of Lemma 3 is a discrete version of the nonstationarity condition $v(t) \not\equiv v_0$ of Corollary 1. Using standard techniques (see also Linz, 1985, sec. 9.2) one can show that the discretization error

$$\varepsilon^{(h)} := \max_{1 \leq i \leq n} | x_i - \bar{x}_i |$$

tends to zero whenever Ω is sufficiently smooth.

Theorem 1: Let $x \in C^2 [0,T]$, $\Omega \in C^2 [0,T]$,

$$|\Omega(t) - \Omega(0)| \geq \mu t \qquad\qquad (\mu > 0, \ 0 \leq t \leq t_0 \leq T)$$

and $\quad |\Omega''(t_1) - \Omega''(t_2)| \leq L |t_1 - t_2| \quad (t_1, t_2 \in [0,T]).$

Then, for a sufficiently small $h > 0$, the problem (PD) is uniquely solvable and we have

$$\varepsilon^{(h)} = \mathcal{O} (h^2) \tag{2.4}$$

□

Proof: Hofmann et al., 1990

If $\dfrac{|\Omega(t) - \Omega(0)|}{t}$ is not bounded below by a positive constant, we have in general a convergence rate of $\varepsilon^{(h)} = \mathcal{O} (h)$ for sufficiently smooth functions Ω.

3   The experimental data problem

In practice we have noisy data vectors
$\hat{v} = (\hat{v}_1, \hat{v}_2, \ldots, \hat{v}_n)^T$ and $\hat{p} = (\hat{p}_1, \hat{p}_2, \ldots, \hat{p}_n)^T$ with

$$\hat{v}_1 = v_1 + \eta_1 \quad , \quad \hat{v}_i = \hat{v}_{i-1} + (v_i - v_{i-1}) + \eta_i \tag{3.1}$$

$$(i = 2, 3, \ldots, n)$$

and

$$\hat{p}_i = p_i + \zeta_i \qquad\qquad (i = 1,2,\ldots,n) \qquad (3.2)$$

That means, the pressures are observed with a random perturbation where we will assume

$$E\,\zeta = 0 \quad , \quad \text{cov}\,(\zeta) = \sigma_p^2\,I \;. \qquad\qquad (3.3)$$

On the other hand, volume measurements come from increment observations, such that

$$E\,\eta = 0 \quad , \quad \text{cov}\,(\eta) = \sigma_v^2\,I \qquad\qquad (3.4)$$

seems to be realistic. Moreover, we assume ζ and η to be
stochastically independent. If we write the system (2.1) of
linear equations in a matrix form

$$A\,x \;=\; b\,, \qquad\qquad (3.5)$$

we can compare the experimental data version

$$\hat{A}\,\hat{x} \;=\; \hat{b} \qquad\qquad (3.6)$$

with

$$\hat{A} \;=\;
\begin{pmatrix}
\hat{a}_1 & & & & \\
\hat{a}_2 & \hat{a}_1 & & & \\
\cdot & \cdot & \cdot & & \\
\cdot & \cdot & & \cdot & \\
\cdot & \cdot & & & \cdot \\
\hat{a}_n & \hat{a}_{n-1} & \cdot & \cdot & \hat{a}_2 & \hat{a}_1
\end{pmatrix}$$

$$\hat{a}_1 \;:=\; \frac{\hat{v}_1}{\hat{p}_1} - \frac{v_0}{p_0}\;,$$

$$\hat{a}_i \;:=\; \frac{\hat{v}_i}{\hat{p}_i} - \frac{\hat{v}_{i-1}}{\hat{p}_{i-1}}$$

$$(i = 2,3,\ldots,n)$$

$$\hat{b}_j \;:=\; \hat{p}_j - p_0 \qquad\qquad (j = 1,2,\ldots,n)$$

Substituting $\hat{v}_i$ and $\hat{p}_i$ by v_i and p_i, respectively, $\hat{a}_i$ carries
over to a_i forming the matrix

$$
A = \begin{bmatrix}
a_1 & & & & \\
a_2 & a_1 & & & \\
\cdot & \cdot & \cdot & & \\
\cdot & \cdot & & \cdot & \\
\cdot & \cdot & & & \cdot \\
a_n & a_{n-1} & \cdot & \cdot & a_2 & a_1
\end{bmatrix}
$$

In the same way, $\hat{b}_j$ carries over to the original right-hand side b_j. The experimental data problem of (PC) is for (3.6):

(PE)   Find $\hat{x} \in \mathbb{R}^n$ when $\hat{p} \in \mathbb{R}^n$ and $\hat{v} \in \mathbb{R}^n$ are given !

From the standard arguments of linear algebra we have

$$
\| \hat{x} - x \|_2 \leq \| \hat{A}^{-1} \|_2 \, (\| \hat{b} - b \|_2 + \| \hat{A} - A \|_2 \, \| x \|_2) \tag{3.7}
$$

($\| x \|_2$ Euclidean norm, $\| A \|_2$ spectral norm). Due to $E \| \hat{b} - b \|_2^2 = n \, \sigma_p^2$ the error term $\| \hat{b} - b \|_2$ of (3.7) tends to zero in the mean as σ_p tends to zero. The same tendency shows $\| \hat{A} - A \|_2$ when both variances σ_p and σ_v tend to zero. The significant error influence comes from $\| \hat{A}^{-1} \|_2$.

Lemma 4: Let $\hat{a}_1 \neq 0$. Then we have

$$
\frac{1}{|\hat{a}_1|} \leq \| \hat{A}^{-1} \|_2 \leq \frac{1}{|\hat{a}_1|} \, (1 + \frac{\| \hat{a} \|_\infty}{|\hat{a}_1|})^{n-1} \tag{3.8}
$$

($\| a \|_\infty$ _maximum norm_). $\square$

If the data are free of noise, then $|a_1| = \Omega(\frac{T}{n}) - \Omega(0)$ tends to zero as $n \longrightarrow \infty$, since $\Omega(t)$ is a continuous function. Therefore, $\| A^{-1} \|_2$ may attain very large values if the used grid is fine. This is the usual situation for an ill-posed problem. If now the data are randomly perturbed, it also cannot be excluded that $|\hat{a}_1|$ takes on very small values. However, note that large norms $\|\hat{A}^{-1}\|_2$ mainly result from the ill-posedness of equation (1.1). Thus in general this norm does not decrease if the data error variances decrease.

4   A lognormal distribution study

Since the error norm bounds (3.7), (3.8) can only give a rough insight into relations between data errors and ill-posedness effects, we should consider the errors in a stronger statistical sence. For simplicity we will do that only for the first component x_1 of the solution vector. In some situations, for instance in the case of monotonically decreasing functions x, this first component error expectation

$$e_1 := \sqrt{E\,(x_1 - \hat{x}_1)^2} \qquad (4.1)$$

essentially charakterizes the total error behaviour.

Now let $\Omega_0 := 0$ and $\hat{p}_1$, $\hat{v}_1$ lognormally distributed with

$$\left. \begin{array}{l} E\,\hat{p}_1 = p_1 > 0,\ E\,\hat{v}_1 = v_1 > 0 \\[2mm] D^2\,\hat{p}_1 = \sigma_p^2,\ D^2\,\hat{v}_1 = \sigma_v^2 \end{array} \right\} \qquad (4.2)$$

Then one can express e_1 in an explicit manner.

Theorem 2: Let $f_p := 1 + \dfrac{\sigma_p^2}{p_1^2}$ and $f_v := 1 + \dfrac{\sigma_p^2}{v_1^2}$. _Under the assumptions stated above we then have_

$$e_1 = \frac{p_1}{v_1} \{(p_1 - p_o)^2 - 2(p_1 - p_o)(p_1 f_p - p_o)f_v +$$

$$+ f_v^3 (f_p^6 p_1^2 - 2f_p^3 p_o p_1 + f_p p_o^2)\}^{1/2} \qquad (4.3)$$

□

This theorem emphasizes the superposition of ill-posedness effects expressed by the first factor $\frac{P_1}{V_1}$ and random data effects expressed by the terms within braces. Using linear approximations for the relative variance powers we may rewrite (4.3) as

$$e_1 = \frac{P_1}{V_1} \left\{ \frac{\sigma_v^2}{V_1^2}(P_1-P_0)^2 + \frac{\sigma_p^2}{P_1^2}(4P_1(P_1-P_0) + P_0^2) + \right.$$

$$\left. + \frac{\sigma_p^2}{P_1^2}\frac{\sigma_v^2}{V_1^2}(16P_1(P_1-P_0) + 3P_0^2)\right\}^{1/2} \qquad (4.4)$$

An intolerable error propagation may be associated with the solution of the stochastic system (3.6). Even if n attains small or moderate values, oscillating solutions are obtained, where $|\hat{x}_i|$ tends to grow extremely as i approaches n. Numerical examples for demonstrating this phenomenon are given by Hofmann et al., 1990. Consequently, a regularization method is required to stabilize the solution of (3.6). For reviews of available regularization methods we refer for example to Louis, 1988 and Hofmann, 1986. This spectrum of methods is considered from a statistical point of view by Titterington, 1985 and O'Sullivan, 1986.

.5 Regularization by linear inequalities

We now restrict our consideration to linear inequalities related to the shape constraints of domain $\mathfrak{D}_x$. The first version of a regularization by linear constraints is

(V1): $\| \hat{A} x - \hat{b} \|_2 \longrightarrow$ min , subject to $x \geq 0$,

using only nonnegativity constraints, wheares

(V2): $\| \hat{A} x - \hat{b} \|_2 \longrightarrow$ min , subject to $Gx \geq 0$,

(for G see (2.2)) is dealt with monotonicity and convexity

constraints. Even if the amount of computational work is somewhat higher for the solution of the constrained least squares
·problem (V2) compared to the NNLS problem (V1), the full-con-
straints version should be preferred. Version (V1) is in general not able to depress oscillations. Empirical results (see
Hofmann et al., 1990, also for algorithmic details concerning
the numerical solution of (V2)) show these considerable differences. Here, we would like to present a theoretical argument. Theorem 3 tries to compare the diameters of compact sets
into which the solutions are embedded by (V1) and (V2), respectively. It becomes clear that monotonicity substantially reduces that diameter and consequently the associated worst case
error.

Theorem 3: Let $\hat{x} = (\hat{x}_1, \hat{x}_2, \ldots, \hat{x}_n)^T$ *be a solution of (V1). Then
we have for* $\hat{a}_1 \neq 0$

$$0 \leq \hat{x}_1 \leq \frac{2 \, \| \hat{b} \|_2}{| \hat{a}_1 |} \; , \quad 0 \leq \hat{x}_j \leq \frac{2 \, \| \hat{b} \|_2 + \| \hat{a} \|_2 \left[\sum_{i=1}^{j-1} \hat{x}_i \right]}{| \hat{a}_1 |}$$

$$(\, j = 2, 3, \ldots, n \,) \tag{5.1}$$

□

On the other hand, for a solution of (V2) we find

$$0 \leq \hat{x}_i \leq \frac{2 \, \| \hat{b} \|_2}{| \hat{a}_1 |} \qquad (\, i = 1, 2, \ldots, n \,) \tag{5.2}$$

There is a tendency of error growth for the solution components of (V1) having large index i. By monotonicity and convexity this error growth is a priori avoided. Moreover, (V2)
has solutions which are not very sensitive to the chosen number n of discretization level.

    If we study the error expectation

$$E \left(\max_{1 \le i \le n} | \hat{x}_i - \bar{x}_i | \right) = e \left(\sigma_v, \sigma_p \right)$$

of (V2) depending on standard deviations σ_v and σ_p by Monte-Carlo simulation (cf. for similar experiments Stoyan, 1979), we obtain

$$e \left(0, \sigma_p \right) = \mathcal{O} \left(\sigma_p^{0.6} \right) \; , \; e \left(\sigma_v, 0 \right) = \mathcal{O} \left(\sigma_v^{0.6} \right) \tag{5.3}$$

This also shows an error reduction of (V2) compared to the solution of (3.6). From formula (4.4) we would derive

$$e_1 \left(0, \sigma_p \right) = \mathcal{O} \left(\sigma_p \right) \; , \; e_1 \left(\sigma_v, 0 \right) = \mathcal{O} \left(\sigma_v \right) \tag{5.4}$$

Finally, note that in the case of simultaneous data errors in p and v empirical results give an estimation of the form

$$\max \left(e \left(\sigma_v, 0 \right) + e \left(0, \sigma_p \right) \right) \le e \left(\sigma_v, \sigma_p \right) \le$$
$$\le e \left(\sigma_v, 0 \right) + e \left(0, \sigma_p \right).$$
$$\tag{5.5}$$

Solutions to (V2) may have piecewise linear or piecewise constant regions. If this shape of solutions is not desired, a Tikhonov regularization version

(V3) $\| \hat{A} x - \hat{b} \|_2 + \alpha \phi(x) \longrightarrow \min$, subject to $Gx \ge 0$

with small $\alpha > 0$ and smoothing term $\phi(x)$ may help to make the solution more attractive. An error reduction, however, cannot be ensured.

References

Berg, L., 1974: Operatorenrechnung Bd. 2 - Funktionentheoretische Methoden. Dt. Verlag d. Wissenschaften, Berlin.
Coats, K.H. et al., 1964: Determination of aquifer influence functions from field data. J. Petroleum Tech. 16, 1417 - 1424.
Gorenflo, R., 1979: Numerical treatment of Abel integral equa-

tions. Inverse and Improperly Posed Problems in Differential Equations. Akademie-Verlag, 125 - 133.

Hofmann, B., 1986: Regularization for Applied Inverse and Ill-Posed Problems (Teubner-Texte zur Mathematik Bd. 85), Teubner, Leipzig.

Hofmann, B., 1988: On the analysis of a particular Volterra-Stieltjes convolution integral equation. Z. Analysis und Anw. 7, 247 - 257.

Hofmann, B., R. Hausding and R. Wolke, 1990: Regularization of a Volterra integral equation by linear inequalities. Computing 43, 361 - 375.

Linz, P., 1985: Analytical and Numerical Methods of Volterra Equations. SIAM, Philadelphia.

Louis, A., 1988: Inverse und schlecht gestellte Probleme. Teubner, Stuttgart.

O'Sullivan, F., 1986: A statistical perspective on ill-posed inverse problems. Statistical Science 1, 502 - 527.

Rutman, R.S., 1988: Use of exogenous information for the regularization of the Fredholm inverse problem. Model Optimization in Exploration Geophysics 2. Vieweg, Braunschweig.

Stoyan, G., 1979: Identification of a spatially varying coefficient in a parabolic equation. Inverse and Improperly Posed Problems in Differential Equations. Akademie-Verlag, Berlin, 249 - 258.

Tikhonov, A.N. et al., 1983: Regularizing Algorithms and Apriori Information (in Russian), Nauka, Moscow.

Titterington, D.M., 1985: Common structure of smoothing techniques in statistics. Int. Statist. Review 53, 151 - 170.

Approximating the Solution to the Cauchy Problem and the Boundary Value for the Laplace Equation

T. D. Van[1], D. N. Hào[1], R. Gorenflo[2]

[1] Institute of Mathematics, Box 631, Bo Ho, Hanoi, Vietnam
[2] Freie Universität Berlin, Fachbereich Mathematik, Institut für Mathematik I, Research Group „Regularization", Arnimallee 2–6, D-1000 Berlin 33, Germany

Abstract. In this paper it is proved that the Cauchy problem and a boundary value problem for the Laplace equation is well-posed provided the data belong to a suitable function space. Explicit numerical procedures are described for approximating the solutions to these problems.

Introduction.

In recent decades great attention has been given to ill-posed problems for partial differential equations in general and to the Cauchy problem for the Laplace equation in particular. As it is well known, the solution of the last problem does not depend continuously on the data, that means the Cauchy problem for the Laplace equation is ill-posed in the sense of Hadamard. Nevertheless, considerable efforts have been devoted to studying various aspects of this ill-posed problem. (See [1-4] for an exhaustive bibliography).

In this paper we apply the theory of pseudodifferential operators with real analytic symbols ($\Psi DOAS$) to establish well-defined solvability of the Cauchy problem for the Laplace equation in the function space $W^{+\infty}_{R^N_\xi}(I\!\!R^N_x) = \{f \in A'(I\!\!R^N), \ supp\ \hat{f}\ is\ compact\}$ and in the space $W^{-\infty}_{R^N_\xi}(I\!\!R^N_x) \equiv (W^{+\infty}_{R^N_\xi}(I\!\!R^N_x))^*$ of generalized functions (see the exact definitions of $W^{\pm\infty}_{R^N_\xi}(I\!\!R^N_x)$ in Section 2 below). We also consider a boundary value problem for the Laplace equation in the strip $(a, b) \times I\!\!R^N_x$ when the boundary conditions belong to function spaces of finite smoothness. We give also approximation methods for solving these problems based on the technique of the $\Psi DOAS$ and on the approximation theory of functions of real variables.

The paper is organized as follows. In the first section we review the results of [5] about the function spaces $W^{\pm\infty}_{R^N_\xi}(I\!\!R^N_x)$ and the differential operators of infinite order (DOIO) which are local representative of $\Psi DOAS$ as far as they are needed in the present paper. In sections 2 and 3 we prove the existence and uniqueness of a solution of the Cauchy problem and of a boundary value problem for Laplace equations. Section 4 is devoted to approximation methods and illustrative examples. We remark also that in writing our paper we were greatly influenced by Dubinskii's works [6,7].

This paper was written during the stay of the first and second authors at the Free University of Berlin. They thank the Humboldt Foundation and the German Academic Exchange Service for supporting their stay in Berlin, the Mathematics Department of the Free University of Berlin for providing good working conditions. Thanks are also due to Mr. F. Haber who has carried out all numerical examples in this paper on computer.

1. The function spaces $W_{R_\xi^N}^{\pm\infty}(\mathbb{R}_x^N)$ and the pseudodifferential operators with real analytic symbols.

Let $x \in \mathbb{R}_x^N$, $N \geq 1$, and $\xi \in R_\xi^N$ be real variables, $D^\alpha = D_1^{\alpha_1} \ldots D_N^{\alpha_N}$, $D_j = -i\partial/\partial x_j$, $j = 1, \ldots, N$, $\alpha = (\alpha_1, \ldots, \alpha_N)$, $|\alpha| = \alpha_1 + \ldots + \alpha_N$. Assume that $f(x) : \mathbb{R}_x^N \to \mathbb{C}^1$, that is, $f(x)$ is a function defined on $\mathbb{R}_x^N$ taking complex values, in general.

Definition 1.1. *The space* $W_{R_\xi^N}^{+\infty}(\mathbb{R}_x^N)$ *is the set of functions* $f(x)$, $x = (x_1, \ldots, x_N)$, *satisfying the following conditions:*
$f(x)$ admits analytic continuation as an entire functions to $\mathbb{C}^N$ and for each $\varepsilon > 0$ there exist constants $r < +\infty$ and C_ε such that

$$|f(x + iy)| \leq C_\varepsilon \, exp(r|y| + \varepsilon|x|), \quad x + iy \in \mathbb{C}^N.$$

The numbers r and C_ε may depend on f.

We list here some classes of functions which belong to $W_{R_\xi^N}^{+\infty}(\mathbb{R}_x^N)$. They are: all functions $f(x) \in L_2(\mathbb{R}_x^N)$, the support of whose Fourier transform $\hat{f}(\xi)$ is compact [6], all functions in $\mathfrak{M}_{\nu p}$, $1 \leq p \leq +\infty$, $\nu < +\infty$ [8], all functions in $W^{+\infty}(\mathbb{R}^N)$ [9]. From the Paley-Wiener theorem it follows that a functions $f(x)$ belongs to $W_{R_\xi^N}^{+\infty}(\mathbb{R}_x^N)$ if and only if its analytic continuation $f(z)$ is the Fourier-Laplace transform of an analytic functional u with compact support, that is

$$\check{f} = u, \quad \hat{u} = f.$$

Definition 1.2. *A sequence of functions* $f_n(x) \in W_{R_\xi^N}^{+\infty}(\mathbb{R}_x^N)$ *is said to converge to $f(x) \in W_{R_\xi^N}^{+\infty}(\mathbb{R}_x^N)$ if and only if: For each $\varepsilon > 0$ there exists a constant $r < \infty$ such that (with $z = x + iy$, $x \in \mathbb{R}^N$, $y \in \mathbb{R}^N$)*

$$\sup_{z \in \mathbb{C}^N} |f_n(z) - f(z)| exp(-r|y| - \varepsilon|x|) \to 0, \quad n \to \infty.$$

We see by virtue of the Paley-Wiener theorem: The sequence (f_n) converges to $f(x)$ in $W_{R_\xi^N}^{+\infty}(\mathbb{R}_x^N)$ if and only if there exists a compact set $K \subset R_\xi^N$ such that $\check{f}_n \to \check{f}$ in $A'[K]$, where $A'[K]$ denotes the space of analytic functionals with support in K (Proposition 1.6 [5]).

Let now the entire function $A(\xi)$, $\xi \in R_\xi^N$, be expanded into the Taylor series

$$A(\xi) = \sum_{|\alpha|=0}^{\infty} a_\alpha \xi^\alpha, \quad a_\alpha = (iD)^\alpha A(0)/\alpha! \tag{1.1}$$

and assuming that this series converges for $\xi \in R_\xi^N$. We consider the action of the DOIO

$$A(D) = \sum_{|\alpha|=0}^{\infty} a_\alpha D^\alpha \tag{1.2}$$

in the space $W_{R_\xi^N}^{+\infty}(I\!R_x^N)$. From Theorem 2.1 in [5] we have: The DOIO (1.2) with the symbol (1.1) analytic in R_ξ^N acts invariantly and continuously in $W_{R_\xi^N}^{+\infty}(I\!R_x^N)$. Further we denote by $W_{R_\xi^N}^{-\infty}(I\!R_x^N)$ the space of all continuous linear functionals defined on $W_{R_\xi^N}^{+\infty}(I\!R_x^N)$. Let $h \in W_{R_\xi^N}^{-\infty}(I\!R_x^N)$ and let $A(D)$ be a DOIO whose symbol is analytic in R_ξ^N . Then

$$< A(D)h(x), \varphi(x) > \overset{def}{=} < h(x), A(-D)\varphi(x) >, \ \forall \varphi \in W_{R_\xi^N}^{+\infty}(I\!R_x^N) .$$

We call elements from $W_{R_\xi^N}^{-\infty}(I\!R_x^N)$ *generalized functions.* The structure theorem for genera-lized functions was established in Theorem 4.3 in [5]: Every generalized function $h \in W_{R_\xi^N}^{-\infty}(I\!R_x^N)$ can be represented in the form

$$h(x) = A(D)(2\sqrt{\pi})^{-N} exp(-x^2/4),$$

where $A(D)$ is a pseudodifferential operator with symbol $A(\xi) = \hat{h}(\xi) \cdot exp(\xi^2)$ analytic in R_ξ^N . As a consequence of Theorem 5.2 [5] we have

Theorem 1.1. *The set of operators $A(D)$ with symbols $A(\xi)$ analytic in the whole space R_ξ^N and defined on $W_{R_\xi^N}^{\pm\infty}(I\!R_x^N)$ forms an operator algebra which is isomorphic to the algebra of functions analytic in R_ξ^N . This isomorphism is defined by the correspondence $A(D) \leftrightarrow A(\xi)$. Here*

$$\alpha A(D) \pm \beta B(D) \ \leftrightarrow \ \alpha A(\xi) \pm \beta B(\xi), \ \alpha, \beta \in \mathbb{C},$$
$$A(D) \cdot B(D) \ \leftrightarrow \ A(\xi) \cdot B(\xi).$$

In particular, if $A^{-1}(\xi)$ is also analytic in R_ξ^N , then $B(D) = I/A(D)$ is the operator inverse to $A(D)$. For any function $A(\xi)$ analytic in R_ξ^N the maps

$$A(D) : \ W_{R_\xi^N}^{\pm\infty}(I\!R_x^N) \ \to W_{R_\xi^N}^{\pm\infty}(I\!R_x^N)$$

are continuous. Here " $+$ " corresponds to " $+$ " and " $-$ " to " $-$ ".

For later use we recall a few simple facts about approximation of functions of real variables. At first we give some notations. Suppose $h = (h_1, \ldots, h_n) \in I\!R^N$ is any vector and $f(x)$ is a function on $I\!R^N$. Let $\Delta_h^0 f = f$, $\Delta_h^1 f = \Delta_h f$,

$$\Delta_h f = \Delta_h f(x) = f(x + h) - f(x).$$

Then

$$\Delta_h^k f = \Delta_h^k f(x) = \Delta_h \Delta_h^{k-1} f(x).$$

Let h be a unit vector. Denote by

$$w^k(\delta) \ = \ w_h^k(f, \delta) \ = \ \sup_{|t| \le \delta} \| \Delta_{th}^k f(x) \|_{L_p(R^N)},$$
$$w(\delta) \ = \ w_h(f, \delta) \ = \ w_h^1(f, \delta).$$

The expression $w_h^k(f, \delta)$ is called the modulus of continuity of order k of the function f in the metric of L_p along the direction h. It is well known that if $f \in L_p(I\!R^N)$ and $1 \le p < \infty$ then

$$\lim_{\delta \to 0} w(\delta) = 0.$$

Let

$$\Omega_{I\!R^N}^k(f, \delta)_p = \sup \{w_h^k(f, \delta)_p | \ h \in I\!R^N, \ |h| = 1\}.$$

Then we have

Proposition 1.1. ([8],§5.2). *Suppose f has derivatives of order s, $|s| \le \rho$, lying in $L_p(I\!R^N)$, $k = l - \rho$. Then there exists an entire function f_ν of spherical type ν such that*

$$\|D^s f_\nu - D^s f\|_{L_p(R^N)} \le \frac{1}{\nu^{\rho-|s|}} \sum_{|s|=\rho} \Omega_{R^N}^k(D^s f, \ 1/\nu).$$

Proposition 1.2. *Let $f \in C^m(I\!R^N)$. Then for any compact set $K \subset I\!R^N$ there exists a sequence of entire functions g_{ν_k} of exponential type ν_k such that*

$$\|f - g_{\nu_k}\|_{C^m(K)} \to 0 \ as \ k \to \infty.$$

These facts give us a possibility to use the $\Psi DOAS$ and $W_{R_\xi^N}^{\pm \infty}(I\!R_x^N)$ -space technique to approximate the solution of the boundary value problem for the Laplace equation in the space of finite smoothness.

2. The Cauchy Problem.

In the halfspace $I\!R_+^{N+1} = \{(t, x), \ t > 0, \ x \in I\!R^N\}$, we consider the Cauchy problem for the Laplace equation

$$\frac{\partial^2 u}{\partial t^2} + \Delta u = 0, \tag{2.1}$$

$$u(0, x) = \varphi_1(x), \tag{2.2}$$

$$\frac{\partial u}{\partial t}(0, x) = \varphi_2(x), \tag{2.3}$$

where $\Delta u = \partial^2 u/\partial x_1^2 + \ldots + \partial^2 u/\partial x_N^2$, $\varphi_1(x)$ and $\varphi_2(x)$ are given functions. The symbol of Δ is $-|\xi|^2 = -\sum_{j=1}^{N} \xi_j^2$.

To solve this problem we set $p = \sqrt{\Delta}$ and, regarding p as a parameter, (2.1) implies

$$\frac{\partial^2 u}{\partial t^2} + p^2 u = 0.$$

Consequently,

$$u(t, x) = e^{itp} c_1(x) + e^{-itp} c_2(x),$$

where $c_1(x)$, $c_2(x)$ are arbitrary functions. These functions are determined by the conditions (2.2) and (2.3), which give the following system of algebraic equations:

$$c_1(x) + c_2(x) = \varphi_1(x), \qquad (2.4)$$
$$ip\, c_1(x) - ip\, c_2(x) = \varphi_2(x). \qquad (2.5)$$

Solving (2.4) and (2.5), we see that $u(t,x)$ is expressed by the formula

$$u(t,x) = \frac{e^{itp} + e^{-itp}}{2}\varphi_1(x) + \frac{e^{itp} - e^{-itp}}{2ip}\varphi_2(x).$$

Replacing in the last formula the symbol p by $\sqrt{\Delta}$ we find that the desired solution has the form

$$u(t,x) = \frac{e^{-it\sqrt{\Delta}} + e^{it\sqrt{\Delta}}}{2}\varphi_1(x) + \frac{e^{it\sqrt{\Delta}} - e^{-it\sqrt{\Delta}}}{2i\sqrt{\Delta}}\varphi_2(x),$$

or

$$u(t,x) = \cos(t\sqrt{\Delta}\,)\varphi_1(x) + \frac{\sin(t\sqrt{\Delta}\,)}{\sqrt{\Delta}}\varphi_2(x)\ . \qquad (2.6)$$

Let us elucidate what the Cauchy data $\varphi_1(x)$ and $\varphi_2(x)$ should be in order that formula (2.6) has a nonformal sense. We see that the operators $\cos(t\sqrt{\Delta}\,)$ and $\sin(t\sqrt{\Delta}\,)/\sqrt{\Delta}$ have the symbols $\cos(it|\xi|)$ and $\sin(it|\xi|)/(it|\xi|)$, analytic in R_ξ^N for any $t > 0$, and we have the representations

$$\cos(t\sqrt{\Delta}\,) \equiv \sum_{n=0}^{\infty} \frac{(-1)^n t^{2n}\Delta^n}{(2n)!}, \qquad (2.7)$$

$$\frac{\sin(t\sqrt{\Delta}\,)}{\sqrt{\Delta}} \equiv \sum_{n=0}^{\infty} \frac{(-1)^{n-1} t^{2n-1}\Delta^{n-1}}{(2n-1)!}. \qquad (2.8)$$

According to Theorem 1.1 the operators (2.7) and (2.8) act invariantly and continuously in the space $W_{R_\xi^N}^{\pm\infty}(I\!R_x^N)$ for any $t > 0$. Thus, for any initial data $\varphi_1(x)$, $\varphi_2(x)$ from $W_{R_\xi^N}^{+\infty}(I\!R_x^N)$ the Cauchy problem (2.1)-(2.3) is well-posed, i.e. for any fixed $t > 0$, the change in $u(t,x)$ is small in the sense of the space $W_{R_\xi^N}^{+\infty}(I\!R_x^N)$, if the changes in the initial functions $\varphi_1(x)$ and $\varphi_2(x)$ are small enough. In the particular case $N = 1$, taking into account that any function $\varphi \in W_{R_\xi^N}^{+\infty}(I\!R_x^N)$ is an entire function and the operator $exp\,(it\partial/\partial x)$ acts as a translation operator:

$$e^{it\partial/\partial x}\varphi_1(x) = \varphi_1(x + it),$$

and putting

$$(\partial/\partial x)^{-1}\varphi_2(x) = \int_0^x \varphi_2(\xi)d\xi + c,$$

we get

$$u(t,x) = \frac{\varphi_1(x+it) + \varphi_1(x-it)}{2} + \frac{1}{2i}\int_{x-it}^{x+it} \varphi_2(\xi)d\xi.$$

As a consequence of Theorem 1.1 we obtain

Theorem 2.1. *Let $\varphi_1(x)$, $\varphi_2(x) \in W_{R_\xi^N}^{\pm\infty}(I\!R_x^N)$. Then there exists a unique solution of the Cauchy problem (2.1)-(2.3) in the space $C^2(I\!R_+^1, W_{R_\xi^N}^{\pm\infty}(I\!R_x^N))$, and*

$$u(t,x) = \sum_{n=0}^{\infty} \frac{(-1)^n t^{2n} \Delta^n \varphi_1(x)}{(2n)!} + \sum_{n=1}^{\infty} \frac{(-1)^{n-1} t^{2n-1} \Delta^{n-1} \varphi_2(x)}{(2n-1)!} \ . \tag{2.9}$$

For any $t > 0$, the right-hand of (2.9) converges in the sense of $W_{R_\xi^N}^{\pm\infty}(I\!R_x^N)$.

3. A boundary value problem.

Let $G = (a,b) \times I\!R^N$ be a strip of the variables (t,x). We consider the following boundary value problem

$$\frac{\partial^2 u}{\partial t^2} + \Delta u = 0 \in G, \tag{3.1}$$

$$u(a,x) = \varphi_1(x), \tag{3.2}$$

$$\frac{\partial u}{\partial t}(b,x) = \varphi_2(x), \tag{3.3}$$

where $\varphi_1(x), \varphi_2(x)$ are known functions. Applying the method from §8 [5], setting $p^2 = -\Delta$ we consider two boundary value problems for the ordinary differential equation with parameter p,

$$v_j''(t,p) - p^2 v_j(t,p) = 0, \ j = 1,2,$$

under the conditions

$$v_1(a,p) = 1, \ v_1'(b,p) = 0;$$

$$v_2(a,p) = 0, \ v_2'(b,p) = 1.$$

After elementary calculations we get the formula

$$v_1(t,p) \ = \ \frac{e^{(t-b)p} + e^{-(t-b)p}}{e^{(b-a)p} + e^{-(b-a)p}},$$

$$v_2(t,p) \ = \ \frac{e^{(t-a)p} - e^{-(t-b)p}}{p(e^{(b-a)p} + e^{-(b-a)p})}.$$

Thus the solution of the problem (3.1)-(3.3) takes the form

$$u(t,x) \ = \ V_1(t, -i\sqrt{\Delta})\varphi_1(x) + V_2(t, -i\sqrt{\Delta})\varphi_2(x)$$

$$= \ \frac{\cos[(t-b)\sqrt{\Delta}]}{\cos[(b-a)\sqrt{\Delta}]}\varphi_1(x) + \frac{\sin[(t-a)\sqrt{\Delta}]}{\sqrt{\Delta}\cos[(b-a)\sqrt{\Delta}]}\varphi_2(x), \tag{3.4}$$

where $v_1(t,p)$ and $v_2(t,p)$ are the symbols of the operators V_1 and V_2, respectively. Furthermore, we can write (3.4) in the form

$$u(t,x) = \cos[(t-b)\sqrt{\Delta}]u_1(x) + \frac{\sin[(t-a)\sqrt{\Delta}]}{\sqrt{\Delta}}u_2(x), \ x \in I\!R^N, \tag{3.5}$$

where the functions $u_1(x), u_2(x)$ are the solutions of the equations

$$\cos[(b-a)\sqrt{\Delta}\,]u_j(x) = \varphi_j(x), \ j = 1, 2, \ x \in \mathbb{R}^N, \tag{3.6}$$

or, equivalently,

$$
\begin{aligned}
u(t,x) &= \sum_{n=0}^{\infty} \frac{(-1)^n (t-b)^{2n}\sqrt{\Delta}^{\,n}\, u_1(x)}{(2n)!} + \\
&+ \sum_{n=0}^{\infty} \frac{(-1)^{n-1}(t-a)^{2n-1}\sqrt{\Delta}^{\,n}\, u_2(x)}{(2n-1)!},
\end{aligned}
\tag{3.7}
$$

$$\sum_{n=0}^{\infty} \frac{(-1)^n (b-a)^{2n}\sqrt{\Delta}^{\,n}\, u_j(x)}{(2n)!} = \varphi_j(x), \ j = 1, 2, \ x \in \mathbb{R}^N. \tag{3.8}$$

Let us consider equations (3.8). The symbol of the operator in the left side of (3.8) is analytic in the whole space R_ξ^N and is not equal to zero. Therefore, from Theorem 1.1 we have, if the boundary value functions $\varphi_j(x)$ belong to $W_{R_\xi^N}^{\pm\infty}(\mathbb{R}^N)$, that there exists a unique solution $u(t,x)$ of the problem (3.1)-(3.3) in the space $C^2((a,b), W_{R_\xi^N}^{\pm\infty}(\mathbb{R}^N_x))$.

Let now $\varphi_1(x), \varphi_2(x)$ belong to $L_2(\mathbb{R}^N_x)$. It is easy to see that $W_{R_\xi^N}^{\pm\infty}(\mathbb{R}^N_x) \cap L_2(\mathbb{R}^N_x)$ is dense in $L_2(\mathbb{R}^N_x)$. In this case the formulas (3.5), (3.6) (or (3.7),(3.8)) give a classical solution of the boundary value problem (3.1)-(3.3).

4. Approximation methods.

In this section we shall use our general approximation methods described in [10] to solve the concrete problem (2.1)-(2.2). In explicit form we have

$$
\begin{aligned}
u(t,x) &= \sum_{n=0}^{\infty} \frac{(-2)^n (t)^{2n}\sqrt{\Delta}^{\,n}\, \varphi_1(x)}{(2n)!} + \\
&= \cos(t\sqrt{\Delta}\,)\varphi_1(x) + \frac{\sin(t\sqrt{\Delta}\,)}{\sqrt{\Delta}}\varphi_2(x).
\end{aligned}
\tag{4.1}
$$

The operator $\cos(t\sqrt{\Delta}\,)$ has the symbol $\cos(it|\xi|) = \cosh(t\xi)$, whereas the operator $\sin(t\sqrt{\Delta}\,)/\sqrt{\Delta}$ has the symbol $\sin(it|\xi|)/i|\xi| = \sinh(t\xi)/\xi$. Now, following our general scheme [10], we approximate these symbols by Taylor series.

4.1. Approximating the symbols by Taylor series.

Theorem 4.1. *Let $\varphi_1(x), \varphi_2(x)$ be elements of $W_{R_\xi^N}^{+\infty}(I\!\!R_x^N)$. Then*

$$u_m(t,x) = \sum_{n=0}^{m} \frac{(-1)^n (t)^{2n} \sqrt{\Delta}^{\,n}}{(2n)!} \varphi_1(x) + \sum_{n=1}^{m} \frac{(-1)^{n-1} (t)^{2n-1} \sqrt{\Delta}^{\,n-1}}{(2n-1)!} \varphi_2(x) \tag{4.2}$$

converges to $u(t,x)$ in the $W_{R_\xi^N}^{+\infty}(I\!\!R_x^N)$ -topology. Furthermore, if $\varphi_1(x) \in \mathfrak{M}_{\nu_1 p}(I\!\!R_x^N)$, $\varphi_2(x) \in \mathfrak{M}_{\nu_2 p}(I\!\!R_x^N)$, where $\nu_1 = (\nu_{11}, \ldots, \nu_{1N})$, $\nu_2 = (\nu_{21}, \ldots, \nu_{2N})$, ν_{1k}, $\nu_{2k} > 0$, then

$$\|u_m(t, \cdot) - u(t, \cdot)\|_p \leq \frac{(t\bar{\nu}_1)^{2m+2}}{(2m+2)!} \cosh(t\bar{\nu}_1) \|\varphi_1\|_p$$

$$+ \frac{(t\bar{\nu}_2)^{2m+1}}{(2m+1)!} \frac{\sinh(t\bar{\nu}_2)}{\nu_2} \|\varphi_2\|_p, \tag{4.3}$$

where $\bar{\nu}_1 = \nu_{11} \ldots \nu_{1N}$, $\bar{\nu}_2 = \nu_{21} \ldots \nu_{2N}$.

Proof: See [10].

From this theorem we see that if φ_1 and φ_2 are algebraic or trigonometric polynomials, then $u_m(t, x)$ is very easily to be calculated and so $u_m(t, x)$ would be a good and comfortable approximation to $u(t, x)$.

Numerical example. Consider the following problem in $I\!\!R_+^2 = \{(t, x), \ t > 0, \ x \in I\!\!R\}$:

$$\begin{aligned} u_{tt} + u_{xx} &= 0, \\ u(0, x) &= \varphi_1(x) = \cos(\omega x), \\ u_t(0, x) &= 0, \end{aligned} \tag{4.4}$$

where $\omega > 0$. It is clear that

$$\begin{aligned} u(t, x) &= \sum_{n=0}^{\infty} \frac{(-1)^n t^{2n}}{(2n)!} \frac{d^{2n} \cos(\omega x)}{dx^{2n}} \\ &= \sum_{n=0}^{\infty} \frac{t^{2n} \omega^{2n}}{(2n)!} \cos(\omega x) \\ &= \cosh(tw) \cos(\omega x). \end{aligned} \tag{4.5}$$

Taking

$$u_k(t, x) = \sum_{n=0}^{m} \frac{(-1)^n t^{2n}}{(2n)!} \frac{d^{2n} \cos(\omega x)}{dw^{2n}} \tag{4.6}$$

as an approximation to $u(t, x)$, then from Theorem 4.1 we obtain

$$\|u_m(t, \cdot) - u(t, \cdot)\|_{L_\infty} \leq \frac{(tw)^{2m+2}}{(2m-2)!} \cosh(tw). \tag{4.7}$$

Here we present error estimates for $t = 1$ and various ω and m.

ω \\ k	0,1	0,2	0,3	0,4	0,5
1	$4.2E-6$	$6.8E-5$	$3.5E-4$	$1.2E-3$	$2.9E-3$
2	$1.4E-9$	$9.1E-8$	$1.1E-6$	$6.2E-6$	$2.4E-3$
3	$2.5E-13$	$6.5E-11$	$1.7E-9$	$1.8E-8$	$1.1E-7$
4	$2.8E-17$	$2.9E-14$	$1.7E-12$	$3.1E-11$	$3.0E-10$
5	$2.1E-21$	$8.7E-18$	$1.2E-15$	$3.8E-14$	$5.7E-13$
6	$1.2E-25$	$1.9E-21$	$5.7E-19$	$3.3E-17$	$7.9E-16$
7	$4.8E-30$	$3.2E-25$	$2.2E-22$	$2.2E-20$	$8.2E-19$
8	$1.6E-34$	$4.2E-29$	$6.3E-26$	$1.2E-23$	$6.7E-22$
9	$4.1E-39$	$4.4E-33$	$1.5E-29$	$4.9E-27$	$4.4E-25$
10	$8.9E-44$	$3.8E-37$	$2.9E-33$	$1.7E-30$	$2.4E-28$

ω \\ k	1	2	3	4	5
1	$6.4E-2$	$2.5E0$	$3.4E1$	$2.9E2$	$1.9E3$
2	$2.1E-3$	$3.3E-1$	$1.0E1$	$1.6E2$	$1.6E3$
3	$3.8E-5$	$2.4E-2$	$1.6E0$	$4.4E1$	$7.2E2$
4	$4.3E-7$	$1.1E-3$	$1.6E-1$	$7.9E0$	$2.0E2$
5	$3.2E-9$	$3.2E-5$	$1.1E-2$	$9.6E-1$	$3.8E1$
6	$1.8E-11$	$7.1E-7$	$5.5E-4$	$8.4E-2$	$5.2E0$
7	$7.4E-14$	$1.2E-8$	$2.1E-5$	$5.6E-3$	$5.4E-1$
8	$2.4E-16$	$1.5E-10$	$6.1E-7$	$2.9E-4$	$4.4E-2$
9	$6.3E-19$	$1.6E-12$	$1.4E-8$	$1.2E-5$	$2.9E-3$
10	$1.4E-21$	$1.4E-14$	$2.8E-10$	$4.3E-7$	$1.6E-4$

Table 1: Values of $\frac{(t\omega)^{2k+2}}{(2k+2)!}\cosh\omega$ for $t = 1$ and various ω and k.

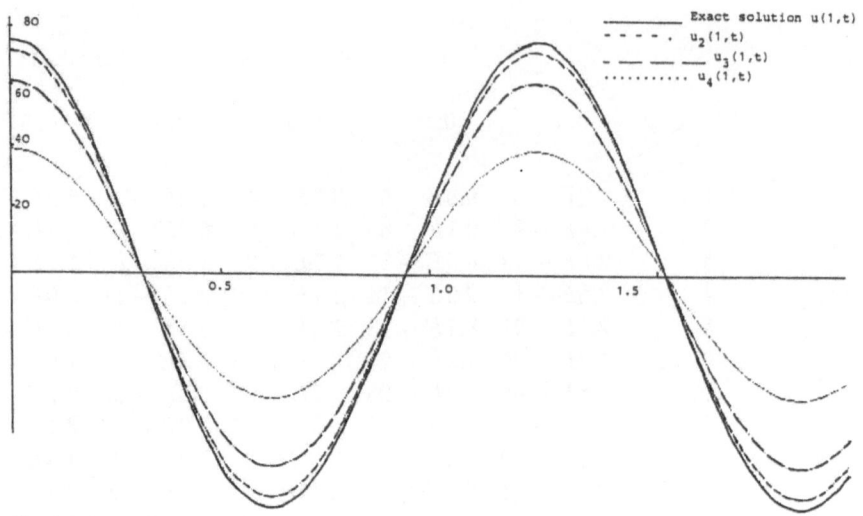

Fig. 1 Approximating the symbols by Taylor series.
Numerical example (4.4) for $\omega = 5$, $t \doteq 1$, $k = 2, 3, 4$.

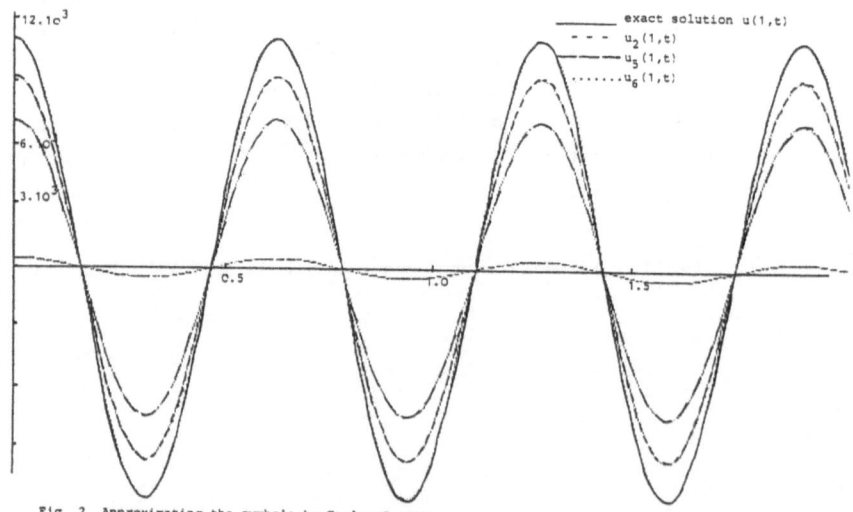

Fig. 2 Approximating the symbols by Taylor Series.
Numerical example (4.4) $\omega = 10$, $k = 2, 5, 6$.

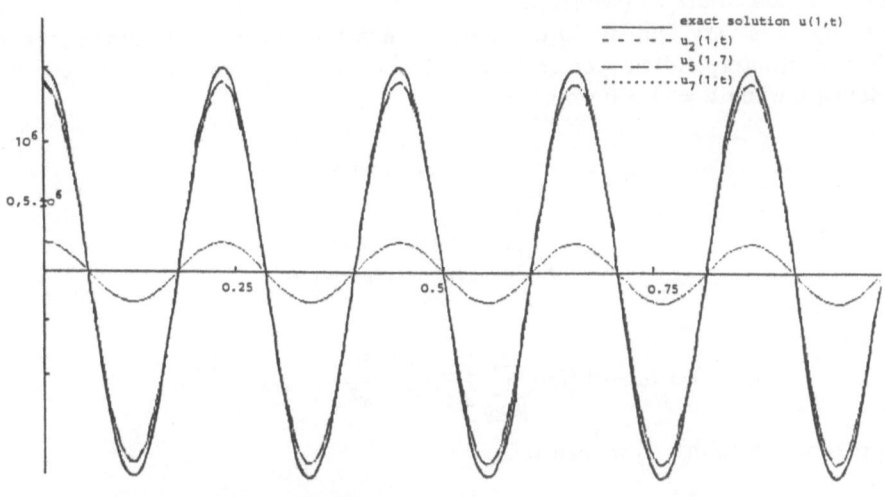

Fig. 3 Approximating the symbols by Taylor series
Numerical example (4.4) for ω = 15, k = 2, 5, 7.

4.2. Approximating the symbols by trigonometric polynomials.

For the sake of simplicity, let $\varphi_2 \equiv 0$. Then $u(t,x) = \cos(t\sqrt{\Delta}\,)\varphi_1(x)$. Let $\varphi_1(x) \in \mathfrak{M}_{\nu p}(\mathbb{R}^N)$, where $\nu = (\nu_1, \nu_2, \ldots, \nu_N)$, $\nu_k > 0$. The operator $\cos(t\sqrt{\Delta}\,)$ has a symbol $\cosh(t\xi)$ which can be expanded in Fourier series in $\Box_\nu = \{\xi \mid |\xi_i| \leq \nu_i,\ i = 1, \ldots, N\}$,

$$\cosh(t\xi) = \sum_{|k| \geq 0} A_k e^{-i\pi \cdot k \cdot \frac{\xi}{\nu}}, \tag{4.8}$$

where $\frac{\xi}{\nu} = (\frac{\xi_1}{\nu_1}, \ldots, \frac{\xi_N}{\nu_N})$, $k = (k_1, \ldots, k_N)$,

$$A_k = \frac{1}{2^N \nu_1 \ldots \nu_N} \int_{\Box_\nu} \cosh(t\xi) e^{-i\pi \cdot k \cdot \frac{\xi}{\nu}} d\xi. \tag{4.9}$$

It is easily seen that the series (4.8) converges absolutely. Now we have

$$u(t,x) = \cos(t\sqrt{\Delta}\,)\varphi_1(x)$$

$$= \sum_{|k| \geq 0} A_k \varphi_1(x + \frac{kx}{\nu}), \tag{4.10}$$

and so we can take

$$u_m(t,x) = \sum_{m \geq |k| \geq 0} A_k \varphi_1(x + \frac{kx}{\nu}) \tag{4.11}$$

as an approximation to $u(t, x)$ (see [10]).

Generally, it is very difficult to calculate A_k exactly and for multidimensional cases one should use an interpolation process as in [10] to approximate $\cosh(t\xi)$, but for the one-dimensional case when $\nu = \omega > 0$ we have

$$A_k = \frac{1}{2\omega} \int_{-\omega}^{\omega} \cosh(t\xi) e^{-i\pi \cdot k \cdot \frac{\xi}{\omega}} d\xi$$

$$= (-1)^{|k|} \frac{t\omega \cosh(t\omega)}{2(t^2\omega^2 + k^2\pi^2)}.$$

Thus,

$$u(t, x) = t\omega \cosh(t\omega) \sum_{|k| \geq 0} \frac{(-1)^{|k|}}{2(t^2\omega^2 + k^2\pi^2)} \varphi_1\left(x - \frac{k\pi}{\omega}\right).$$

As an approximation to $u(t, x)$ we can take

$$u_m(t, x) = t\omega \cosh(t\omega) \sum_{m \geq |k| \geq 0} \frac{(-1)^{|k|}}{2(t^2\omega^2 + k^2\pi^2)} \varphi_1\left(x + \frac{k\pi}{\omega}\right).$$

Numerical example: We consider again problem (4.4), and here we draw some pictures for various ω and m.

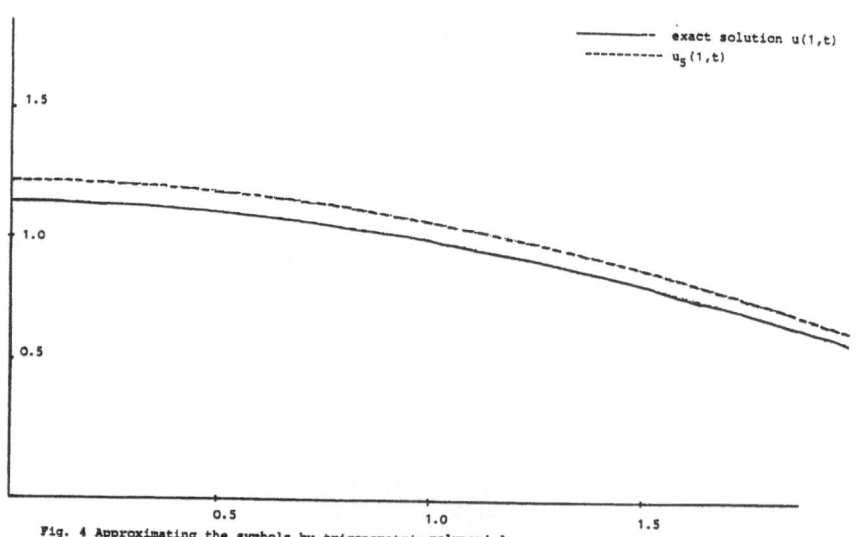

Fig. 4 Approximating the symbols by trigonometric polynomials
Numerical example (4.4) for $\omega = 0,5$, $m = 5$.

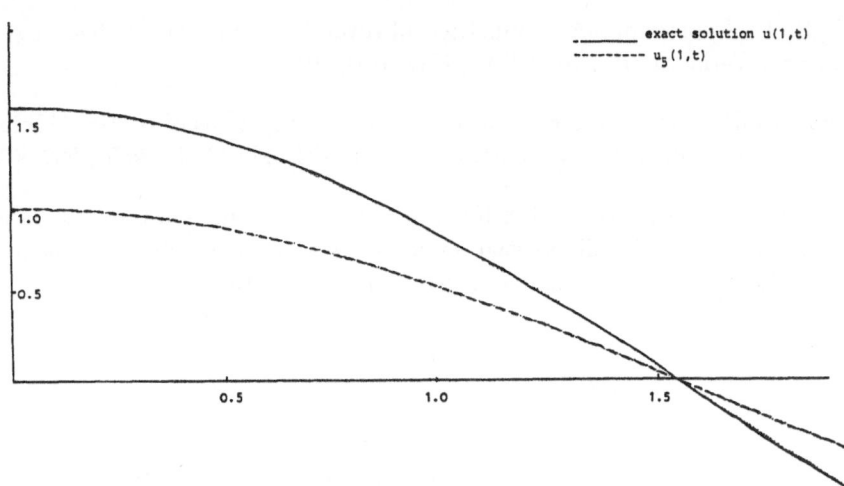

Fig. 5 Approximating the symbols by trigonometric polynomials
Numerical example (4.4) ω = 1, m = 5.

References

1. Tikhonov, A.N., Arsenin, V.Y.: Solutions of ill-posed problems. Winston-Wiley, Washington, 1977.

2. Lavrentev, M.M., Romanov, V.G., Shishatskii, S.P.: Ill-posed problems in mathematical physics and analysis. Transl. of Math. Monographs, Vol. 64, AMS, Providence, Rhode Island, 1986.

3. Cannon, J.R., DuChateau, P.: Approximating the solution to the Cauchy problem for Laplace's equation. SIAM J. Numer. Anal., 14, 3 (1977), 473-483.

4. Medeiros, L.A.: Remarks on a non-well posed problem. Proceedings of the Royal Society of Edinburgh, 102A (1986), 131-140.

5. Tran Duc Van: On the pseudodifferential operators with real analytic symbols and their applications. J. Fac. Sci. Univ. Tokyo, IA Mathematics, 36, 3 (1989), 803-825.

6. Dubinskii, Yu.A.: The algebra of pseudodifferential operators with analytic symbols and its applications to mathematical physics. Russian Math. Surveys, 37 (1982), 109-153.

7. Dubinskii, Yu.A.: Sobolev spaces of infinite order and differential equations. Teubner-Texte zur Mathematik, Bd. 87, Leipzig, 1986.

8. Nikolskii, S.M.: Approximation of functions of several variables and imbedding theorems. Springer-Verlag, Berlin-Heidelberg-New York, 1975.

9. Trinh Ngoc Minh, Tran Duc Van: Cauchy problems for systems of partial differential equations with a distinguished variable. Soviet Math. Dokl. 32, 2 (1985), 562-565.

10. Tran Duc Van, Dinh Nho Hào, Trinh Ngoc Minh, Gorenflo, R.: On the Cauchy problems for systems of partial differential equations with a distinguished variable. To appear in J. "Numerical Functional Analysis and Optimization."

On Numerical Integration Differentiation of Fractional Order: a Systems Theory Approach

R. S. Rutman

Department of Electrical and Computer Engineering, Southeastern Massachusetts University, North Dartmouth, MA 02747, USA

Abstract

Inversion of the fractional integration and of the Abel transform require fractional differentiation. In the introduction to the paper, fractional integration is presented in an informal way as a dynamical system. Then the detailed dynamical models are presented for all the operations involved, on the basis of a developed inversion method. In each case the model reduces to a linear filter combined with trivial transformations. Regularization of inverse operations is analysed in the frequency domain. Numerical examples are presented with the comparison to the existing methods.

1. INTRODUCTION

Consider n cascaded integrators as shown in Figure 1

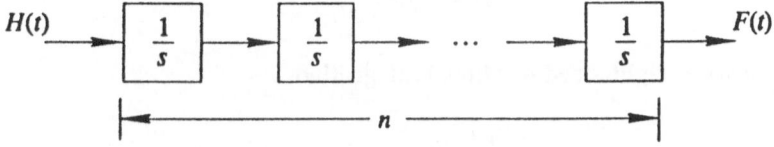

Figure 1. Cascaded integrators.

where $1/s$ stands for the transfer function of an individual integrator. Thus, the impulse response of the cascade is:

$$g(t) = \mathcal{L}^{-1}\left\{\frac{1}{s^n}\right\} = \begin{cases} \dfrac{t^{n-1}}{(n-1)!} & , t \geq 0 \\ \\ 0 & , t < 0 \end{cases} \qquad (1)$$

Replacing the cascade by a single block with the impulse response $g(t)$ (whose Laplace transform is denoted by $\hat{g}(s)$ in Figure 2), we get a causal convolution formula,

$$F(t) = \int_0^t g(t - \tau)H(t)d\tau = \frac{1}{(n-1)!}\int_0^t (t - \tau)^{n-1}H(t)dt \qquad (2)$$

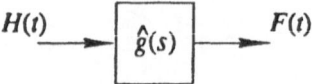

$$H(t) \longrightarrow \boxed{\hat{g}(s)} \longrightarrow F(t)$$

Figure 2.

The above is a systems theory interpretation of the well-known representation for the n-fold integration, which is presented below in a more general form:

$$\underbrace{\int_a^t dt \int_a^t dt \dots \int_a^t H(t)dt}_{n} = \frac{1}{(n-1)!}\int_a^t (t - \tau)^{n-1}H(t)dt \qquad (3)$$

Since $(n-1)!$ can be replaced by the gamma function $\Gamma(n)$, the right side of equation (3) may also be defined for non-integer n. This reasoning attaches rationality to the definitions of non-integer (traditionally called **fractional**) **left-sided integration** (Oldham, Spanier, 1974; Samko et al., 1983),

$$(\mathcal{I}_{a+}^\alpha H)(x) := \frac{1}{\Gamma(\alpha)}\int_a^x \frac{H(r)}{(x - r)^{1-\alpha}}dr = F(x), \quad x > a \qquad (4)$$

and, by extension, **right-sided fractional integration**

$$(\mathcal{I}_{b-}^\alpha H)(x) := \frac{1}{\Gamma(\alpha)}\int_x^b \frac{H(r)}{(r - x)^{1-\alpha}}dr = F(x), \quad x < b \qquad (5)$$

of order α, $\alpha > 0$. In a more rigorous setting, the class of admissible functions $H(r)$ can be defined as $L_1(a, b)$, i.e.

$$\int_a^b |H(r)|\,dr < \infty$$

Operators inverse to $\mathscr{I}^{\alpha}_{a+}$ and $\mathscr{I}^{\alpha}_{b-}$ define **fractional differentiation** of order α, $0 < \alpha < 1$:

left-sided

$$(\mathscr{D}^{\alpha}_{a+}F)(r) = \frac{1}{\Gamma(1-\alpha)} \frac{d}{dr} \int_a^r \frac{F(x)}{(r-x)^{\alpha}} dx = H(r) \qquad (6)$$

and **right-sided**

$$(\mathscr{D}^{\alpha}_{b-}F)(r) = -\frac{1}{\Gamma(1-\alpha)} \frac{d}{dr} \int_r^b \frac{F(x)}{(x-r)^{\alpha}} dx = H(r) \qquad (7)$$

Assuming $F(x)$ to be differentiable, we can write

$$(\mathscr{D}^{\alpha}_{a+}F)(r) = \frac{1}{\Gamma(1-\alpha)} \left[\frac{F(a)}{(r-a)^{\alpha}} + \int_a^r \frac{F'(x)dx}{(r-x)^{\alpha}} \right] \qquad (8)$$

$$(\mathscr{D}^{\alpha}_{b-}F)(r) = -\frac{1}{\Gamma(1-\alpha)} \left[\frac{F(b)}{(b-r)^{\alpha}} - \int_r^b \frac{F'(x)dx}{(x-r)^{\alpha}} \right] \qquad (9)$$

Extension of fractional differentiation to $\alpha > 1$ is straightforward.

The abbreviations FI and FD will hereafter denote fractional integration and differentiation, and the letters L and R will mean left-sided or right-sided, respectively.

We will consider formulas (8) and (9) (for $F(a) = 0$ or $F(b) = 0$) as the basic formulas of fractional differentiation, and (6) - (7) as **alternative** formulas, which will be reflected in the notations as FDLA and FDRA.

Remarks:

1. Fractional integrals and derivatives as defined above are also called **Riemann - Liouville** fractional integrals and derivatives.

2. Left-sided fractional integration and differentiation are represented by causal, and right-sided by anticausal, convolution integrals.

3. Numerous important applications of fractional integrals and derivatives in various fields of physics, biology, astronomy, medicine, etc. are known. Some are referred to in Gorenflo, Vessella, 1990.

4. Some numerical methods available to date for FI and FD can be found in Oldham, Spanier, 1974, and Gorenflo, 1979.

By the techniques similar to those developed in Rutman, Estes, 1989 and Rutman, 1989 (see Appendix 1) and utilizing the above relations, we reduce formulas (4) - (9) of fractional integration/differentiation essentially to a **filtering** process:

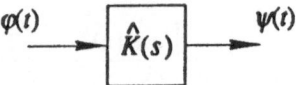

Figure 3. Fractional Integration/Differentiation as a Filtering Process.

a) The input, $\varphi(t)$, of the filter is formed by replacing the physical variable (r for FI and x for FD) by the corresponding substitutions (10) - (13) in the function $\Phi(r)$ or $\Phi(x)$ in Table 1, column a.

b) The transfer function, $\hat{K}(s)$, of the filter is taken from column b of Table 1. In practical realization, the truncated transfer functions, $\hat{k}_{n,1-\alpha}(s)$ or $\hat{k}_{n,\alpha}(s)$ are used. See formulas (20) and (21).

c) The output of the filter, $\psi(t)$, is returned into the physical domain r or x, for FD or FI respectively, by the corresponding substitution inverse to (10) - (13), and adjusted to produce the final result in column c, Table 1.

Table 1.

	a) Input $\Phi(r)$ or $\Phi(x)$	b) Transfer function of the filter, $\hat{K}(s)$	c) Solution
FIL	$\Phi(r) = \dfrac{1}{\Gamma(\alpha)} H(r) \cdot r$	$\hat{k}_{1-\alpha}(s)$	$F(x) = \psi(t) x^{\alpha-1}$
FDL	$\Phi(z) = \dfrac{1}{\Gamma(1-\alpha)} F(x)$	$s\,\hat{k}_\alpha(s)$	$H(r) = \psi(t) r^{-\alpha}$
FDLA	$\Phi(x) = \dfrac{1}{\Gamma(1-\alpha)} F(x) \cdot x$	$(s - \alpha)\hat{k}_\alpha(s)$	$H(r) = \psi(t) \cdot r^{-(\alpha+1)}$
FIR	$\Phi(r) = \dfrac{1}{\Gamma(\alpha)} H(r) r^\alpha$	$\hat{k}_{1-\alpha}(s)$	$F(x) = \psi(t)$
FDR	$\Phi(x) = \dfrac{1}{\Gamma(1-\alpha)} \dfrac{F(x)}{x^\alpha}$	$(s - \alpha)\hat{k}_\alpha(s)$	$H(r) = \psi(t)$
FDRA	$\Phi(x) = -\dfrac{1}{\Gamma(1-\alpha)} F(x) x^{1-\alpha}$	$s\,\hat{k}_\alpha(s)$	$H(r) = \psi(t) \cdot \dfrac{1}{r}$

Table 1. Summary of relations for fractional integration/differentiation.

2. SYSTEMS THEORY MODELS

The method for representing Abel-type integrals as systems theory models (filters) has been developed in Rutman, Estes, 1989; Rutman, 1989; Rutman 1990. Here, the results and computer simulations are presented in detail as applied to integrals (4) - (9). For a detailed derivation of each case, see Appendix 1. The essential variable transformations are:

$$\text{FIL} \qquad\qquad x = ae^{t} \quad r = ae^{\tau} \qquad\qquad (10)$$

$$\text{FDL and FDLA} \qquad x = ae^{\tau} \quad r = ae^{t} \qquad\qquad (11)$$

$$\text{FIR} \qquad\qquad x = be^{-t} \quad r = be^{-\tau} \qquad\qquad (12)$$

$$\text{FDR and FDRA} \qquad x = be^{-\tau} \quad r = be^{-t} \qquad\qquad (13)$$

Two major reference formulas utilized for the derivation are:

$$\int_{0}^{\infty} \left(1 - e^{-\frac{x}{\beta}}\right)^{\nu-1} e^{-\mu x}\, dx = \beta B(\beta\mu, \nu) \qquad\qquad (14)$$

$$(\text{Re } \beta > 0,\ \text{Re } \nu > 0,\ \text{Re } \mu > 0)$$

see Goldshteyn, Ryzhik, 1980, 3.312, p. 305, and

$$(x + y + 1)B(x + 1, y + 1) = \prod_{k=1}^{\infty} k \frac{x + y + k}{(x + k)(y + k)} \qquad\qquad (15)$$

ibid, 8.383, p. 950, where B is the **beta function**.

Define

$$k_{\alpha}(t) := \left(1 - e^{-t}\right)^{-\alpha} \qquad\qquad (16)$$

so that

$$k_{1-\alpha}(t) = \left(1 - e^{-t}\right)^{\alpha-1} \qquad\qquad (17)$$

Using (14) at $\beta = 1$, $x = t$, and $\nu = \alpha$ or $\nu = 1 - \alpha$, we Laplace-transform (17) and (16):

$$\mathcal{L}\{k_{1-\alpha}(t)\} = \hat{k}_{1-\alpha}(s) = B(s, \alpha)$$

$$\mathcal{L}\{k_{\alpha}(t)\} = \hat{k}_{\alpha}(s) = B(s, 1 - \alpha)$$

Equation (15) is then used to obtain

$$\hat{k}_{1-\alpha}(s) = \frac{1}{\alpha} \cdot \frac{1}{s} \prod_{k=2}^{\infty} \frac{k}{k-1+\alpha} \frac{s+k-2+\alpha}{s+k-1} \qquad (18)$$

and

$$\hat{k}_{\alpha}(s) = \frac{1}{1-\alpha} \cdot \frac{1}{s} \prod_{k=2}^{\infty} \frac{k}{k-\alpha} \frac{s+(k-1)-\alpha}{s+(k-1)} \qquad (19)$$

Truncation of (18) and (19) yields, respectively:

$$\hat{k}_{n,1-\alpha}(s) = \frac{1}{\alpha} \cdot \frac{1}{s} \prod_{k=2}^{n} \frac{k}{k-1+\alpha} \frac{s+k-2+\alpha}{s+k-1} \qquad (20)$$

and

$$\hat{k}_{n,\alpha}(s) = \frac{1}{1-\alpha} \cdot \frac{1}{s} \prod_{k=2}^{n} \frac{k}{k-\alpha} \frac{s+(k-1)-\alpha}{s+(k-1)} \qquad (21)$$

Observe that $\hat{k}_{n,1-\alpha}(s)$ is obtained from $\hat{k}_{n,\alpha}(s)$ by replacement of α with 1 - α, and vice-versa.

3. POLE / ZERO SPREADING: MOTIVATION AND IMPLEMENTATION

We will show below some preliminary computer results in order to illustrate the above points and a necessity for further improvement.

As an example, consider a case of FDL. The Bode plots of the filter are shown in Figure 4a, for $\alpha = 1/2$ and $n = 8$. We can see that the basic slope of the gain plot which, at high frequencies, should be α times 20 dB/decade = 10 dB/decade (Rutman, Estes, 1989) is kept at such a value roughly up to $\omega = 4$. Figure 4b displays the result of the differentiation of the function $F(x) = x - a$ ($a = 0.001$). Comparison with the exact result

$$\frac{2}{\sqrt{\pi}}(x - a)^{\frac{1}{2}}$$

illustrates the low accuracy exhibited at such a value of n. Along the frequency axis, the pole and zero locations are also marked, poles as crosses as zeros as circles. From this example it is clear that although the developed algorithm enables us to obtain results, its accuracy is limited by the necessity of using a filter of too high an order.

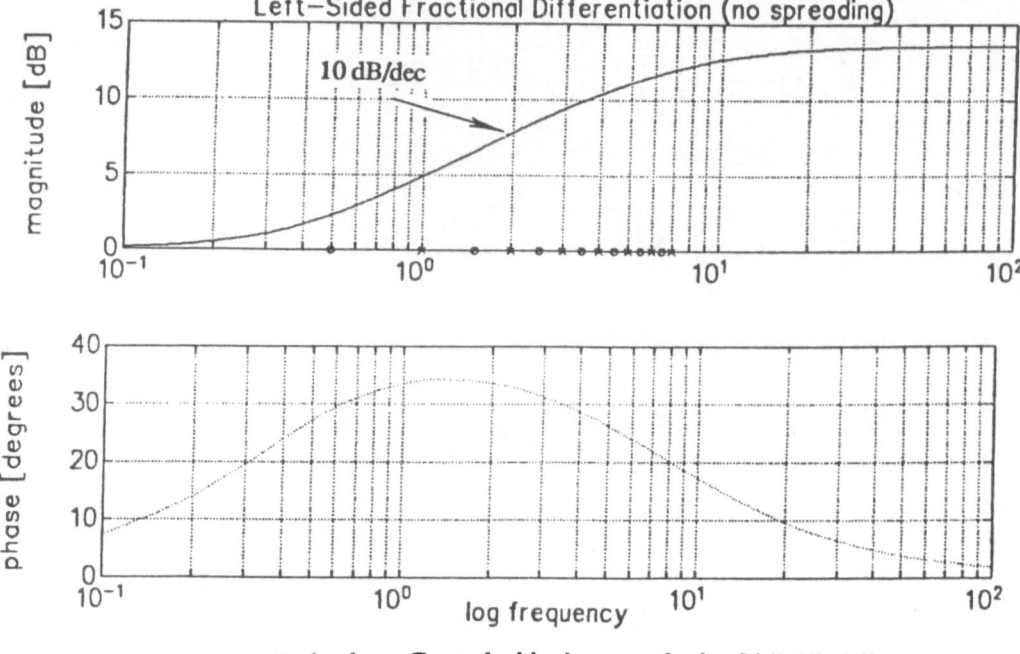

a) Bode plots. Created with: [response] = freqDL(8,1/2,-1,2)

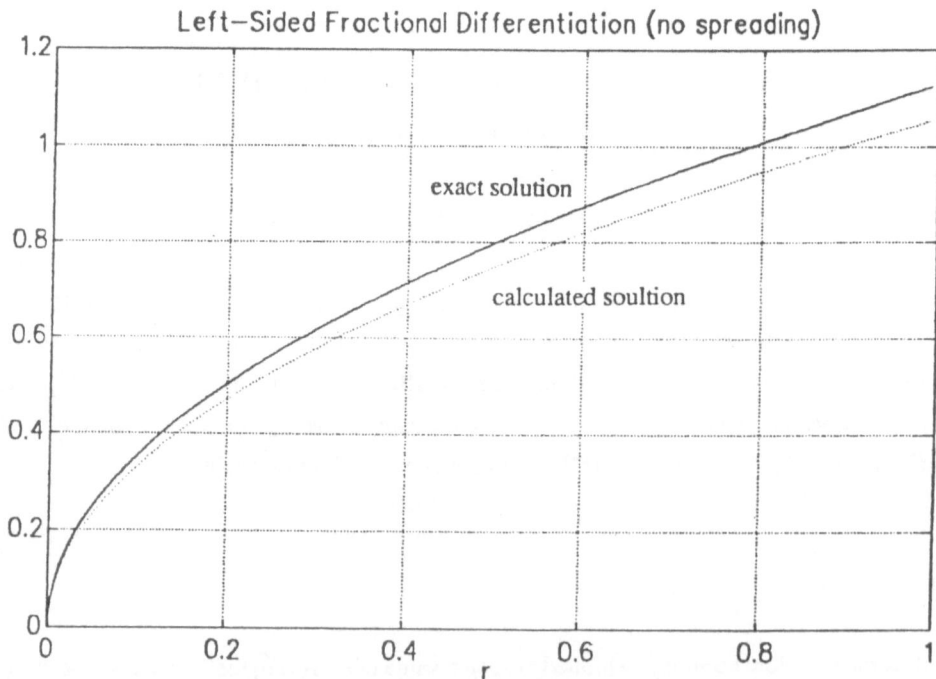

b) The exact solution and calculated half-derivative of $F(x) = x - a$ $(a = 0.001)$.
Created with: [filtout,H] = outDL(8,0,log(1000),0.01,1/2,0.001,1)

Figure 4. FDR with $n = 8$, $\alpha = 1/2$.

Indeed, the poles λ_k and the zeros β_k obey an arithmetic progression:

$$\lambda_k = -(k-1) \qquad (22)$$

$$\beta_k = -(k-2+\alpha) \qquad (23)$$

for $\hat{k}_{n,1-\alpha}(s)$ and

$$\lambda_k = -(k-1) \qquad (24)$$

$$\beta_k = -(k-1-\alpha) \qquad (25)$$

for $\hat{k}_{n,\alpha}(s)$.

Therefore, in order to extend the straight line portion of the Bode gain plot over a couple of decades, one would need hundreds of poles and zeros, as they tend to cluster together ever closer on the logarithmic frequency axis as $\omega \to \infty$. Figure 5 illustrates this limitation by showing that with $n = 50$ the straight line portion of the Bode gain plot barely encompasses a single decade! Note also the clustering of the poles and zeros on the logarithmic frequency axis. Obviously, the error of the method is inversely related to the bandwidth of the gain plot which is understood here as the upper border of the frequency interval where the gain plot

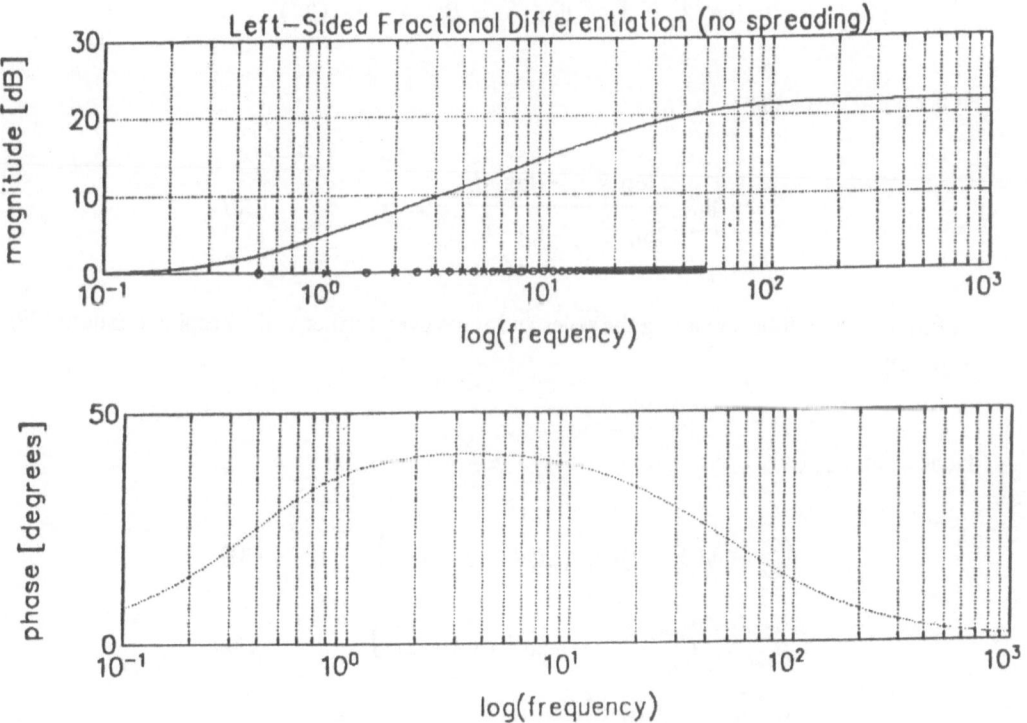

Figure 5. Illustration of pole/zero clustering in FDR with $n = 50$, $\alpha = 1/2$.

has the desired slope of α times 20 dB/dec. These simple considerations check with the analysis in Rutman, 1989, where it was shown that the asymptotic accuracy of the computation is determined, first of all, by the value of the **regularization constant**. In that early version presented heretofore, the value of the regularization constant is

$$\sigma = \frac{1}{n+1} \qquad (26)$$

The following improvement is presented in this paper. A better use of a fixed order n filter can be made by **spreading** the values of the poles and zeros over the desired bandwidth so that they are spaced evenly over the logarithmic scale. To this end one can use:

$$\hat{\lambda}_{k,\alpha} = -(k-1)\exp\{\sigma(k-2+\alpha)\} \qquad (27)$$

$$\hat{\beta}_{k,\alpha} = -(k-1-\alpha)\exp\{\sigma(k-2)\} \qquad (28)$$

for $\hat{k}_{n,\alpha}(s)$, and the formulas for $\hat{k}_{n,1-\alpha}(s)$ can be obtained from those for $\hat{k}_{n,\alpha}(s)$ by replacement of α with $1 - \alpha$. As $n \to \infty$, $\sigma \to 0$, and

$$\lim_{\sigma \to 0} \hat{\lambda}_k = \lambda_k, \quad \lim_{\sigma \to 0} \hat{\beta}_k = \beta_k \qquad (29)$$

It is easy to check that:

$$\lim_{k \to \infty} \frac{\log \hat{\beta}_{k,\alpha} - \log \hat{\beta}_{k,\alpha}}{\log \hat{\lambda}_{k,\alpha} - \log \hat{\lambda}_{k-1,\alpha}} = \alpha \qquad (30)$$

A further step will be spreading the poles and zeros even further while keeping relations (29) and (30).

Hence, for $\hat{k}_{n,\alpha}(s)$,

$$\hat{\lambda}_{k,\alpha} = -(k-1)\exp\{\sigma(k-2+\alpha)^2\} \qquad (31)$$

$$\hat{\beta}_{k,\alpha} = -(k-1-\alpha)\exp\{\sigma(k-2)^2\} \qquad (32)$$

It is easy to see that the properties of (29) and (30) have been maintained. The improvement in accuracy obtained by the widening of the filter bandwidth for the given n vastly overcomes the adverse effect of thinning out the poles and zeros at high frequency ranges. The regularization constant here should be considered the value

$$\hat{\sigma} = \frac{1}{2 - \lambda_n}$$

Figure 6 shows an example of a system (FDL, $\alpha = 1/2$, $n = 8$) with pole/zero spreading. Compare the Bode gain plot of Figure 6a with that of Figure 4a, and note how the straight line portion of the plot has been extended to the upper border of $\omega = 700$. Comparison of Figure 6b with Figure 4b illustrates how pole/zero spreading reduces error.

To perform the numerical integration/differentiation, a software package has been developed. The code has been implemented on PC MATLAB with the Control Systems Toolbox. All the results below will be presented for the algorithms with pole/zero spreaching only.

To illustrate the simulation results, the fractional differentiation of the function

$$F(x) = \sqrt{\pi}\left(\frac{x - a}{\lambda}\right)^{\alpha - \frac{1}{2}} J_\nu(\lambda\sqrt{x - a}) Y_\nu(\lambda\sqrt{x - a})$$

(Fig. 7a ; $\alpha = 3/5$) is presented.

Here $Y_\nu(z)$ is the Bessel function of the second kind (the Neumann function)

$$Y_\nu(z) = \csc \nu\pi [J_\nu(z) \cos \nu\pi - J_{-\nu}(z)]$$

and $J_\nu(z)$ is the Bessel function of the first kind.

The analytical result

$$\mathcal{D}^\alpha_{a+}(F)(r) = (r - a)^{\frac{1}{2}\alpha - \frac{3}{4}} Y_\nu(2\lambda\sqrt{r - a})$$

is compared with the result of the numerical fractional differentiation on Fig. 7b.

More examples of numerical integration differentiation, block-diagrams of simulation and detailed software can be found in Rutman, Medeiros (1990).

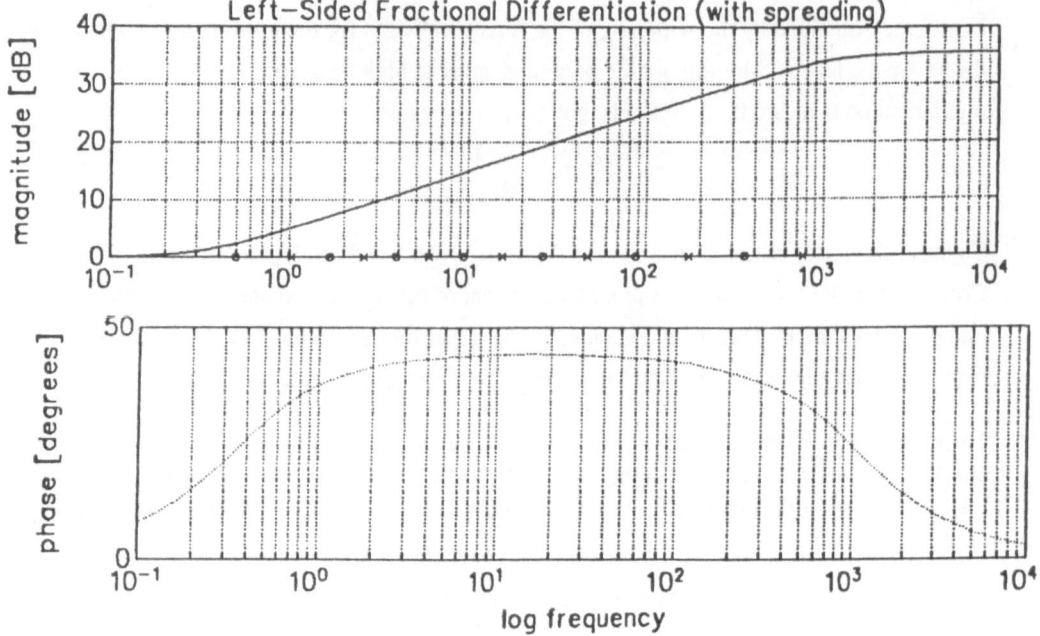

a) Bode plots. Created with: [response] = freqDLs(8,2,1/2,-1,4)

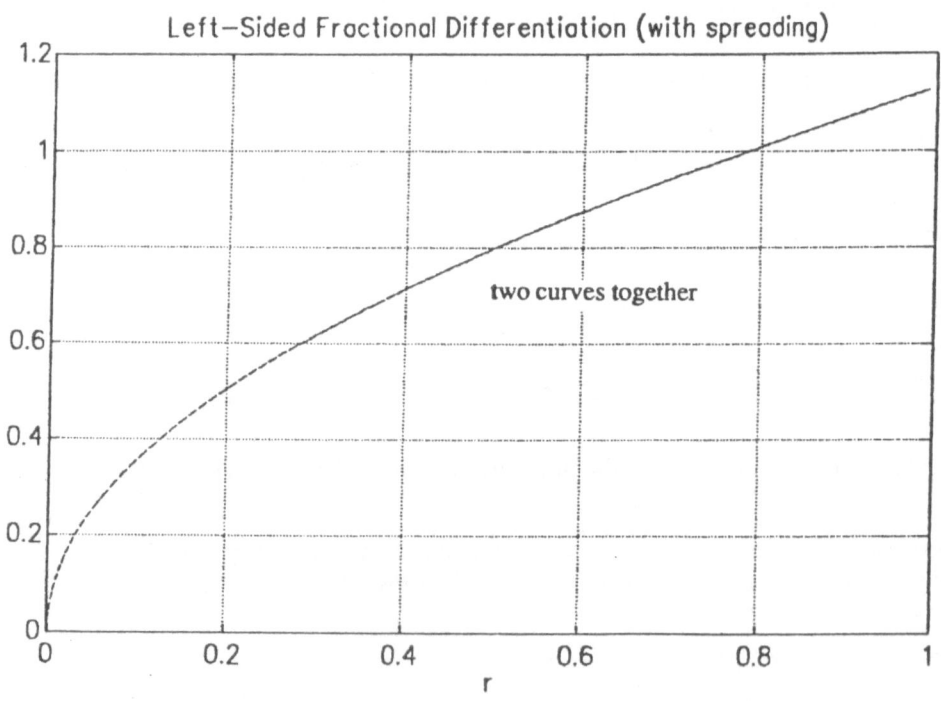

b) The exact solution and calculated half-derivative of $F(x) = x - a$ ($a = 0.001$).
Created with: [filtout,H]=outDLs(8,2,log(1000),0.01,1/2,0.001,1)

Figure 6. FDR with spreading, $n = 8$, $\alpha = 1/2$.

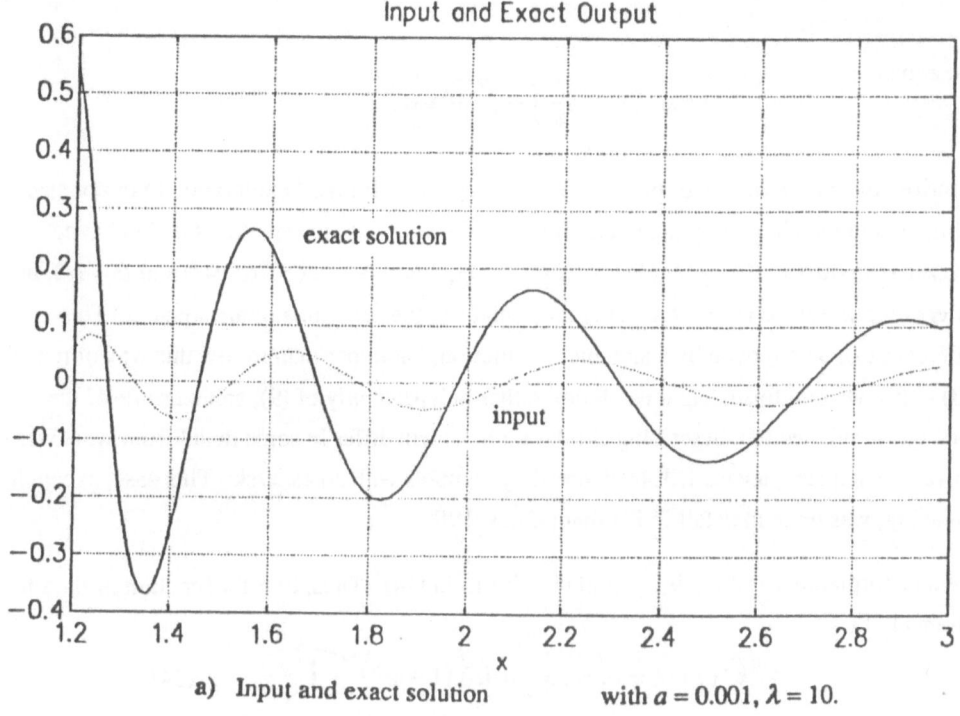

a) Input and exact solution with $a = 0.001$, $\lambda = 10$.

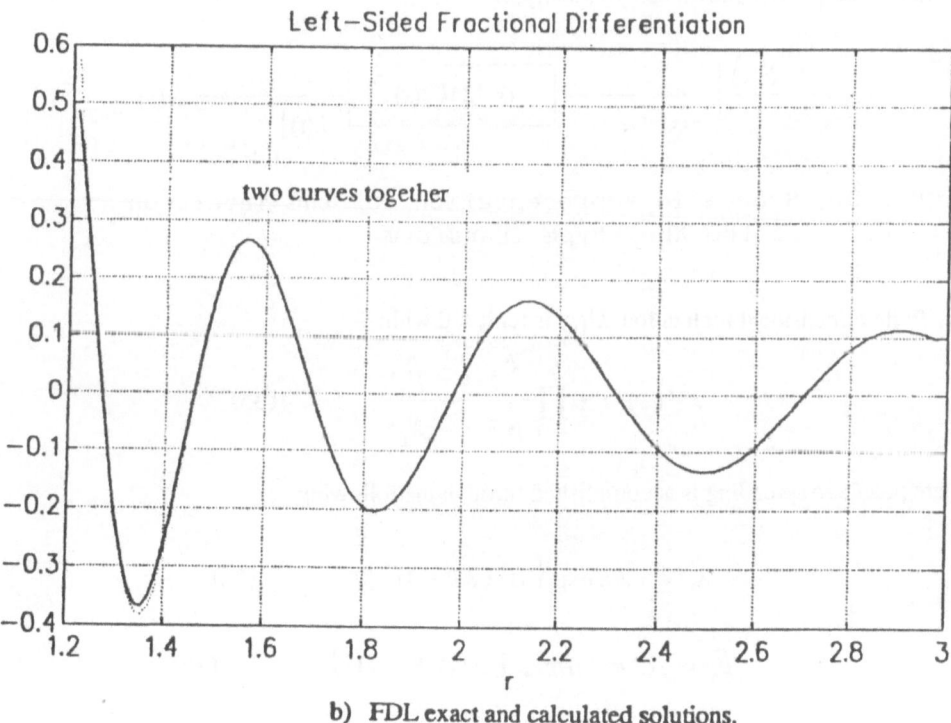

b) FDL exact and calculated solutions.

Figure 7. FDL output simulation with $n = 8$, $\alpha = 3/5$.

6. MODIFICATION: AN ABEL-TYPE INVERSE TRANSFORM

The equation

$$H(r) = -\frac{1}{\pi} \int_r^b \frac{F'(x)}{\sqrt{x^2 - r^2}} dx \qquad (33)$$

describes the whole variety of physical problems ranging from oil exploration to tumor size estimation to metallography to astronomy, etc. We will call this equation the **Abel-type inverse transform**. Equation (33) is similar to equation (9) when $F(b) = 0$ and $\alpha = 1/2$, but differs (except for a constant factor) in that x and r in the denominator are squared. This difference can be removed by a suitable substitution, but rather than follow this way of reducing (33) to (9), we will formulate the solution rather independently of (9), so that some of the notations in this section, describing similar notions, will differ in some detail from what was above. Such discrepancies will be denoted by symbols with an asterisk. This case, except for spreading, was done in detail in Rutman, Estes, 1989.

The transformation to take this integral remains as in (14). Thus, transfer function of the filter is found to be:

$$\hat{k}^*(s) = \frac{1}{2} B\left(\frac{s}{2}, \frac{1}{2}\right) = L\left\{(1 - e^{2t})^{-1/2}\right\} \qquad (34)$$

and the solution $H(r)$, shown schematically, is

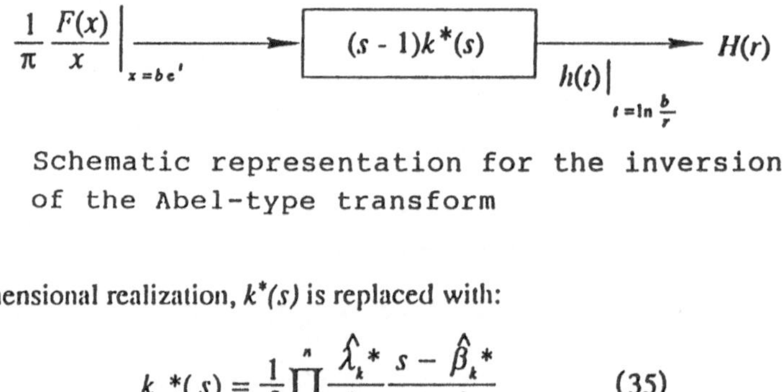

$$\left.\frac{1}{\pi} \frac{F(x)}{x}\right|_{x = be^t} \longrightarrow \boxed{(s - 1)k^*(s)} \longrightarrow H(r)$$
$$h(t)\big|_{t = \ln \frac{b}{r}}$$

Figure 8. Schematic representation for the inversion of the Abel-type transform

For finite-dimensional realization, $k^*(s)$ is replaced with:

$$k_n^*(s) = \frac{1}{s} \prod_{k=2}^n \frac{\hat{\lambda}_k^* s - \hat{\beta}_k^*}{\hat{\beta}_k^* s - \hat{\lambda}_k^*} \qquad (35)$$

where pole/zero spreading is accomplished through the following:

$$\hat{\lambda}_k^* = -2 k \exp\left\{\sigma^*(2k - 1)^4\right\} \qquad (36)$$

$$\hat{\beta}_k^* = -(2k - 1)\exp\left\{\sigma^*(2k - 2)^4\right\} \qquad (37)$$

$$\sigma^* = \frac{1}{2n - 1} \qquad (38)$$

The spreading constant d has been introduced in (36) - (37) instead of the value 2 as in (31) - (32), in order to provide further leverage on adjustment.

The computer simulation was performed for two types of input signal:

Curve A:

$$F(x) = \begin{cases} -\frac{2}{3}\left(\frac{1}{4} - x^2\right)^{1/2}(1 + 8x^2) + \frac{4}{3}(1 - x^2)^{1/2}(1 + 2x^2) \\ \\ -4x^2\ln\left\{\dfrac{1 + (1 - x^2)^{1/2}}{1/2 + (1/4 - x^2)^{1/2}}\right\}, \end{cases} \quad 0 \le x \le 1/2 \quad \text{(39 a)}$$

$$F(x) = \frac{4}{3}(1 - x^2)^{1/2}(1 + 2x^2) - 4x^2\ln\left\{\frac{1 + (1 - x^2)^{1/2}}{x}\right\}, \quad 1/2 \le x \le 1 \quad \text{(39b)}$$

and Curve B:

$$F(x) = \frac{\sqrt{\pi}}{\beta}(1 - x^2)^{-1/2}\exp\left[\beta^2\left(1 - \frac{1}{1 - x^2}\right)\right] \quad \text{where} \quad \beta = 1.1 \quad \text{(40)}$$

The answers are known to be:

Curve A:

$$H(r) = \begin{cases} 1 - 2r^2, & 0 \le r \le 1/2 \\ 2(1 - r)^2, & 1/2 \le r \le 1 \end{cases} \quad \text{(41)}$$

and Curve B:

$$H(r) = (1 - r^2)^{-3/2}\exp\left[\beta^2\left(1 - \frac{1}{1 - r^2}\right)\right] \quad \text{(42)}$$

Such names are given to these functions in the paper by Minerbo, Levi, 1969, and these examples served as test cases for a whole number of researchers to compare the accuracies of their methods.

Figure 9 shows the Bode plots for such a filter. Note how similar these plot are to those for fractional differentiation.

Figure 10 displays the results of the output simulation for Curves A and B. As one can easily see, there is very little error between the exact output and the computed output for both Curve A and Curve B. Table 3 presents a **quantitative** comparison between the above method and previous methods. In addition to the root-mean-square error reported and compared in the literature sources, the **maximum error** 3.4 * 10^{-4} and the **average error** 2.6 * 10^{-4} were computed in this work for Curve A.

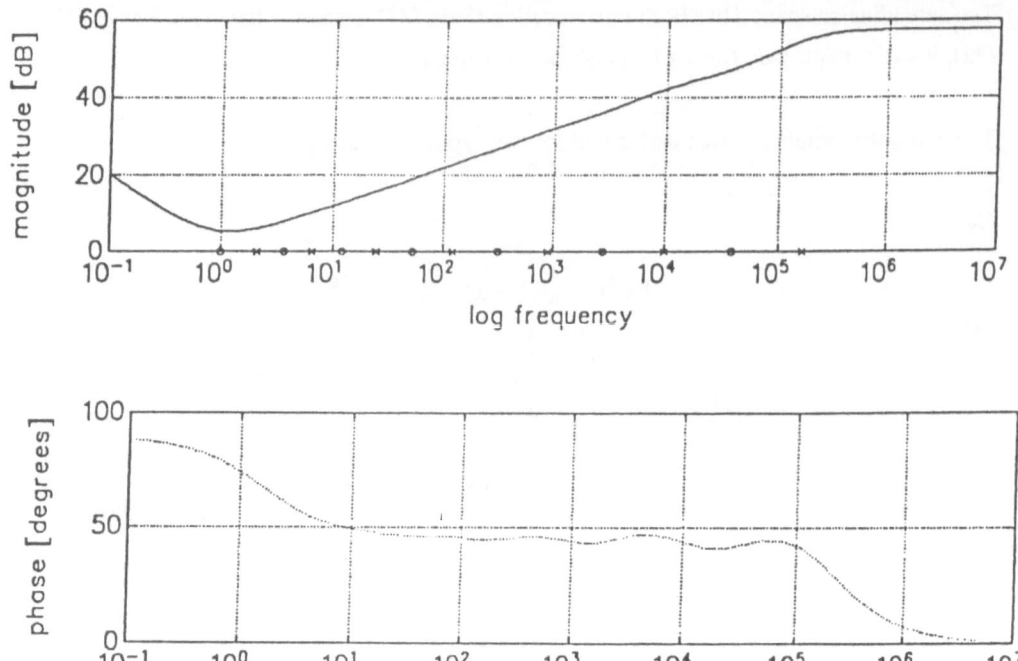

Figure 9.   Bode plots for the modified filter

Table 2. Comparison of the root-mean-square (RMS) error from different sources.

	Curve A	Curve B
This work	$2.4 * 10^{-4}$	$5 * 10^{-4}$
Andersen, 1976	$3.2 * 10^{-4}$	$1.67 * 10^{-4}$
Minerbo - Levy, 1969	$11 * 10^{-4}$	$7.8 * 10^{-4}$
Maldonado et al., 1965	$11 * 10^{-4}$	$82 * 10^{-4}$
Bokasten, 1961	$12 * 10^{-4}$	$28 * 10^{-4}$
Nestor - Olsen, 1960	$51 * 10^{-4}$	$118 * 10^{-4}$

The plots in Figure 10 and the RMS errors presented in Table 2 were created with a special simulation program using **optimal values** for both the time increment between samples, and the spreading exponent d, which controls the degree of pole/zero spreading. Figure 11 presents sample results of the two programs that were used to determine the optimal values of the increment and d.

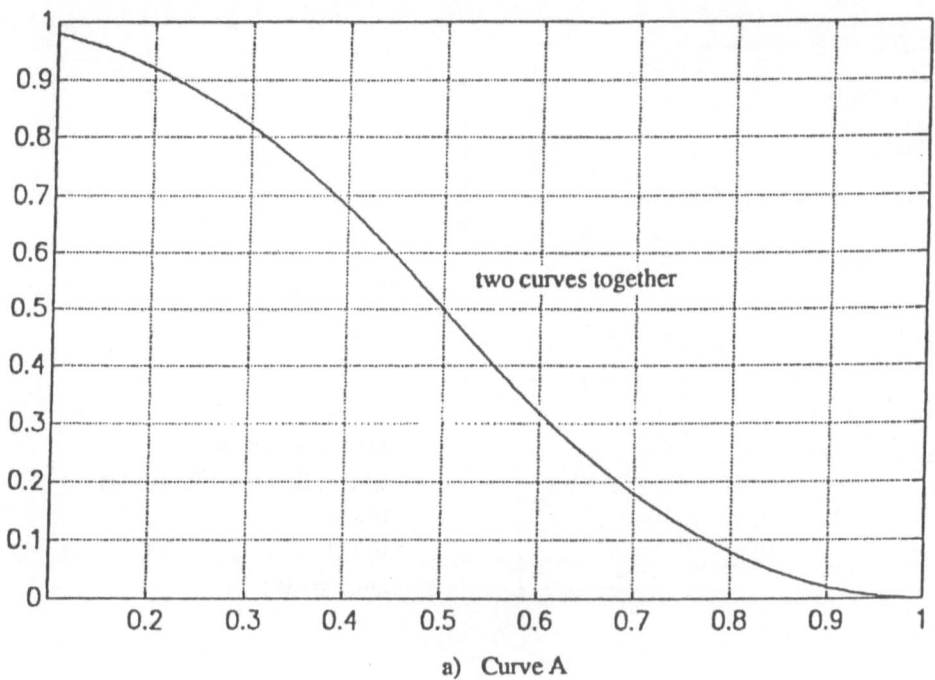

two curves together

a) Curve A

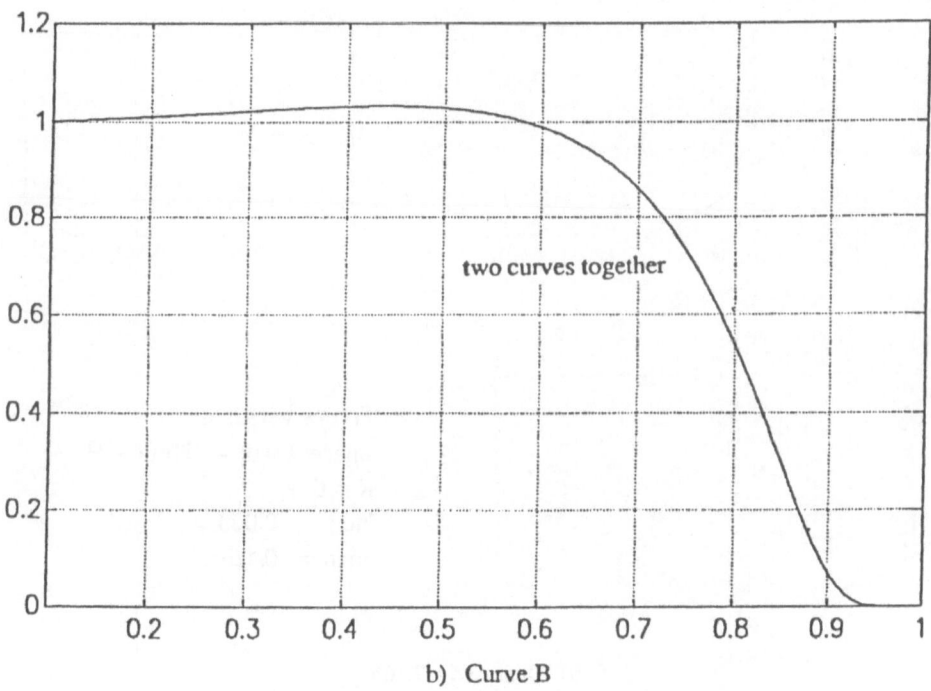

two curves together

b) Curve B

Figure 10.   Output simulations with $n = 17$.

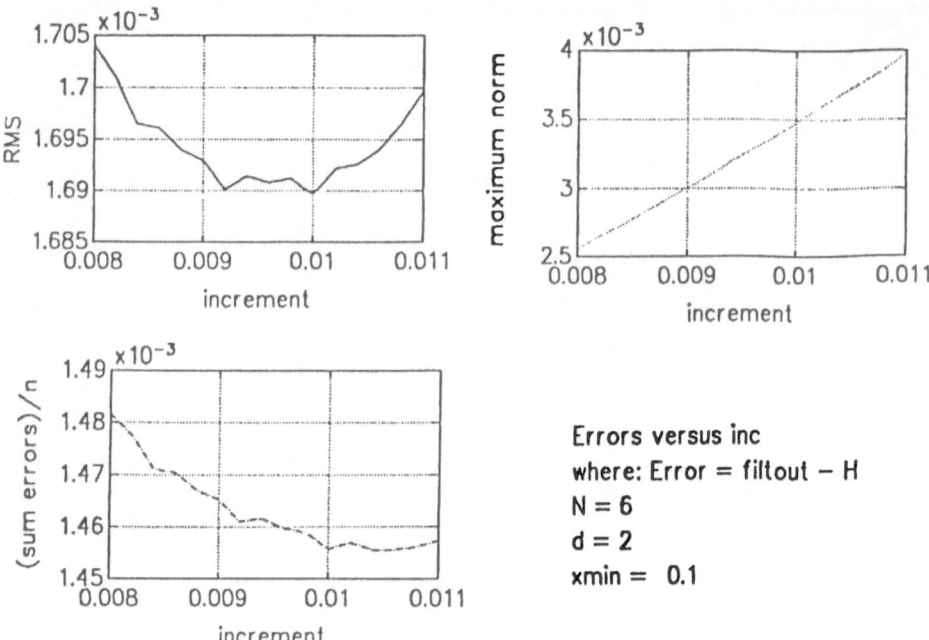

Errors versus inc
where: Error = filtout − H
N = 6
d = 2
xmin = 0.1

a) Optimal increment = 0.01.

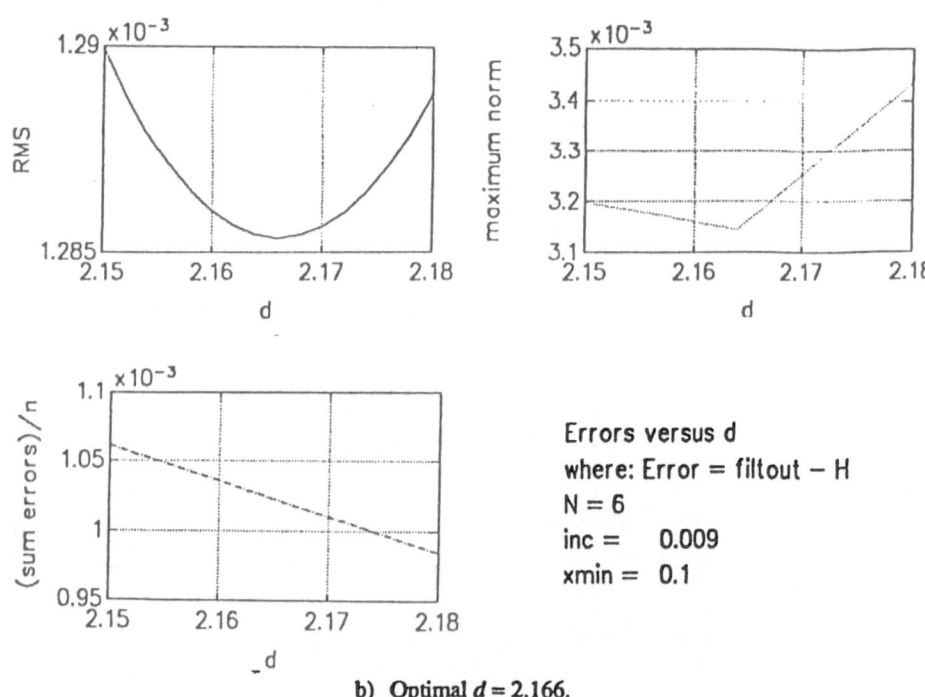

Errors versus d
where: Error = filtout − H
N = 6
inc = 0.009
xmin = 0.1

b) Optimal d = 2.166.

Figure 11. Results of programs used to determine the optimal increment and d (Curve B, n = 6).

Figure 12 demonstrates the dependence of the RMS error on n. This dependence reflects the fact that the regularization constant depends inversely on n. Recall the major regularization property

$$\lim_{\sigma \to 0} \|\varepsilon\| = 0 \qquad (43)$$

where ε is the difference between the exact (ideal) solution and a regularized solution.

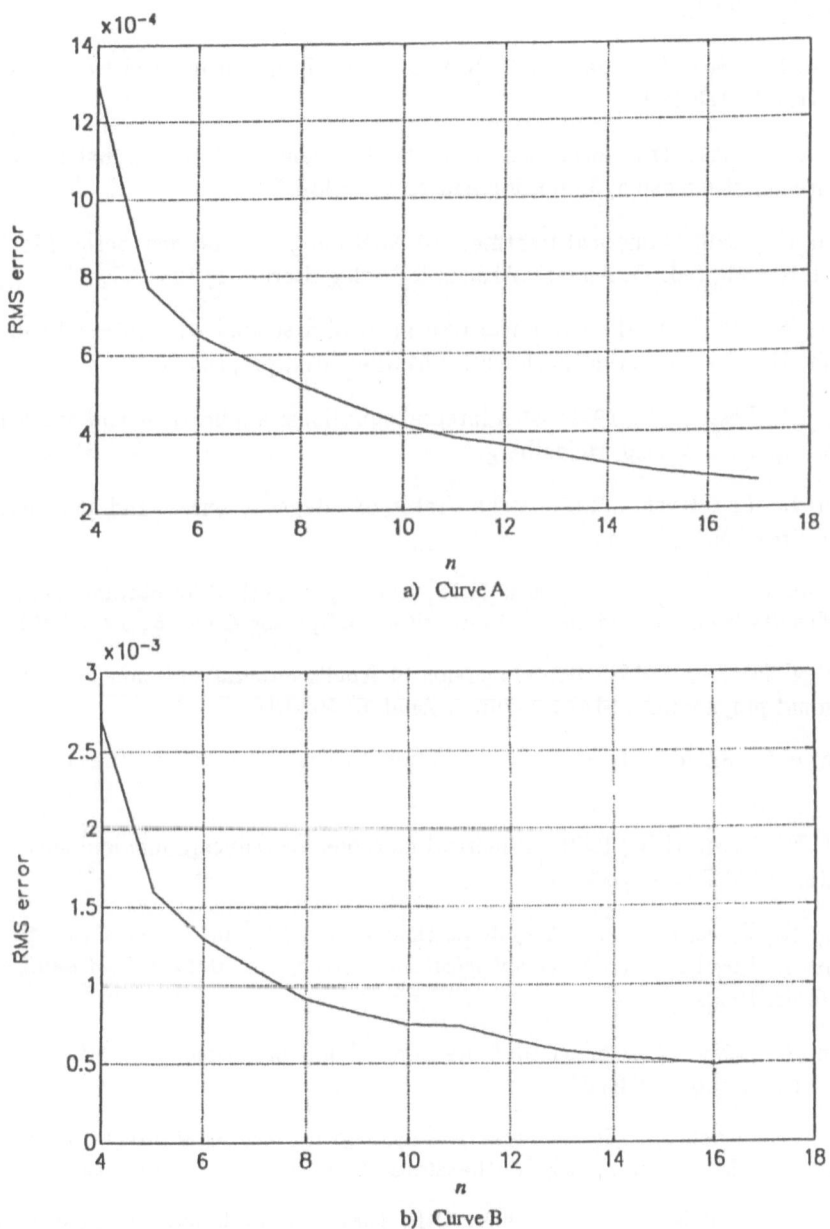

a) Curve A

b) Curve B

Figure 12. Illustration of the dependence of the RMS error on n.

Acknowledgement

The author is acknowledging the valuable help by Michael R. Medeiros in developing the software and numerical simulation.

BIBLIOGRAPHY

[1] Anderson,R.S.,1976: Stable procedures for the inversion of Abel's equation, J.Inst.Math. Its Appl.,**17**, 329-342.

[2] Bateman, H., 1954: Bateman manuscript project: Tables of integral transforms. McGraw-Hill, **2**, 185-199.

[3] Bokasten, K., 1961: Transformation of observed radiances into radial distribution of the emission of a plasma, J.Opt.Soc.Amer., **51**, 943-947.

[4] Gorenflo, R., 1979: Numerical treatment of Abel integral equations, Series Mathematical Research, (ed. G. Anger), Akademie-Verlag, Berlin, **1**, 125-133.

[5] Gorenflo, R., 1987: On the numerical treatment of first kind Abel integral equations, Freie Universität Berlin, Fachbereich Mathematik, preprint No. 258.

[6] Gorenflo, R., Vessella, S., 1991: Abel integral equations: applications and analytic properties, Springer-Verlag, Heidelberg.

[7] Gradshteyn, I.S., Ryzhik, I.M., 1980: Table of integrals, series, and products, Academic Press, N.Y., London.

[8] Maldonado, C.D., Caron A.P., Olsen, H.N., 1965: New method for obtaining emission coefficients from emitted spectral intensities, J.Opt.Soc.Amer.,**55**,1247-1254.

[9] Minerbo, G.N., Levy, M.E.,1969: Inversion of Abel's integral equation by means of orthogonal polynomials, SIAM J.Numer.Anal.,**6**, 598-616.

[10] Oldham, K.B., Spanier, J., 1974: The fractional calculus, Academic Press, N.Y., London.

[11] Olsen, O.H., Olsen, H.N., 1960: Numerical methods for reducing line and surface probe data, SIAM Rev.**2**, 200-207.

[12] Rutman, R., Estes, L., 1989: A systems theory model for inversion of the Abel transform. A. Vogel et al.(ed). *Geophysical data inversion: Methods and applications.* Vieweg, 1990.

[13] Rutman, R., 1989: A regularized finite-dimensional inversion of an Abel-type equation, SMU Report ECE-TR-12.

[14] Rutman, R., Medeiros, M.R.: A numerical method for fractional integration and differentiation. SMU-ECE-TR-16, Southeastern Mass.Univ., N. Dartmouth: 1990

[15] Samko, S.G., Kilbas, A.A., Marichev, O.I., 1987: Integrals and derivatives of fractional order with some applications (in Russian), Science and Engineering Publishing, Minsk.

FIL: $\quad x = ae^{t}, r = ae^{\tau}, h(\tau) = H(ae^{\tau}), dr = r\, d\tau$

$$F(x) = (\mathcal{J}_{a+}^{\alpha} H)(x) = \frac{1}{\Gamma(\alpha)} \int_{a}^{x} \frac{H(r)}{(x-r)^{1-\alpha}}\, dr$$

$$= \frac{1}{\Gamma(\alpha)} \int_{0}^{t} \frac{h(\tau)\, r\, dt}{x^{1-\alpha}\left(1 - \frac{r}{x}\right)^{1-\alpha}} = x^{\alpha-1} \frac{1}{\Gamma(\alpha)} \int_{0}^{t} \frac{h(t)\, r\, d\tau}{[1 - e^{-(t-\tau)}]^{1-\alpha}}$$

$$= \frac{1}{\Gamma(\alpha)} [h(t)\, ae^{t} * k_{1-\alpha}(t)] \cdot x^{\alpha-1}$$

FDL: $\quad x = ae^{\tau}, r = ae^{t}, F'(x)\, dx = f'(\tau)\, d\tau, f(\tau) = F(ae^{\tau})$

$$H(r) = (\mathcal{D}_{a+}^{\alpha} F)(x) = \frac{1}{\Gamma(1-\alpha)} \int_{a}^{r} \frac{F'(x)}{(r-x)^{\alpha}}\, dx$$

$$= \frac{1}{\Gamma(1-\alpha)} \cdot \frac{1}{r^{\alpha}} \int_{a}^{r} \frac{F'(x)\, dx}{\left(1 - \frac{x}{r}\right)^{\alpha}} = \frac{1}{\Gamma(1-\alpha)} \cdot \frac{1}{r^{\alpha}} \int_{0}^{t} \frac{f'(\tau)\, d\tau}{(1 - e^{-(t-\tau)})^{\alpha}}$$

$$= \frac{1}{\Gamma(1-\alpha)} [f'(t) * k_{\alpha}(t)] \cdot \frac{1}{r^{\alpha}}$$

The derivative accounts for the cancellation of the $1/s$ term in $\hat{K}(s)$.

FDLA : $x = ae^{\tau}, r = ae^{t}, dx = x\, d\tau$

$$H(r) = (\mathcal{D}_{a+}^{\alpha} F)(x) = \frac{1}{\Gamma(1-\alpha)} \frac{d}{dr} \int_{a}^{r} \frac{F(x)}{(r-x)^{\alpha}}\, dx$$

$$= \frac{1}{\Gamma(1-\alpha)} \frac{d}{dr} \int_{0}^{t} \frac{F(x) x\, d\tau}{r^{\alpha}\left(1 - \frac{x}{r}\right)^{\alpha}} = \frac{1}{\Gamma(1-\alpha)} \frac{d}{dr} r^{-\alpha} \int_{0}^{t} \frac{f(\tau)\tau\, d\tau}{(1 - e^{-(t-\tau)})^{\alpha}}$$

$$= \frac{1}{\Gamma(1-\alpha)} \frac{d}{r\, dt}\left[(ae^{t})^{-\alpha} y(t)\right] \qquad (y(t) = f(t) * k_{\alpha}(t))$$

$$= \frac{1}{\Gamma(1-\alpha)} \frac{1}{r}\left[(-\alpha)(ae^{t})^{-\alpha} y(t) + (ae^{t})^{-\alpha} y'(t)\right]$$

$$= \frac{1}{\Gamma(1-\alpha)} \frac{1}{r}\left[(-\alpha) r^{-\alpha} y(t) + r^{-\alpha} y'(t)\right]$$

$$= \frac{1}{\Gamma(1-\alpha)} r^{-(1+\alpha)}\left[y'(t) - \alpha\, y(t)\right]$$

The operations in the brackets account for the extra s - α factor in $\hat{K}(s)$.

FIR: $x = be^{-t}, r = be^{-\tau}, H(be^{-\tau}) = K(\tau), dr = -r\, d\tau$

$$F(x) = (\mathcal{J}_{b-}^{\alpha} H)(x) = \frac{1}{\Gamma(\alpha)} \int_{x}^{b} \frac{H(r)}{(r-x)^{1-\alpha}}\, dr$$

$$= -\frac{1}{\Gamma(\alpha)} \int_{t}^{0} \frac{H(r) r\, d\tau}{r^{1-\alpha}\left(1 - \frac{x}{r}\right)^{1-\alpha}} = \frac{1}{\Gamma(\alpha)} \int_{0}^{t} \frac{h(\tau) r^{\alpha}\, d\tau}{[1 - e^{-(t-\tau)}]^{1-\alpha}}$$

$$= \frac{1}{\Gamma(\alpha)}\left[K(t)(be^{-t})^{\alpha} * k_{1-\alpha}(t)\right]$$

FDR: $x = be^{-\tau}, r = be^{-t}, F'(x)\,dx = f'(\tau)\,d\tau$

$$H(r) = (\mathcal{D}^\alpha_{b-} F)(x) = -\frac{1}{\Gamma(1-\alpha)} \int_r^b \frac{F'(x)\,dx}{(x-r)^\alpha} = -\frac{1}{\Gamma(1-\alpha)} \int_t^0 \frac{f'(\tau)\,d\tau}{x^\alpha(1-e^{-(t-\tau)})^\alpha}$$

$$= \frac{1}{\Gamma(1-\alpha)} \int_0^t \frac{f'(\tau)/(be^{-\tau})^\alpha\,d\tau}{(1-e^{-(t-\tau)})^\alpha}$$

$$= \frac{1}{\Gamma(1-\alpha)}[f'(t)(be^{-t})^{-\alpha} * k_\alpha(t)]$$

But since

$$\frac{d}{dt}[e^{\alpha t} f(t)] = \alpha e^{\alpha t} f(t) + e^{\alpha t} f'(t)$$

we get

$$\frac{1}{b^\alpha} e^{\alpha t} f'(t) = \frac{d}{dt}\left[\frac{f(t)}{(be^{-t})^\alpha}\right] - \alpha \frac{f(t)}{(be^{-t})^\alpha}$$

This accounts for the extra $s - \alpha$ factor in $\hat{K}(s)$.

FDRA: $x = be^{-\tau}, r = be^{-t}, dx = -x\,d\tau, dr = -r\,dt, F(be^{-t}) = f(t)$

$$H(r) = (\mathcal{D}^\alpha_{b-} F)(x) = -\frac{1}{\Gamma(1-\alpha)} \frac{d}{dr} \int_r^b \frac{F(x)\,dx}{(x-r)^\alpha}$$

$$= \frac{1}{\Gamma(1-\alpha)} \frac{d}{dt}\left[f(t)(be^{-t})^{1-\alpha} * k_\alpha(t)\right] \cdot \frac{1}{r}$$

The derivative accounts for the cancellation of the $1/s$ term in $\hat{K}(s)$.

Recent Results in the Study of the Moment Problem

G. Inglese

IAGA-CNR, via S. Marta 13/a, 50139 Firenze, Italy

Abstract. Accuracy and stability results are collected for the approximate solutions of finite Hausdorff Moment Problem; a numerical example is discussed in the light of classical approximation theory.

1 Introduction and brief history

A classical set of inverse (and ill-posed) problems is constituted by problems of moments. The goal is to find a positive function u in (a,b) $a,b \in \mathbb{R} \cup \{-\infty, \infty\}$ such that its "moments"

$$\int_a^b x^{j-1} u(x) dx, \quad j = 1,2,\ldots$$

(this word, taken from mechanics, was used first by Stieltjes) have the prescribed values μ_j.

Generally the following classification is used :

$a = 0,\ b = \infty$	Stieltjes	1895 ;
$a = -\infty, b = \infty$	Hamburger	1920 ;
$a < b\ ;\ a,b \in \mathbb{R}$	Hausdorff	1923 .

The above mentioned mathematicians solved the respective problems proving existence and, when possible, uniqueness of the solutions. Strictly related to the problem of the moments are some papers by Tchebischev (1855 and after) that laid the foundations of the general theory of orthogonal polynomials. Other mathematicians, like Markoff and Heine, gave their contributions in the study of this problems. A complete survey in which are specified the theorems and pointed out the techniques is Shoat et al.,'43 .

We will consider in more detail the Hausdorff Moment Problem (HMP) in a Hilbert spaces framework.

Let $u \in L^2(0,1)$ be the unknown function, μ be the sequence of the moments. The linear operator $A: L^2 \to \ell^2$ maps a function into the sequence of its moments. Let us list the classical results that solve the HMP :

(a) A is injective (unicity of the solution).

Proof: it derives from the density in L^2 of the linear combinations of $1, x, x^2, \ldots$.

(b)　The ℓ^2 sequence μ is in Range(A) <u>if and only if</u> the following condition is true:

$$(n+1) \sum_{k=0}^{n} \binom{n}{k}^2 (\Delta^{n-k}\mu_k)^2 \leq \text{constant} \quad \text{for } n=0,1,2,\dots \qquad \text{where } \Delta \text{ is the difference}$$

operator such that $\Delta^m \mu_k = \sum_{k=0}^{m} \binom{m}{k}(-1)^i \mu_{k+i}$.

Proof: this is esentially the result by Hausdorff; see for example Askey et al.,'82 .

(c)　$\|Au\|_{\ell^2} < \pi \|u\|_{L^2}$; hence A is a continuous operator.

Proof: see Hardy et al..,'64 . It is a Fejer-Riesz's theorem .

Now we can define the operator A^{-1} on Range(A). The problem of stability (related to the norm $\|A^{-1}\|$) was disregarded up to recent years. It is a fact that numerical analysts have known for a lot of time that there are big difficulties, due to the severe ill-conditioning, in using moments as data; moreover geometrical considerations in moment space suggest that the solution of HMP cannot be stable.

A rigorous argument proving the unboundedness of A^{-1} is due to Talenti : a sequence of functions u_p is exibited such that $\|Au_p\|_{\ell^2}$ is uniformly bounded while $\|u_p\|_{L^2}$ is arbitrarily large. Let $u_p = x^{-1/2 - (1/p)\ln x}$, $p=1,2,\dots$; we have

$$\mu_k = \int_0^1 x^{k-1/2} x^{-(1/p)\ln x} \, dx =$$

$$= \int_0^\infty e^{-t(k+1/2) - t^2/p} \, dt \leq \int_0^\infty e^{-t(k+1/2)} \, dt = \frac{1}{k+1/2} ,$$

and so

$$\|Au_p\|_{\ell^2} \leq \left(\sum_{k=0}^{\infty} \frac{1}{(k+1/2)^2} \right)^{1/2} \quad \text{for all } p .$$

On the other hand we have

$$\|u_p\|_{L^2} = \left(\int_0^1 x^{-1 - (2/p)\ln x} \, dx \right)^{1/2} =$$

$$= \left(\sqrt{\frac{p}{2}} \int_0^\infty e^{-y^2} dy \right)^{1/2} \to \infty \quad \text{for } p \to \infty .$$

Analogous examples show the unboundedness of A^{-1} in L^p too, for $p \geq 1$. Because of the just shown instability we can affirm that HMP is an ill–posed problem in the sense of Hadamard.

2 Finite HMP. Stability estimates and discretization.

We are interested in the numerical study of the HMP: in this case the problem has to be reformulated opportunely. Our interest is not only theoretical, but it is motivated by the close relation with the inversion of the Laplace transform (with discrete equally spaced data); see Bertero et al.,'85. Let us recall that the Laplace tranform of a function f on $(0,\infty)$ is $F(p) =$

$$\int_0^\infty e^{-pt} f(t)dt \ , \quad p \in \mathbb{R}^+ .$$

If we use a uniform sampling of p, as $p = 0,1,2,...$ and change the variable $x=e^{-t}$, we have

$$F(p) = \int_0^1 x^p u(x)dx \ \text{ where } u(x) = f(\ln\frac{1}{x}) .$$

Thus we can state the finite problem:

Let u be a real continuously differentiable function on $(0,1)$ and let ε, E, μ_j $(j=1,...,n)$ be real numbers such that

(i) $\qquad\qquad \sum_{j=1}^n | \int_0^1 x^{j-1}u(x)dx - \mu_j |^2 \leq \varepsilon^2$

(ii) $\qquad\qquad \int_0^1 (u')^2 dx \leq E^2 \ .$

Approximate u in some m-dimensional space $X_m \subset L^2(0,1)$.

Remarks. Condition (i) means that the available data are near to the moments of the unknown function; the distance is estimated by ε. Condition (ii) gives us global 'a priori information about u' and can be used to restrict the set of the possible solutions; in fact if S is the set of the functions satisfying (i), we have (see Talenti,'87)

$$\sup_S \|u_1-u_2\|_2^2 \leq \text{ Const. } \left(e^{3.5n} \varepsilon + \frac{1}{2n^2} \int_0^1 (u')^2 dx \right) ;$$

such an upper bound is finite if (ii) holds.

In the following v will be the approximation of u in X_m. The operative choice of X_m provides the definition of a linear space of functions and a basis $\{b_k\}_{k=1}^m$; hence we write

$v = \sum_{k=1}^m v_k b_k(x)$. The moments can be represented in the following way

$$\int_0^1 x^{j-1} v(x)dx = \sum_{k=1}^m a_{jk} v_k \quad \text{with} \quad a_{jk} = \int_0^1 b_k(x) x^{j-1} dx . \tag{1}$$

The matrix $A_X = (a_{jk})$ $(j=1,\ldots,n \; ; \; k=1,\ldots,m)$ is the discretization of the operator A and defines a linear system that relates moments and elements of X_m (regarded as m-dimensional vectors). For simplicity we limit ourselves to the case m=n.

Ex. 2a. Approximation by polynomials. If $X_n=\{$polynomials of degree lower than n$\}$ with basis $1,x,x^2,\ldots,x^{n-1}$, A_X is the Gram matrix of such a basis and coincides with the nth segment H_n of the Hilbert matrix; if we use orthogonal polynomials (Legendre) as a basis of X_m , we obtain that A is a triangular non-singular matrix such that $A_X^* A_X = H_n$ so that

$k_2(A_X) = \sqrt{k_2(H_n)}$ (k_2 indicates the spectral condition number); see for example Talenti,'87 .

Ex. 2b. Approximation by piecewise constant functions . Once fixed the nodes

$x_0=0 < x_1 < \ldots < x_{n-1} < x_n=1$ and $b_k(x) = \chi_{(x_{k-1}, x_k)}(x)$, we have $a_{jk} = \int_{x_{k-1}}^{x_k} x^{j-1} dx = \dfrac{x_k^j - x_{k-1}^j}{j}$.

Thus $A_X = \text{Diag}(1/i) \, V_n \, \text{Diag}(x_i) \, B$ where V_n is a Vandermonde matrix and $B=(b_{jk})$ is such that $b_{jk}=1$ for j=k and $b_{jk}=-1$ for j=k–1.

Hence in this case too A_X is invertible and we can estimate its condition number (see Inglese,'88).

Ex. 2c. Approximation by continuous piecewise linear functions . In this case we consider the following representation : $v(x) = \alpha - \int_x^1 \sum_{j=1}^{n-1} \chi_{(x_{j-1}, x_j)} m_j$. The approximation is now determined

by its value α for x=1 and by its angular coefficients m_j j=1,...,n–1 .

We obtain $a_{j1} = \dfrac{1}{j}$ $(j=1,\ldots,n)$, $a_{j\,k+1} = -\dfrac{x_k^{j+1} - x_{k-1}^{j+1}}{j(j+1)}$ $(j=1,\ldots,n \; ; \; k=1,\ldots,n-1)$. This matrix too is invertible and we have the following estimates for the condition numbers (for the proof, see Appendix) :

76

$$\min_{x_1,\dots,x_{n-2}} k_2(A_X) \geq \frac{2^{n-2}}{n^5} \quad,$$

$$k_2(A_X) \leq n^3 \ln n \ 2^{3n} \quad \text{(equally spaced nodes)}.$$

With respect to the three examples above we have the following theorem:

Theorem. Let v be the (unique) function in X_n having $\mu_1,\dots,\mu_n$ as first moments. Suppose moreover that $\dfrac{\varepsilon}{\|\mu\|_2} < 1$. We have the following estimate for the accuracy of the approximation:

$$\frac{\|u-v\|_2}{\|u\|_2} \leq \text{Constant} \left\{ \frac{E}{\|\mu\|_2} \gamma(n) + \delta(n) \frac{\varepsilon}{\|\mu\|_2} \right\} \theta(\varepsilon) \ ,$$

where

$$\gamma(n) = \begin{cases} \displaystyle\max_{1 \leq k \leq n} (x_k - x_{k-1}) & \text{ex. 2b} \\[2ex] \displaystyle\max_{1 \leq k \leq n-1} (x_k - x_{k-1}) & \text{ex. 2c} \\[2ex] \dfrac{1}{n} & \text{ex. 2a} \end{cases} \quad,$$

$$\delta(n) = k_2(A_X)$$

and

$$\theta(\varepsilon) = \frac{\sqrt{\pi}}{\sqrt{1 - \dfrac{\varepsilon^2}{\|\mu\|_2^2}}} \quad.$$

Proof. Let $P_n u$ be the function in X_n having exactly the same first n moments as u. We remark that these moments differ from the μ_j's at most by ε in euclidean norm. Hence we have

$$\|u-v\|_2 \leq \|u-P_n u\|_2 + \|v-P_n u\|_2 \ .$$

The first term on the right hand side (discretization error) is lower than (or equal to) $(1+\|P_n\|_2)\cdot$ $d(u,X_n)$; in the case 2.a, thanks to the orthogonality of the basis, it is simply lower than $d(u,X_n)$.

Such a distance can be evaluated (see Talenti,'87 for 2.a and Inglese,'88 for 2.b; 2.c is analogous to 2.b) and it turns out to be finite because of the limitation (ii) : $d(u,X_n) \leq$ Constant E $\gamma(n)$. The second term is the error on the data magnified by the condition number of A_X ; we gave some estimate about it in the previous examples.

Remark 1. The discretization error converges to zero when n goes to infinity; our estimates give a rate like $\frac{1}{n}$. The second term increases exponentially in n reflecting in this way the severe ill-posedness of the HMP. Hence the sum of the two terms takes a minimum for a value n_{opt} (see for example Talenti,'87); numerical tests confirm the behaviour but, fortunately, with an optimal value larger than n_{opt}.

Remark 2. If one uses splines, a suitable choice of the nodes can improve the approximation. In the next section we will analyze this fact and describe numerical tests; we will use only step functions but a further developement of the research might consider the use of higher order splines. The regularization role of high order splines is studied in Natterer,'77.

3   Step functions approximation: optimal collocation of the nodes

Using splines we can try to improve the approximation by moving the nodes. Two strategies, not necessarily compatible, are the following:

(i)   minimize the discretization error ,

(ii)   minimize the condition number of the matrix A .

We can't expect the conditioning to be improved radically because of the just seen lower bounds for $k_2(A_X)$. However it is possible to compute the minimum of $k_2(A_X)$ as a function of $x_1,...,x_n$; we used MINUIT package of CERN library for $n=1,...,15$. The optimal $x_{opt,i}$'s approximately behave like the zeros of Legendre polynomials . Moreover we used $y_k=x_{opt,k}$, $y_0=0$, $y_1=1$ as nodes for the step function approximation ; convergence to a quite precise approximation was observed for different test functions with exact data . Let us show in Figure 1 the beaviour of such an approximation for $u(x) = \frac{1}{(x+.1)^2} - \frac{4}{x+.1}$. We remark that in this example the approximation is particularly good (see Inglese,'88 for further details). In

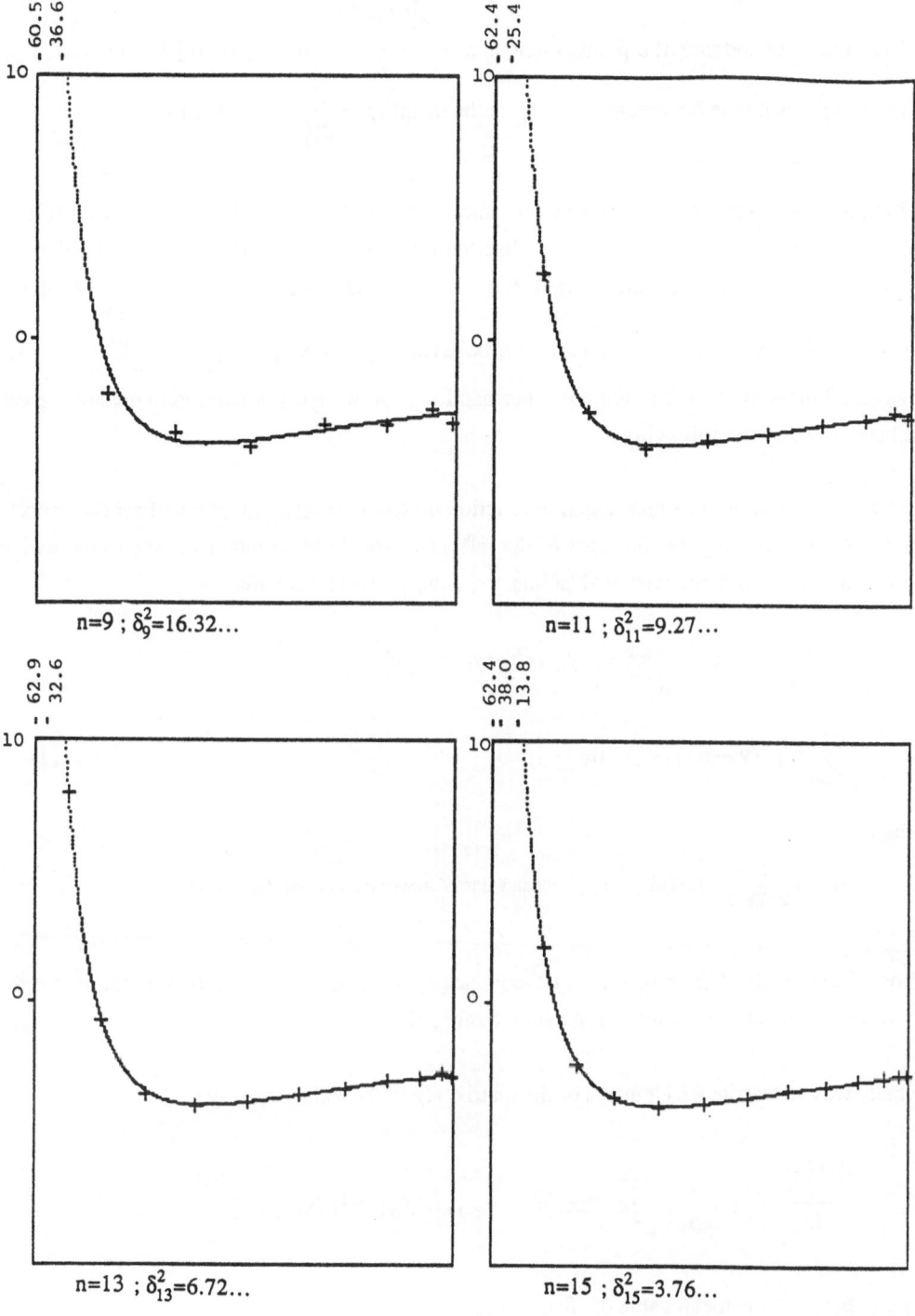

Figure 1. Approximation of $u(x)=1/(x+.1)^2 - 4/(x+.1)$ using 9,11,13,15 moments and optimal configurations of the nodes

Figure 1 crosses indicate the points (z_k, v_k), $z_k = \dfrac{x_{k-1}^{(opt)} + x_k^{(opt)}}{2}$ and v_k as in §2, for $k=1,\ldots,n$;

the continuous line is the graph of u. δ_n^2 is the quantity $\dfrac{1}{n} \sum_{k=1}^{n} |v_k - u(z_k)|^2$.

Though a theoretical explanation of this phenomenon is still far, there are some results in approximation theory related to our observations. First let us recall that, provided not restrictive conditions for the weight function u, the zeros of the relative orthogonal polynomials have a common asymptotic behaviour (precisely $\eta_k = \dfrac{1 + \cos\dfrac{\pi k}{n+1}}{2}$; see Szegö,'39 theorem 12.7.2). On the other hand we can formally relate moments and "good" values of v, by the following

Proposition. Let u be a non-negative function on $(0,1)$. Let $\mu_1,\ldots,\mu_n$ be its first (Hausdorff) moments and $\xi_1,\ldots,\xi_n$ be the zeros of the nth orthogonal polynomial (relative to the weight function u). Then there exist $n-1$ points $y_1,\ldots,y_{n-1} \in (0,1)$ such that

$$y_0 \equiv 0 < \xi_1 < y_1 < \xi_2 < \ldots < \xi_{n-1} < y_{n-1} < \xi_n < y_n \equiv 1 \quad \text{and}$$

$$\sum_{k=1}^{n} \xi_k^{j-1} (y_k - y_{k-1}) \, v_k = \mu_j \tag{2}$$

where

$$v_k = \frac{1}{y_k - y_{k-1}} \int_{y_{k-1}}^{y_k} u(t)\,dt = \bar{u}_k \quad \text{(mean value of u over the interval (y_{k-1}, y_k)).}$$

Proof. This result is an immediate consequence of a classical theorem of separation for the zeros of ortogonal polynomials; see again Szegö,'39, theorem 3.41.1 .

Finally we can rewrite the elements of the matrix A_X in the following way :

$$\frac{y_k^j - y_{k-1}^j}{j} = \left(\frac{1}{y_k - y_{k-1}} \int_{y_{k-1}}^{y_k} x^{j-1} dx \right)(y_k - y_{k-1}) = E_k[x^{j-1}] \, (y_k - y_{k-1})$$

where $E_k[f]$ is the mean value of f over (y_{k-1}, y_k), and

$$\sum_{k=1}^{n} E_k[x^{j-1}] \, (y_k - y_{k-1}) \, v_k = \mu_j \qquad\qquad (2bis)$$

It might be interesting to compare (2bis) and (2) and try to understand if the analogies are only formal.

Finally, the above arguments suggest a question : can we restrict the class of the unknown functions of finite HMP (with exact data) in a way that the stepwise constant approximation with nodes x_{opt} (or η?) is rapidly covergent to $\bar{u}_k$?

Some final remarks : because of the severe ill-conditioning of the problem, a regularization method is necessary when data are affected by noise . In this case it seems that accuracy and stability of the approximation are independent on the choice of the nodes. It was foreseeable in the light of our previous considerations: in fact the improvement due to a good collocation of the x_i's is observed essentially in the discretization term. Recent applications of a regularization Tikhonov method to HMP are due to Frontini et al.,'89 ; they show a successful implementation of an algorithm for finite HMP is succesful. Moreover, in Amato,'89 regularization by maximum entropy method is studied .

4   Appendix.

Proposition 1. The matrix A_x in 2.c is invertible.

Proof. We recall the following properties of determinants (see for example Bini et al.,'88):

(a) $\det B = \det A$ if B is obtained from A adding to a column another column times a number.

(b) $\det B = \gamma \det A$ if B is obtained from A multiplying a column times γ.

Substituting column j^{th} (j=2,n) with the sum of the column from j to n, we obtain a matrix A' with the first column equal to the first column of A_x, whilst the jth column is $(-\dfrac{1-x_{j-1}^{i}}{i})$ where i=2,...,n+1 ($x_0=0$). Again subtracting the second column from the third one, then from the fourth one and so on we obtain

$$A'' = \begin{bmatrix} 1 & -1/2 & x_{j-1}^{2}/2 & \cdots & x_{n-1}^{2}/2 \\[2mm] 1 & -1/3 & x_{j-1}^{3}/3 & \cdots & x_{n-1}^{2}/3 \\[2mm] \cdots & \cdots & \cdots & \cdots & \cdots \\[2mm] 1 & \dfrac{-1}{(n+1)} & \dfrac{x_{j-1}^{n+1}}{(n+1)} & \cdots & \dfrac{x_{n-1}^{n+1}}{(n+1)} \end{bmatrix}$$

From the recalled properties of determinants it follows that if A'' is invertible, then A_X is invertible too. Let us suppose now (by contrapposition) that the <u>rows</u> of A'' are linearly dependent; in this case there is a non zero vector $(c_1,...,c_n)$ such that $\sum_{i=1}^{n} c_i = 0$ and

$$\sum_{i=1}^{n} c_i \frac{x_{j-1}^{i+1}}{i+1} = 0 , \quad j=2,..., n.$$

Let $p(x) = \sum_{i=1}^{n} c_i \frac{x^{i+1}}{i+1}$; $p(x)$ has a zero of 2nd order in $x=0$. Moreover from

$$p'(1) = \sum_{i=1}^{n} c_i = 0,$$ it follows that $x=1$ is another zero of 2nd order. On the other hand, the points x_j $j=1,...,n-2$ are zeroes of $p(x)$ by hypotesis. But p is a polynomial of maximum degree $n+1$ and so it cannot have $n+2$ zeroes without being identically zero. Finally the c_i's are all zero and the rows of A'' are independent.

Proposition 2. $\quad \min_{x_1,...,x_{n-2}} \|A_X^{-1}\|_\infty \geq \frac{2^n}{5n^4}$.

Proof. Let us consider the matrix $A' = \text{Diag}[i]\,\text{Diag}[i+1]_{i=1,...,n}\,A_X$. The elements of the first column are $2,3,...,n$; we name the others $-a_{ij}$ where $a_{ij}=(x_j^{i+1} - x_{j-1}^{i+1})$ $(i=1,...,n ; j=1,...,n-1)$.

Let B be the auxiliary matrix

$$\begin{bmatrix} 0 & 0 & \cdots & 0 & 1 \\ -b_{11} & -b_{12} & \cdots & -b_{1,n-1} & 0 \\ \cdots & \cdots & \cdots & \cdots & \cdots \\ -b_{1,n-1} & -b_{2,n-1} & \cdots & -b_{n-1,n-1} & 0 \end{bmatrix}$$

where $\sum_{k=1}^{n-1} b_{ik} a_{kj} = \delta_{ij}$ (Kronecker δ) . Then we consider the product $A'\,B$ and obtain the $n \times n$ matrix

$$L = \begin{bmatrix} 1 & 0 & \cdots & 0 & 2 \\ 0 & 1 & \cdots & 0 & 3 \\ \cdots & \cdots & \cdots & \cdots & \cdots \\ 0 & \cdots & \cdots & 1 & n \\ c_1 & c_2 & \cdots & c_{n-1} & n+1 \end{bmatrix}$$

where $c_k = \sum_{l=1}^{n-1} a_{n,l} b_{lk}$. Now let B' and $\bar{A}'$ the $(n-1) \times (n-1)$ matrices respectively defined by

the b_{ij}'s and by the a_{ij}'s. Observe that $\bar{A}' = V_{n-1} \text{Diag}[x_i^2] B_{n-1}$ and $\bar{A}' B' = I_{n-1}$.

Moreover the row $a_{n,1}, \ldots, a_{n,n-1}$ is the $(n-1)$th row of the matrix $A'' = V_{n-1} \text{Diag}[x_i^3] B_{n-1}$.

Hence every c_k is equal to the $(n-1,k)$ element of the matrix $A^\circ \equiv A''B' = V_{n-1} \text{Diag}[x_i] V_{n-1}^{-1}$.

Let us recall that $L = A'B$ and so $B = A'^{-1} L$ (A' is invertible from proposition 1); since $\|A'^{-1}\|_\infty \geq \frac{\|B\|_\infty}{\|L\|_\infty}$ and $\|B\|_\infty > \frac{2n-2}{n-1}$ (see Gautschi et al.,'88) only remains to estimate $\|L\|_\infty$.

It is easy to check that $\|L\|_\infty = (n+1) + \sum_{k=1}^{n-1} |c_k| \leq n+1 + \|A^\circ\|_\infty$; A° is similar to $\text{Diag}[x_i]$ by means

of the similitude matrix V_{n-1}. Then (see for example Bini et al.,'88) the x_i's are eigenvalues of A° whilst the columns w_i of V_{n-1} are linearly independent eigenvectors; moreover they are normalized to have ∞–norm equal to 1.

Let $u = \sum_{k=1}^{n-1} \hat{u}_k w_k \in \mathbb{R}^{n-1}$ be a vector of ∞–norm equal to 1 ; then

$$\|A^\circ\|_\infty = \max \|A^\circ u\|_\infty \leq \max \sum_{k=1}^{n-1} |\hat{u}_k| x_k \leq n-1.$$

and we have $\|L\|_\infty \leq \frac{3}{2}(n-1) + \frac{5}{2}$ and $\|A'^{-1}\|_\infty \geq \frac{2^n}{5n^2}$; finally $\|A_X^{-1}\|_\infty \geq \frac{2^n}{5n^4}$.

The result for the euclidean norm comes immediately from the well–known relations between norms.

Proposition 3. $\|A_X^{-1}\|_\infty \leq n^2 \ln n \ 2^{3n+1}$ when the nodes are equally spaced.

Proof. We observe that A" (only changing the sign of the second column) is a confluence of the nth to the 1st column of the matrix $W_n = V_n(x) \text{Diag}(x_i^2)$. In Gautschi,'62 an estimate is

given for confluent Vandermonde matrices ; the same calculations with W_n instead of V_n lead

us to prove the statement.

References

Amato,U. 1989: Regularization by entropy for first kind Fredholm integral equations, Rapp.Tecnico IAM, 53/89 .

Askey R.,Schoenberg I.J. and Sharma A. 1982: Hausdorff Moment Problem and Expansion in Legendre Polynomials , J.Math.Anal.Appl., 86.

Bertero M., De Mol C. and Pike R. 1985: Linear inverse problems with discrete data I , Inverse Problems, 1.

Bini D.,Capovani M. and Menchi O. 1988: Metodi numerici per l'algebra lineare, Zanichelli, Bologna .

Frontini M.,Rodriguez G.,Seatzu S. 1989: An Algorithm for Computing Minimum Norm Solutions of Finite Moment Problems , in Algorithms for Approximation, J.C.Mason   and M.G.Cox eds., Chapman-Hall .

Gautschi W. 1962: On Inverses of Vandermonde and Confluent Vandermonde Matrices, Numer.Math. 4.

Gautschi W. and Inglese G. 1988: Lower Bounds for the Condition Number of Vandermonde Matrices , Numer.Math.,52.

Hardy G., Littlewood J.E. and Polya G. 1964: Inequalities, Cambridge U.P., Cambridge.

Inglese G. 1988: Approximate solutions for a finite moment problem, Calcolo, vol.25 n.3.

Natterer F. 1977: Regularisierung schlecht gestellter Probleme durch Projektionsverfahren, Numer.Math.28.

Shoat J.A. and Tamarkin J.T. 1943: The problem of the moments, AMS Math. Survey.

Szegö G. 1939: Orthogonal Polynomials, AMS Coll.Publ. vol XXIII , New York .

Talenti G. 1987: Recovering a function from a finite number of moments, Inverse Problems, 3, 501-518 .

On the Structure of Uniqueness in Linear Inverse Source Source Problems

L. Ballani, D. Stromeyer

Central Institute for Physics of the Earth, Telegrafenberg, 1561 Potsdam, Germany

Abstract

The formulation and solution of inverse problems require a functional analytic embedding in each case. As an important example for a linear inverse source problem, the inverse gravimetric problem is considered in the frame of Hilbert function space. Different ways are demonstrated how the uniqueness and nonuniqueness of the solvability of this inverse problem can be characterized. For this purpose, bases functions being linearly independent or orthogonal as well as trial functions, e.g. point masses, are efficient tools. They permit description of mass distributions with zero potential in the outer space and give the possibility to determine and include additional conditions which can lead to uniqueness to a certain degree if the solution belongs to a corresponding model class. Especially for the model class of finite point mass systems, the unique solvability of the inverse gravimetric problem is proved.

1. Introduction

The inverse gravimetric problem is a classical topic in theoretical geophysics as well as an exciting question again and again. Recently its study is going on more intensively. The importance has still increased, especially by the closer connection between geodesy and geophysics, also based on modern geodetic measuring techniques, and last but not least because of sophisticated geodynamic theories developing quickly. The main point is - as stressed e.g. in *Anger* (1990) - to find clear mathematical structures, in our case for the sources represented by the right-hand sides (inhomogeneities) of linear differential equations.

In the following pages some essential mathematical statements on the uniqueness and nonuniqueness of the inverse gravimetric problem shall be discussed. The connection with other geophysical fields concerning the effect of nonuniqueness from the physical point of view was already impressively demonstrated by *Buchheim* (1974).

Some simple principles (like a "philosophy") are necessary to guarantee a successful mathematical handling of inverse problems and thus their practical interpretation as well:
- Mathematical embedding of the problem
- Investigation of the mathematical structures
    - Study of unique solvability/ search for equivalent descriptions/zero space
    - Decomposition of the source by means of bases or trial functions
- Truncation to get finite systems under the aspect of spectral, functionally decomposed source characteristics
- Analysis concerning the inclusion of additional conditions to restrict the set of solutions/choice of "model classes"
- Solving procedure on the basis of a well-fitted stabilizing algorithm ("regularization")

The last aspect of regularization will be not included here, but for instance *Oganesjan* 1981 is related to our frame of Hilbert function space. Recently, *Seidman and Vogel* 1989 have shown that in the uniqueness case the true solution can be reached by regularization.

This paper can be considered as the continuation of *Ballani and Stromeyer* (1982), *Stromeyer and Ballani* (1984) and *Ballani* (1986). Some important results are presented once more.

2. Mathematical embedding: The inverse gravimetric problem in Hilbert space

The inverse gravimetric problem shall be considered here (cf. *Ballani and Stromeyer* (1982)) in the very useful and convenient frame of Hilbert space theory. The Hilbert space is represented here by the function space $L_2(G)$, $G \subset \mathbb{R}^3$ bounded, containing all real-valued quadratically integrable functions on G.

In detail the inverse gravimetric problem consists in finding a density function $f \in L_2(G)$ from some knowledge of the (outer) potential V in $G_a = \mathbb{R}^3 \backslash \bar{G}$. The quantities V and f are connected by the known formulae (γ - gravitational constant)

$$V(x) = \gamma \int_G \frac{f(x)}{\|x - y\|} dG_y \qquad x \in G_a \qquad \text{(Newtonian potential)} \qquad (2.1)$$

$$\Delta V(x) = \begin{cases} -4\pi\gamma f & x \in G & \text{(Poisson equation)} \\ 0 & x \in G_a & \text{(Laplace equation)} \end{cases} \qquad (2.2)$$

The basis for the following considerations is formed by a theorem of *Weck* (1972) concerning the orthogonal decomposition of $L_2(G)$:

$$L_2(G) = N(\Delta) \oplus \Delta\mathring{H}_2(G) \qquad (2.3)$$

with $N(\Delta) := \{ h: h \in L_2(G), \; \Delta h = 0 \}$ \qquad (2.4)

and $\Delta\mathring{H}_2(G) := \{ g: g = \Delta v, \; v \in \mathring{H}_2(G) \}$ \qquad (2.5)

$\mathring{H}_2(G)$ is the closure of $C_0^\infty(G)$ in the norm of the Sobolev space $W_2^2(G)$. For regular regions $\mathring{H}_2$ is characterized by the equivalence

$$v \in \mathring{H}_2 \iff v \in H_2 \quad \text{and} \quad v\Big|_{\partial G} = \frac{\partial v}{\partial n}\Big|_{\partial G} = 0,$$

where ∂G is the boundary of G.

The second item of the decomposition (2.3) -the space $\Delta\mathring{H}_2$ ="anharmonic rest"- exactly contains those density functions having zero-potential in the outer space G_a. That means the general solution of the inverse gravimetric problem (without additional conditions) for the density as solution has the form

$$f = h + \Delta\mathring{H}_2 \qquad (2.6)$$

with unique h, $h \in N(\Delta)$. For other analogous inverse problems the possibility of the existence of several different geophysical source functions generating zero-potential or zero-field in the outer space ("Nullaffekt") was demonstrated by *Buchheim* (1974) applying directly the integral formulae of Green and Gauss.

3. Equivalent descriptions

A way often used to solve the inverse gravimetric problem is the restriction of all possible density models to certain function classes. In view of the practical determination of "free" model parameters, the question concerning the uniqueness of the inverse problem for a given family of density models ("model class") is of special interest.

In order to characterize the cases of uniqueness and nonuniqueness some simple derivations or modifications of (2.3) and (2.6) will be helpful in the following:

Theorem: The inverse gravimetric problem has at most one solution in the model class $A \subset L_2$ if for all $f, \bar{f} \in A$, $f \neq \bar{f}$

$$f - \bar{f} \notin \Delta \mathring{H}_2$$

holds.

Because of the equivalence of this statement with the relation

$$(f - \bar{f})\big|_{N(\Delta)} \neq 0$$

($g_{N(\Delta)}$ denotes the harmonic part of g), two different functions have also different harmonic parts if the inverse problem is uniquely solvable in the model class A. If A is even a linear space one easily finds the relation

$$f \neq 0, \quad f \in A \iff f\big|_{N(\Delta)} \neq 0.$$

Density functions $f, \bar{f}$, $f \neq \bar{f}$ with identical outer potentials fulfil the relation $(f - \bar{f}) \in \Delta \mathring{H}_2$. Another formulation for this fact is given by the coincidence of their harmonic parts

$$f\big|_{N(\Delta)} = \bar{f}\big|_{N(\Delta)}$$

as well as the difference of their anharmonic rests

$$f\big|_{\Delta \mathring{H}_2} \neq \bar{f}\big|_{\Delta \mathring{H}_2}.$$

Remark: A formulation equivalent to the theorem is the following one:

For the uniqueness of the inverse problem in A, A $\subset$ $\mathbb{L}_2$, it is necessary and sufficient that for f, $\bar{f}$ $\in$ A, f$\neq\bar{f}$ from

$$(f - \bar{f}, h) = 0 \qquad\qquad (3.1)$$

for all h $\in$ $\mathbb{N}(\Delta)$

$$f = \bar{f}$$

follows.

The validity of this statement, however, is not constrained to squared integrable density functions but can also be formulated for arbitrary measures μ, η $\in$ A (for instance with point support = "point masses") as generalized momentum identity (*Anger* (1990), *Schulze and Wildenhain* (1977))

$$\int_G h(x)\mu(x)dx = \int_G h(x)\eta(x)dx \ , \quad h \in \mathbb{N}(\Delta) \qquad (3.2)$$

The two density distributions $\mu(x)$ and $\eta(x)$ have the same Newtonian Potential in the outer space G_a if and only if (3.2) holds for all h $\in$ N(Δ).
If -instead of (3.1)- from (3.2) for all h $\in$ $\mathbb{N}(\Delta)$ $\mu=\eta$ follows, the unique solvability of the inverse gravimetric problem in A is fulfilled (sufficiency criterion).
It is an interesting technical fact that one often even succeeds in constructing special harmonic functions h(x), so that from (3.2) the identity $\mu=\eta$ immediately can be concluded.

4. Inverse gravimetric problem for Point Mass Models

Point mass models are a very useful and efficient tool for the approximation of the gravitational field (*Barthelmes* (1986)). On the other hand they form a model class of mass (density) distributions for the inverse gravimetric interpretation. It is interesting that the property of a finite number of point mass potentials to be linearly independent, which is important for numerical processes, is identical with the unique solvability of the inverse problem in the model class of finite point mass models. The mathematical proof for this unique solvability in the case of two-, three- or n-dimensional space (c.f. *Stromeyer and Ballani* (1984)) is founded on the generalized momentum identity (3.2). In order to illustrate the features here, only the two-dimensonal case is presented.

Let μ be a finite point mass model linearly combined by means of δ-distributions

$$\mu = \sum_{j=1}^{n} m_j \delta_j := \sum_{j=1}^{n} m_j \delta(x-x_j, y-y_j), \qquad (4.1)$$

$m_j \neq 0$, $(x_j, y_j) \neq (x_k, y_k)$ for $j \neq k$, and $\eta=0$ the zero distribution. After having inserted μ and η in (3.2) this identity is transferred to

$$\int_G h(x,y) \sum_{j=1}^{n} m_j \delta_j \, dxdy = 0. \qquad (4.2)$$

Choosing as harmonic functions $h_k(x,y) = (x + iy)^k = z^k$, $i^2 = -1$, $k = 0,1,\ldots n-1$ and taking into account that

$$\int_G z^k \delta_j \, dxdy = (x_j + iy_j)^k = z_j^k$$

holds, then the coefficient matrix D of the linear homogeneous system (4.2) for the unknown quantities m_j has the form

$$D = \begin{bmatrix} 1 & 1 & \ldots 1 \\ z_1^1 & z_2^1 & \ldots z_n^1 \\ \vdots & \vdots & \vdots \\ z_1^{n-1} & z_2^{n-1} & \ldots z_n^{n-1} \end{bmatrix}$$

Its determinant det(D), often named Vandermonde's determinant, has the known analytical expression

$$\det(D) = \prod_{1 \leq k < j \leq n} (z_j - z_k)$$

and can therefore be equal to zero only if one factor $z_j - z_k$ vanishes (two mass points coincide). Hence the system (4.2) has for different mass point locations only the trivial zero solution $m_j = 0$ for $j = 1, \ldots, n$. That means that no finite point mass model can generate zero potential in the .outer space. On the other hand this statement is equivalent with the unique solvability of the inverse gravimetric problem in this model class. The proof in higher dimensions requires the choice of other suited harmonic functions h to be inserted in (4.2) (*Stromeyer and Ballani* (1984)).

5. Bases systems and nonuniqueness

Basing on (2.3) the nonuniqueness of the inverse gravimetric problem in the Hilbert space $L_2(G)$ can be very impressively and systematically described by means of structural statements of the space $\Delta \mathring{H}_2$ containing exactly the density functions with outer zero potentials ("anharmonic rests"). Looking for possibilities to find bases function systems for the zero space $\Delta \mathring{H}_2$, there exist two ways on principle.

5.1 First Possibility ("outer way")

Starting with (2.3) $\Delta \mathring{H}_2$ can be interpreted as "orthogonal difference"

$$\Delta \mathring{H}_2 = L_2(G) \ominus N(\Delta) \tag{5.1}$$

In general this leads to the problem how to split up an orthogonal basis $\{e_i\}$ of L_2 into the harmonic part $\{h_i\} \subset N(\Delta)$ and the anharmonic rest $\{g_i\} \subset \Delta \mathring{H}_2$.

 -The most convenient case is given if the inclusion

$$\{e_i\} \supset \{h_i\} , \qquad \{h_i\} \text{ orthogonal basis in } N(\Delta), \tag{5.2}$$

 exists which immediately results in

$$\{g_i\} = \{e_i\} \setminus \{h_i\}. \tag{5.3}$$

 and thus $\qquad\qquad \{g_i\} \subset \Delta \mathring{H}_2 .$

 ┌──────────┐
 │ Example 1 │ Legendre polynomials $\{P_k\}$ on $G = [-1, 1]$
 └──────────┘
$$\{e_k\} = \{P_k\}, \qquad k = 0, 1, \ldots \qquad \{h_k\} = \{P_0, P_1\}$$
$$\{g_k\} = \{P_2, P_3, \ldots\}$$

-The general case, if an orthogonal harmonic basis $\{h_i\}$ on G is known, can be handled by orthogonal projection

$$g_k = e_k - \sum_j (e_k, h_j) \frac{h_j}{\| h_j \|^2} \qquad (5.4)$$

The function system $\{g_k\}$ constructed in this way is in every case a complete system in $\Delta\overset{\circ}{H}_2$, but is in general no longer linearly independent.

Example 2   For $G = [-\pi, \pi]$, $\{e_k\} = \{1, \sin(kx), \cos(kx), \ldots\}$, $k=1,2,\ldots$ and $\{h_k\} = \{1,x\}$ we get

$$\{g_k\} = \left\{\sin(kx) - \frac{3(-1)^k x}{\pi^2 k}, \cos(kx)\right\} \quad k=1,2,\ldots$$

Example 3   Let $G = K(0,1) \subset \mathbb{R}^2$ be the unit circle. Taking as bases the functions for $\mathbb{L}_2(G)$

$$\{e_{kl}\} = \left\{ r^k \begin{bmatrix} \sin(l\phi) \\ \cos(l\phi) \end{bmatrix} \right\} \quad k,l = 0,1,2,\ldots,$$

and for $\mathbb{N}(\Delta)$

$$\{h_l\} = \left\{ r^l \begin{bmatrix} \sin(l\phi) \\ \cos(l\phi) \end{bmatrix} \right\} \quad l=0,1,2,\ldots$$

the constructive equation (5.4) provides for $\Delta\overset{\circ}{H}_2$

$$\{g_{kl}\} = \left\{ \left[r^k - \frac{2l+2}{l+k+2} r^l \right] \begin{matrix} \sin(l\phi) \\ \cos(l\phi) \end{matrix} \right\}, \quad k \neq l, \quad k,l=0,1,2\ldots$$

In the special case of spherical symmetry ($l=0$) only a constant density h_0 is harmonic and a complete basis system of "sources" with zero potential is

$$\{g_k\} = (r^k - 2/(k+2)), \quad k=1,2,\ldots$$

Example 4   Let $G = K(0,1) \subset \mathbb{R}^3$ the full unit sphere. With the complete system in $\mathbb{L}_2(G)$

$$\{e_{knm}\} = \left\{ r^k P_{nm}(\cos\theta) \begin{bmatrix} \cos(m\lambda) \\ \sin(m\lambda) \end{bmatrix} \right\}, \quad k,n=0,1,2\ldots; \quad m=0,1,\ldots,n.$$

(r,θ,λ) spherical coordinates, P_{nm} associated Legendre polynomials and

$$\{h_{nm}\} = \left\{r^m P_{nm}(\cos\theta)\begin{bmatrix}\cos(m\lambda)\\\sin(m\lambda)\end{bmatrix}\right\}, \quad n=0,1,2\ldots; \quad m=0,1,\ldots,n.$$

in $N(\Delta)$ the zero space $\Delta\mathring{H}_2$ basis

$$\{g_{knm}\} = \left\{(r^k - \frac{2n+3}{k+n+3}\, r^n)P_{nm}(\cos\theta)\begin{bmatrix}\cos(m\lambda)\\\sin(m\lambda)\end{bmatrix}\right\}, \quad \begin{array}{l}k\neq n,\, k,\, n=0,1,2\ldots\\ m=0,1,\ldots,n\end{array}$$

can be constructed.

5.2 Second possibility:

An "inner way" which avoids the necessity to know any harmonic basis $\{h_k\}$ in $N(\Delta)$ uses the direct characterization of the space $\mathring{H}_2$:

Theorem: Let $S(x)$ be a function which defines the boundary ∂G of G (sufficiently smooth) by means of $S(x)=0 \iff x \in \partial G$. Then the equivalent relation

$$v \in \mathring{H}_2 \quad \iff \quad v = S^2 f, \quad f \in L_2(G) \tag{5.5}$$

is valid.

Hence, every density function $g \in \Delta\mathring{H}_2$ must have the form

$$g = \Delta v = \Delta(S^2 f), \quad v \in \mathring{H}_2, \quad f \in L_2(G) \tag{5.6}$$

The property of completeness of the $L_2(G)$-system $\{e_k\}$ is transmitted to both systems $\{v_k\} \subset \mathring{H}_2$ and $\{g_k\} \subset \Delta\mathring{H}_2$ as well, if these systems are generated from $\{e_k\}$ by means of(5.5) and (5.6):

$$g_k = \Delta v_k = \Delta(S^2 e_k)$$

However, the connection between $\{g_k\}$ and $\{e_k\}$ is not so clear as in the case of orthogonal projection (5.4). This is a certain disadvantage in studying linear independence and orthogonality of the derived system $\{g_k\}$.

Example 5 $G = [-\frac{\pi}{2}, \frac{\pi}{2}] \times [-\frac{\pi}{2}, \frac{\pi}{2}]$

$L_2(G)$ system: $\{e_{kl}\} = \{\cos(kx)\cos(ly)\}, \quad k, l = 0, 1, \ldots$

boundary function: $S(x,y) = \cos(x)\cos(y)$,

help-functions:

$$u_k(x) = 0.25(\cos((k-2)x) + 2\cos(kx) + \cos((k+2)x))$$

$$w_k(x) = 0.25((k-2)^2\cos(k-2)x + 2k^2\cos(kx) + (k+2)^2\cos(k+2)x).$$

complete system in $\Delta \mathring{H}_2(G)$:

$$\{g_{kl}(x,y)\} = \{u_k(x)w_l(y) + u_l(y)w_k(x)\}.$$

Example 6

$G = [-1,1] \times [-1,1]$,

$L_2(G)$ system: $\{e_{kl}\} = \{x^k y^l\}$, $k,l = 0,1,2,\ldots$

boundary function $S(x,y) = (1 - x^2)(1 - y^2)$.

The polynomial of lowest degree in $\Delta \mathring{H}_2$ has the form:

$$g_{00}(x,y) = 3x^2 y^2(x^2 + y^2) - 12x^2 y^2 - (x^4 + y^4) + 5(x^2 + y^2) - 2$$

6. Moments

The gravitational potential V of the Earth or any other planetary body is usually given as a series in terms of spherical harmonics $\{H_i\}$ with gravitational coefficients C_i containing the information on shape/topography and density distribution of the interior

$$V(r,\phi,\lambda) = \sum_j C_j H_j(r,\phi,\lambda). \tag{6.1}$$

Assuming complete bases $\{e_i\}$ for $L_2(G)$ and $\{h_j\}$, $\{g_k\}$ for $\mathbb{N}(\Delta)$ and $\Delta\mathring{H}_2$, respectively, the connection between (6.1) and the density development (6.2)

$$f = \sum_i (f,e_i)e_i = \sum_j (f,h_j)h_j + \sum_k (f,g_k)g_k \tag{6.2}$$

can be found by inserting (6.2) into the Newtonian potential and some simple calculations:

$$V(x) = \gamma \int_G \frac{f(y)}{\|x - y\|} dG_y = \gamma \sum_j (f,h_j) \int_G \frac{h_j(y)}{\|x - y\|} dG_y \tag{6.3}$$

The right hand side of (6.3) has exactly the structure of (6.1). The quantity C_j

$$C_j = (f,h_j) = (f|_{\mathbb{N}(\Delta)}, h_j) \tag{6.4}$$

with $$f = f|_{\mathbb{N}(\Delta)} \oplus f|_{\Delta\mathring{H}_2}$$

is usually called "moment" and is of its nature the projection of the (harmonic) density on the single basis function h_j. Hence from C_j only this harmonic part of the density can be uniquely concluded without additional information ("decomposed inverse problem"). The zero space of one moment C_j comprehends thus the whole space $L_2(G)$ reduced by the one-dimensional function space generated by h_j. Of course, the common zero space for all moments remains the space $\Delta \mathring{H}_2$.

Remark: The Earth rotation as an gyroscope motion seen from an Earth fixed position can be described by means of the angular momentum balance. Inserting the angular momentum $\mathfrak{H}$,

$$\mathfrak{H} = I \cdot \omega \quad , \quad \text{I tensor of inertia,} \quad \omega \text{ Earth rotation vector,}$$

in the balance equation the tensor of inertia

$$I = (\theta_{ik}) \quad i,k=1,2,3, \quad \theta_{ik} \text{ moments of inertia}$$

as well as its time derivative

$$\dot{I} = (\dot{\theta}_{ik}) \quad i,k=1,2,3$$

are contained there. Thus not only the moments of the density f but also of its timely variable increments $\dot{f}$ (mass geometry and its changes with time)

$$\left\{ \begin{array}{c} \theta_{ik} \\ \dot{\theta}_{ik} \end{array} \right\} = \int\limits_{\text{Earth}} \left\{ \begin{array}{c} f \\ \dot{f} \end{array} \right\} h_1^{(2)} dG \quad , \quad h_1^{(2)} \quad \begin{array}{l} \text{harmonic polynomial} \\ \text{of second degree} \end{array}$$

are given (c.f. Ballani (1986)). However, these moments are only related to harmonic polynomials of second degree, so that by this Earth rotation modelling (angular momentum approach) from Earth rotation data without additional information, only very confined global density estimations can be concluded: the projections onto these harmonic bases functions (applications, e.g. Jochmann (1984)).

7. Further examples: Additional conditions

The following examples will show how the study of the influence and the knowledge of additional conditions will be useful, particularly in the construction of reliable detailed Earth models under the aspect of the outer gravitational field.

Considering the model class A_R

$$A_R = \{f: f(r,\phi) = R(r)\Phi(\phi)\}, \quad R(r) \text{ known}$$

we are interested in studying uniqueness properties for the inverse gravitational problem. For the unit circle $K(0,1) \subset \mathbb{R}^2$ every density function $f \in A_R$ can be expanded in a Fourier series

$$f(r,\phi) = R(r)\Phi(\phi) = R(r)\sum_k (a_k \cos(k\phi) + b_k \sin(k\phi)).$$

From the remark in section 3, it follows that with harmonic functions

$$h_k = \left\{ r^k \left[\begin{array}{c} \cos(k\phi) \\ \sin(k\phi) \end{array} \right] \right\}$$

the relation $f = \bar{f}$ (with coefficients a_k, b_k, $\bar{a}_k$, $\bar{b}_k$ respectively) holds if and only if the property

$$\int_0^1 R(r)r^{k+1}dr \neq 0, \quad k = 0,1,2\ldots \qquad (6.6)$$

for the given function $R(r)$ is fulfilled (c.f. Zidarov, 1980: $R(r) = r^p$). If there exists one index k for which the relation (6.6) is not fulfilled, then the portion

$$f_k(r,\phi) = R(r)(a_k \cos(k\phi) + b_k \sin(k\phi))$$

is a density function with zero potential (a_k, b_k arbitrary !).

Let the model class A_R be specified in that way that for $R(r)$ a step function

$$R(r) = s_1 \quad \left\{ \begin{array}{l} r_{1-1} \geq r > r_1 \\ 1=1,\ldots,n; \quad r=1; \quad r=0 \end{array} \right.$$

with free parameters (r_1, s_1) is selected. From the uniquely determined harmonic part h of f

$$h = \sum_k (f,h_k)h_k / \|h_k\|^2$$

$$= \sum_k \frac{2k + 2}{k + 2} r^k (a_k \cos(k\phi) + b_k \sin(k\phi))\sum_{1=1}^n s_1 (r_{1-1}^{k+2} - r_1^{k+2})$$

can be concluded that only the composed coefficients A_k, B_k

$$\left\{ \begin{matrix} A_k \\ B_k \end{matrix} \right\} = \left\{ \begin{matrix} a_k \\ b_k \end{matrix} \right\} \sum_{l=1}^{n} s_l (r_{l-1}^{k+2} - r_l^{k+2}), \quad k = 0, 1, \ldots$$

are uniquely connected with a given outer potential and correspond to its expansion coefficients. Thus, the decision on uniqueness or nonuniqueness is transformed into a simple algebraic problem.

8 References

Anger,G. (1990): Inverse problems in differential equations.
London-New York: Plenum; Berlin: Akademie-Verlag.

Ballani,L.,Stromeyer,D.(1982): The Inverse Gravimetric Problem : A Hilbert Space Approach. Proc. Internat. Sympos. "Figure of the Earth, the Moon and other Planets", Prague, Czechoslovakia, 1982, Sept 20-25, Research Institute of Geodesy, Topography and Cartography, ed. P. Holota, Prague 1983, pp. 359-373.

Stromeyer,D.,Ballani,L.(1984): Uniqueness of the Inverse Gravimetric Problem for Point Mass Models. manuscr. geodaet. Vol.9, No 1/2, pp. 125-136.

Ballani,L.(1986): Inverse Problems concerning the Dynamics of the Earth's Rotation. Proc of the Internat. Sympos. "Figure and Dynamics of the Earth, Moon, and Planets", Prague, Czechoslovakia, 1986, Sept 15-20. Astron. Institute of the Czechoslovak Academy of Sciences, ed. P. Holota, Prague 1987, Part 3, pp. 851-878.

Barthelmes,F.(1986): Untersuchungen zur Approximation des äußeren Gravitationsfeldes der Erde durch Punktmassen mit optimierten Positionen. Veröff. d. Zentralinst. für Physik d. Erde Potsdam, Nr. 92.

Buchheim,W.(1974): Zur geophysikalischen Inversionsproblematik. Proc. Internat. Sympos. "Seismology and Solid-Earth-Physics", Jena, GDR, 1974, April 1-6. Veröff. d. Zentralinst. für Physik d. Erde Potsdam, 1975, Nr.31, part 2, pp. 305-310.

Jochmann, H. (1984): Die inverse Lösung der Differentialgleichung
der Erdrotation als Hilfsmittel zur
Interpretation zyklischer geophysikalischer
Ereignisse. Wiss. Z. Univ. Dresden, Vol.33, No.6 ,
pp.98-101.

Jochmann, H. (1987): The detection of global sea level changes by
inverse solution of equations of variations of
the vector of rotation. Gerl. Beitr. Geophysik,
Leipzig, Vol.96, No.3/4, pp.222-229.

Oganesjan, S. M. (1981): Rešenije obratnoj zadači gravimetrii v klasse
$L_2(S)$ raspredelenija plotnostej.
Dokl. Akad. Nauk Ukr. SSR, Ser. B, No.6, pp.39-43.

Schulze, B.-W., Wildenhain, G. (1977): Methoden der Potentialtheorie
für Differentialgleichungen beliebiger Ordnung.
Berlin: Akademie-Verlag; Basel, Stuttgart:
Birkhäuser.

Seidman, T. I., Vogel, C. R. (1989): Well posedness and convergence
of some regularisation methods for nonlinear
ill posed problems.
Inverse Problems, Vol.5, No.2, pp.227-238.

Weck, N. (1972): Zwei inverse Probleme in der Potentialtheorie.
Mitt. Inst. Theor. Geod. Univ. Bonn, Nr.4,
pp.27-36.

Zidarov, D. P. (1980): Some uniqueness conditions for the solution
of the inverse gravimetric problem.
C. R. Acad. Bulgar. Sci., Vol.33, No.7, pp.909-912.

(Contribution of the Central Institute for Physics of the Earth
Potsdam, No.1880)

Fast Algorithm for Solving Potential Field Problems

V. Bezvoda[1], J. Hrabe[1], K. Segeth[2]

[1] Charles University, Faculty of Sciences, Albertov 6, 12843 Prague, Czechoslovakia
[2] Czechoslovak Academy of Sciences, Mathematical Institute, Zitna 25, 11567 Prague, Czechoslovakia

Abstract

The present state of the art in computer science favors the use of fast methods. The direct as well as the inverse problem of potential field is a typical example. The efficient solution of the direct problem based on an application of 2-D FFT was devised by Bhattacharyya's group (Berkeley, California) in the 70's. A new study of this problem has proved the efficiency of the solution of the direct problem and has shown a promising way of solving the inverse problem.

1. Introduction

The importance of the 1-D DFT especially in connection with time series analysis in communication, radio location, etc. leads producers of computers to the construction of special hardware units, which implement the discrete Fourier transform (DFT) in a very effective way. As we deal with a "black box" it is not correct to speak about fast Fourier transform (FFT) — there are also other possibilities to perform the DFT quickly using special hardware approaches (Blahut 1985).

In principle, it is possible to state that multidimensional linear filtering (LF) of fairly large data sets is implementable also on standard PC's now. This fact is important especially for solving potential field problems where the funds are usually limited and, therefore, we must be much more modest in our demands both in hard- and software as compared, for example, with the solution of seismic or telecommunication problems.

2. Solving the direct problem of potential field

It is well known that one can write the potential of gravity in the form

$$(1) \qquad U(r) = \varkappa \int\limits_{V} \frac{\varrho_v(r')}{|r - r'|} \, dV',$$

where $\varrho_v(r)$ is the density of anomalous body, $\varkappa$ is the gravity constant, and the meaning of other symbols as well as the axes orientation are evident from Fig.1.

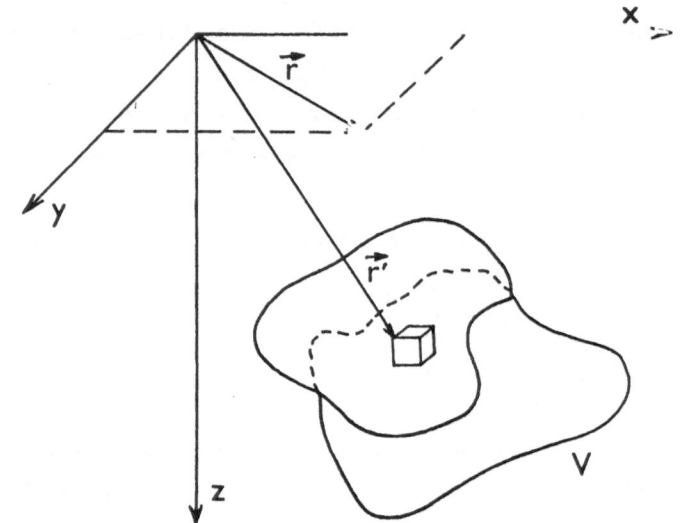

Fig. 1.

Axes orientation and notation used.

The other quantities important for gravity and magnetic fields are the derivatives of formula (1) and they are nothing new in principle.

If we put $\varrho_v(r)$ equal to zero outside the anomalous body, formula (1) transforms into the 3-D convolution

$$(2) \qquad U(r) = \varkappa \int\limits_{-\infty}^{\infty}\!\!\!\int\!\!\!\int \frac{\varrho_v(r')}{|r - r'|} \, dV' = \varkappa \, \varrho_v(r) \overset{3}{*} |r|^{-1},$$

where the index above the asterisk indicates the dimension of convolution.

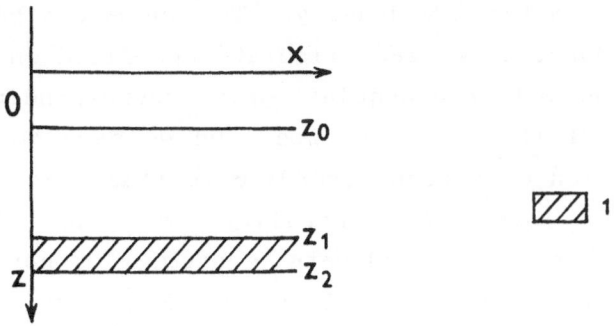

Fig.2. The location of a 2-D body and the plane of field
determination. 1 - region where $\varrho_s > 0$.

In what follows we confine ourselves to the 2-D problem
only. Thus we seek for the field outside a thin layer
$z_1 < z < z_2$ parallel to the xy plane, with the density
$\varrho_s = \varrho_s(x,y)$ - see Fig.2. An arbitrary 3-D body can be
understood as a sum of suitable (sandwich type) bodies
fulfilling the conditions given in Fig.2; especially the
domain where the field is to be computed (3-D is allowed
again) must be outside the anomalous body. As far as the
computation is concerned, 2-D as well as 3-D algorithms can
be used. The problem of the optimal choice depends on the
computer capability.

In this way we seek for the field of a thin layer where
ϱ_s does not depend on z. After differentiation with respect
to z, rearrangement, and eventually by integrating with
respect to z' we obtain for the field strength

$$(3) \quad g(r) = \varkappa \int\int\limits_{-\infty}^{\infty} \varrho_s(x',y') \left\{ [(x-x')^2 + (y-y')^2 + (z-z_1)^2]^{-\frac{1}{2}} - \right.$$

$$\left. - [(x-x')^2 + (y-y')^2 + (z-z_2)^2]^{-\frac{1}{2}} \right\} dS' =$$

$$= \varkappa \, \varrho_s(x,y) \overset{2}{\ast} \left(|r_1|^{-1} - |r_2|^{-1} \right) .$$

As we solve a numerical problem of discrete LF, we must
confine ourselves to a parallelepiped D consisting of basic
prisms of equal size, each of them being homogeneous and

101

possessing an arbitrary density. The anomalous body is then inside it. Moreover we can calculate the field only at nodal points of a 3-D rectangular grid corresponding to the subdivision of the anomalous body and we have to discretize the integration operation according to this grid.

Without loss of generality we can choose the orientation of the coordinate system. Further, we will confine ourselves to the 2-D domain O in the plane $z=z_o$, where the field is computed. In Fig.3 we can see the geometry of the problem as well as the rectangular domain o – projection of the parallelepiped D to the plane $z=z_o$, and the domain O (see Fig.2). Both the domains have the same center.

The formula (3) transforms then to

$$(4) \qquad g(m\Delta x, n\Delta y) = g_{mn} = \varkappa \ \Delta x \Delta y \sum_{j=-\mu}^{\mu} \sum_{k=-\nu}^{\nu} w_{jk} \ \varrho_{jk} \ \alpha_{m-j,n-k} \quad ,$$

where

$$m = -M, \ldots, M \ , \qquad n = -N, \ldots, N \ ,$$

$$\alpha_{pq} = |r_1|^{-1} - |r_2|^{-1} \quad ,$$

$$|r_i| = \sqrt{(p\Delta x)^2 + (q\Delta y)^2 + z_i^2} \quad , \qquad i = 1, 2 \ .$$

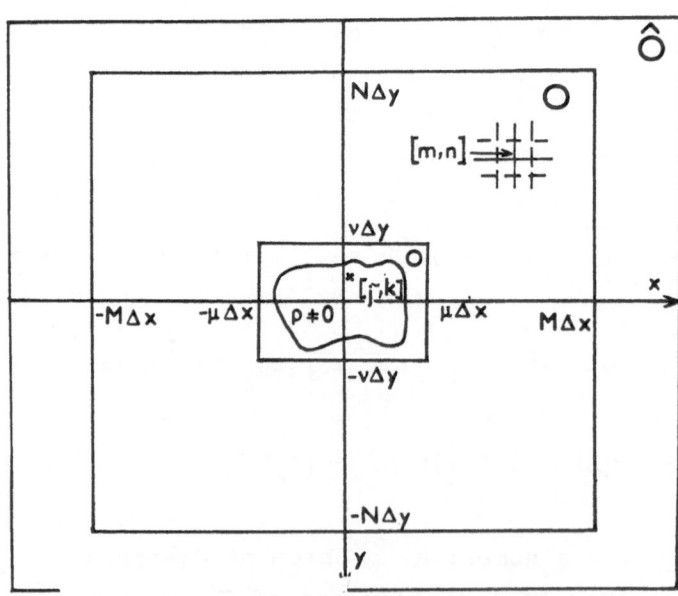

Fig.3. Projection of the parallelepiped D containing the body to the domain O (the plane $z=z_o$).

102

From the numerical point of view it is more suitable and for geophysicists more familiar to apply the other version of the discrete convolution

$$(5) \qquad g_{mn} = \varkappa \, \Delta x \Delta y \sum_{j=m+\mu}^{m-\mu} \sum_{k=n+\nu}^{n-\nu} w_{m-j,n-k} \, \varrho_{m-j,n-k} \, \alpha_{jk} \quad .$$

The geometrical meaning of this formula is as follows: Instead of computing the field at different nodal points of the domain O we compute it in its center. At the same time the anomalous body is not fixed, it is "traveling" through the domain O being rotated by $180°$. It is obvious now that we can precompute all the necessary values of α_{jk}, however, we need them on the domain $\hat{O}$, which is larger than O. In this way we have obtained a standard 2-D LF. Only the situation where o, O, $\hat{O}$ have the same centre and orientation is meaningful.

As far as the evaluation of formula (5) is concerned, there are two equivalent approaches. Priority has to be given to the frequency domain computation that is put into effect according to the scheme

$$(6) \qquad g = \varkappa \, DFT^{-1}[DFT(\varrho_s) \, DFT(\alpha)] \quad .$$

As all the functions here have the same support we must add a proper number of zero values to ϱ_{jk}. The result of the 2-D LF is computed correctly on the domain of the size O but it can be shifted because of the necessity of re-indexing. To avoid this shift we make use of the periodicity of DFT - see Fig.4. Thus we start the numeration in a proper way from zero to $2(M+\mu)$ and $2(N+\nu)$, respectively. Only in case of a very small support - for example 3x3 points - one can efficiently apply the direct approach in the space domain using formula (5). Some computers are furnished with special fast processors, which can perform such a convolution in almost real time. No other error except for round-off errors and that of the quadrature formula used can affect the results. There is thus no reason to apply the philosophy of aliasing , i.e. errors connected with replacing the continuous FT by the discrete FT.

As far as the weight function w_{jk} is concerned, it is possible to choose it in different manner. If we put w=1 everywhere the "effective" support ϱ_s must be taken larger than the smallest region containing all the nodes. That is why it is recommendable to choose w_{jk} in a more sophisticated way. The trapezoidal rule - see Fig.5 - should be preferred.

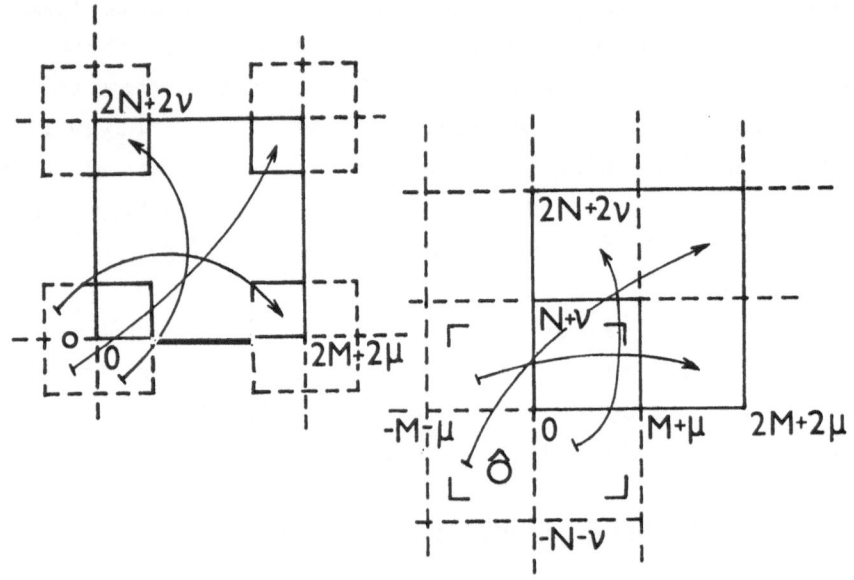

Fig.4. Re-indexing of o and O to the "main" domain of the functions ϱ_{jk} and α_{jk} in the first quadrant.

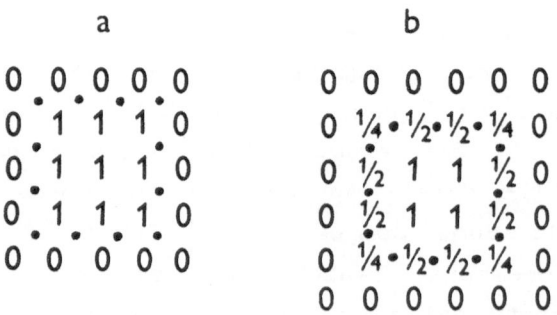

       a   b

```
0 0 0 0 0          0 0 0 0 0 0
0 1 1 1 0          0 ¼ ½ ½ ¼ 0
0 1 1 1 0          0 ½ 1 1 ½ 0
0 1 1 1 0          0 ½ 1 1 ½ 0
0 0 0 0 0          0 ¼ ½ ½ ¼ 0
                   0 0 0 0 0 0
```

• • •1

Fig.5. Choice of the weight function w for:
a) the rectangular rule, b) the trapezoidal rule.
1 - boundary of the region.

Our numerical experiments as well as the theoretical conclusions have proved that the above described approach to the direct problem of any potential field is very effective also in case of the application of a standard microcomputer (i.e. without special hardware unit). In this case - of course - the FFT is applied.

All that has been said was described by Bhattacharyya and his team from Berkeley in the seventies (for a survey see Bhattacharyya 1977). Surprisingly bad numerical results obtained by this group are most probably connected with a bad choice of the weight function w_{jk} (in fact they didn't introduce it) and with little experience with numerical integration. The convolution of two relatively very well-behaved functions ϱ_{jk} and α_{jk} in formula (6), one of them moreover having a finite support, does give good results also in case of a relatively coarse grid.

We have found that for the accuracy of about 1% it is enough to choose $\Delta x = \Delta y = 0.25h$, where h is the smallest orthogonal distance of the anomalous body and the domain O. For the same accuracy, Bhattacharyya's group recommended steps almost 10 times smaller in each dimension. This underestimation in connection with the smaller efficiency and flexibility of the computers in the seventies led to the fact that the method has almost been forgotten.

3. Solving the inverse problem

All that has been said about the fast direct method can be applied to the inverse problem, too. The application of the inverse problem operators in the frequency domain brings technical problems, however, there is another very important aspect of the whole task. The application of fast direct methods in combination with the computer effective enough allows to create detailed 3-D models of the density distribution. Such models represent a new quality in gravimetry and - because of the number of parameters - it is not possible to apply standard inverse methods at any stage of the interpretation. In this case we can rely on the trial and error method only. Of course the interactive way of its application brings also possibilities for a new and more effective implementation of the method.

This concept of the inverse problem may bring new quality into the interpretation of 3-D problems with a relatively well known density distribution. Such a situation is typical for some mining districts and archaeological sites.

References

Bhattacharyya, B. K. and Leu, L. K., 1977: Spectral Analysis of Gravity and Magnetic Anomalies Due to Prismatic Bodies. Geophysics, 42(1), 41-50.
Blahut, R. E., 1985: Fast Algorithms for Digital Signal Processing. Addison-Wesley Publishing Company, Reading, Mass.

2. Gravity and Magnetics

9 Gravity and Magnetics

Solving Inverse Problems for Potential Fields for Nonuniform Data with Error

O. I. Kounchev

Institute of Mathematics, Bulgarian Academy of Sciences, Acad. G. Bonchev, 8, 1113 Sofia, Bulgaria

Abstract. The interpretation of data from potential fields meets substantial difficulties when these data are not uniform. In such a case the nonlinear, and even the linear regressional models do not have good statistical properties if directly applying the data fitting using least squares. To avoid this we propose an "uniformized" data fitting. Then one must use some cubature formula. The choice of this cubature formula is not unique and must be flexible with respect to the addition of information on the properties of the sources of the field. Here we present an idea how to overcome this obstacle which we demonstrate in detail in the case of geomagnetism and paleomagnetism. In a subsequent paper we shall demonstrate the realization of the same idea in the case of gravimetry and magnetics.

1. Introduction

In a previous paper Kounchev (1990) we discussed the necessity of proper formulation of the inverse problems for the case of nonuniformly distributed data obtained from measuring a potential field. Such data are present in the paleomagnetic studies or from satellite observations of the magnetic field of a planet, when the trajectory of the satellite is not regular.

In the present paper we consider the problem of modelling the global magnetic field of a planet. We also make an account of the experience obtained through computer modelling in the cases of paleomagnetism and magnetospheric studies Parkinson (1983), resp. Alfven (1981).

In the following we suppose that we are working in a geocentric Cartesian coordinate system , and every data are represented in it.

The field vector of the magnetic field in the point x outside the planet will be denoted by F(x) ($\in \mathbf{R}^3$) .

2. The model

First, we shall describe the model M through which we are going to approximate the field F(x) . Let us note that we provide here this model for the sake of concreteness and to simplify the comprehension of the idea of what we are doing. In every other case one has more information about the structure of the sources of the field and he may choose another model M .

After that we shall demonstrate the scheme for interpreting the nonuniform data. This scheme does not depend of the concrete model which we choose but it is applicable to a wide variety of models.

The model M which we use is formed by magnetic dipoles. To avoid complicated notations we describe the dipole through the parameters

$\alpha = (\alpha_1, \alpha_2, \alpha_3, \alpha_4, \alpha_5, \alpha_6)$, where $(\alpha_1, \alpha_2, \alpha_3)$ is the centre of the dipole and $(\alpha_4, \alpha_5, \alpha_6)$ is the moment vector.

The field vector generated by this dipole in the point x we denote by F(x, α) ($\in \mathbf{R}^3$) (cf. Smythe (1950) or Tamm (1966) for the exact formula).

The model M_n represents the sets of n independent dipoles, so we shall write

$M_n = \{ (\alpha^{(1)}, \alpha^{(2)}, \ldots, \alpha^{(n)}) : \alpha^{(j)} , j = 1, 2, \ldots, n,$
        is a dipole $\}$

The model M is equal to the infinite union $\bigcup\limits_{n=1}^{\infty} M_n$.

The field generated by the set of dipoles $d \in M_n$ is given by the superposition formula

$$F(x,d) \overset{def}{=} \sum_{k=1}^{n} F(x,\alpha^{(k)}) \quad .$$

3. The data fitting and the cubature formulas

The data which are measured in the points x_j, $j = 1,2,\ldots,N$ will be denoted by $F_j = F(x_j)$.

The following situations of nonuniform data we meet in the practice:

i) The points x_j lie on the surface of the planet which is supposed to be geometrically symmetric and regular, usually sphere or ellipsoid. This may be not the surface of the planet but a surface concentric with the planet centre. This is the case of the paleomagnetic and geomagnetic data.

ii) The points x_j lie between two concentric (with the planet centre) spheres or ellipsoids outside the planet. This is usually tha data obtained from satellites.

In Kounchev (1990) we discussed why the data fitting

$$(1) \qquad \phi(d) = \sum_{j=1}^{N} (F(x_j,d) - F_j)^2 \quad ---> \quad \min_{d \in M_n}$$

(here d in fact varies over the set of n independent dipoles) is not adequate, especially when the points of observation x_j are not uniformly distributed.

We proposed an "uniformized" data fitting which is formulated below in the cases pointed above.

In the case i) let us suppose that the points x_j lie on a sphere or an ellipsoid S which is concentric with the planet centre. Then instead of (1) we consider the uniformized data fitting:

(2) $\qquad \phi(d) = \iint\limits_{S} (F(x,d) - F(x))^2 d\sigma(x) \quad \text{---}> \quad \min\limits_{d \in M_n}$

The integral may be approximated through a cubature formula:

(3) $\quad \phi(d) = \iint\limits_{S} (F(x,d) - F(x))^2 d\sigma(x) \cong \sum\limits_{j=1}^{N} (F(x_j,d) - F_j)^2 q_j$

and problem (2) is reduced to a weighted least squares problem (but the weights q_j may have negative sign!).

The essential problem is what cubature formula (3) to choose which could permit easily to take into account any additional information on the model M.

In the case ii) let us suppose that the points x_j lie between two concentric spheres or ellipsoids (concentric with the planet centre), a set which we shall denote by B.

Just like above we consider the "uniformized" data fitting problem

(4) $\qquad \phi(d) = \iiint\limits_{B} (F(x,d) - F(x))^2 dx \quad \text{---}> \quad \min\limits_{d \in M_n}$

Here the integral may be approximated through the cubature formula

(5) $\quad \phi(d) = \iiint\limits_{B} (F(x,d) - F(x))^2 dx \cong \sum\limits_{j=1}^{N} (F(x_j,d) - F_j)^2 q_j$

which reduces as above problem (4) to a least squares problem with coefficients with a mixed sign.

In Kounchev (1990) we proposed to use cubature formulas based on the spherical functions which are very good for approximating functions on the sphere (their transforms are good for the ellipsoid too). Another reason is that they are harmonic functions just like the coordinates of the magnetic field vector.

Another approach to the cubature formulas (3),(5) is to use the products of spherical functions since in (3) and (5) we have to approximate not the field vector itself but its square which contains products of harmonic functions.

Conceptually, this way of treating the cubature formulas (3),(5) seems to be equivalent to an upward continuation of the magnetic field by using harmonic (spherical) analysis. Through the last we may find a sphere S and observation points y_k, $k=1,2,...,N_1$ which are uniform on S, and after that directly apply the data fitting (1). But the experience shows that this approach fails and the reason is again the insufficiency of the data which spoils the upward continuation of the field through harmonic analysis.

To find the coefficients q_j we need a good base of N functions.

One of the reasonable choices is to use some concrete values of the parameter d, i.e. to fix some concrete sets of dipoles.

This can be formulated as a system of 3 equations:

$$(6) \qquad \iint_S F(x,\tilde{d})\,d\sigma(x) \cong \sum_{j=1}^{N} F(x_j,\tilde{d})q_j$$

for every given $d = \tilde{d}$. Here the left-hand side is computable and the right-hand side too.

4. Cubature formulas through model functions

For finding the coefficients q_j through system (6) we need to specify $p = [N/3] + 1$ sets of dipoles, namely $d_1,d_2,...,d_p$.

The problem is how to choose these parameters in the most reasonable way.

113

Let us suppose that according to some a priori information we know that the possible states of the solution of problem (2) (resp. (4)) is in a set D, $D \subseteq M$.

For solving (2) (resp. (4)) we shall use a cubature formula (3) (resp. (5)) based on (6).

The parameters $d_1, d_2, \ldots, d_p$ should be constructed in a proper way in the set $D \cap M_n$. It seems that the best is when they are chosen as uniform as possible (from geometrical point of view) in the set $D \cap M_n$.

This would guarantee that the value of d which solves problem (2) (resp. (4)) be a convex combination of these concrete values of $d_1, d_2, \ldots, d_p$.

How good is such a choice must be proved of course rigorously as well as practically through a sufficient number of test experiments.

Below we describe the idea of a new algorithm implemented in a computer program for realizing the choice of the values of $d_1, d_2, \ldots, d_3$ in the case of approximation of a planetary magnetic field through the field of one dipole, i.e. problem (2) (resp. (4)) with the model M_1.

The choice is not the optimal one but in many cases it works well.

In the case of the model M_1 we shall consider parameter states d which are represented by one dipole only, i.e. $d = \alpha = (\alpha_1, \alpha_2, \ldots, \alpha_6)$.

We introduce some a priori information in the following way:

The centre $(\alpha_1, \alpha_2, \alpha_3)$ is no more than $R/3$ units far from the planet centre, i.e.

(7) $\qquad \alpha_1^2 + \alpha_2^2 + \alpha_3^2 \leq (R/3)^2$.

Also the latitude of the dipole moment is not less than 60° south, i.e. if we write using spherical coordinates

$$\alpha_4 = I\sin\theta\cos\phi$$

(8) $$\alpha_5 = I\sin\theta\sin\phi$$

$$\alpha_6 = I\cos\theta$$

then

(9) $150^O \leq \theta \leq 180^O$.

So the set $D \cap M_1$ is given through the restrictions (7),(9).

Now the problem is to find a sequence of parameters

$$d_1 = \alpha^{(1)} = (\alpha_1^{(1)}, \alpha_2^{(1)}, \ldots, \alpha_6^{(1)}) \quad , \quad 1 = 1,2,3\ldots$$

so that for every number of data N the set $d_1, d_2, \ldots, d_N$ is geometrically uniform in the set $D \cap M_1$.

A relatively simple way for programming is to use the barycentric subdivision of a simplex which is also quite uniform in the simplex. We shall choose the centres $(\alpha_1, \alpha_2, \alpha_3)$ of the models $d_1, d_2, \ldots, d_p$ in the points of a proper subdivision of simplex which is described below.

The nodes of the primary simplex approximating the set $D \cap M_1$ are given by

$$S_1^O = (-a/2, -\sqrt[2]{b^2 - R^2}, -R/3)$$

$$S_2^O = (a/2, -\sqrt[2]{b^2 - R^2}, -R/3)$$

$$S_3^O = (0, -R/\sqrt{2}, -R/3)$$

$$S_4^O = (0, 0, -2R/3)$$

where $a = R\sqrt{3/2}$, $b = 3/(2\sqrt{2})$.

Shortly speaking, at every step the edges of the simplices are divided in two parts and their centres are connected with the barycentre to create the new simplices .

At the first step, let us denote the barycentre of S^O by

$$CS^O = (S_1^O + S_2^O + S_3^O + S_4^O)/4 \quad .$$

Denote through $S_{ij}^1 = (S_i^O + S_j^O)/2$ for $i \neq j$

the new nodes of the simplices which we count below:

$$[s_1^0, s_{12}^1, s_{13}^1, s_{14}^1], \quad [[cs^0, s_{12}^1, s_{13}^1, s_{14}^1], \quad [[cs^0, s_{23}^1, s_{24}^1, s_{34}^1], \ldots etc.,$$

in total 12 simplices which are obtained through connecting
the nodes s_{ij}^1 and the barycentre cs^0. As a result we
get the new set of nodes s^1 which contains those of s^0
and 6 new. We repeat the same procedure to the simplices of
s^1 and get $s^2 \ldots$ etc.

Let us denote by t_i the number of the new points
added to s^{i-1} in order to obtain s^i. The total number
of points in s^i is $T_i = \sum\limits_{k=0}^{i} t_k$.

When the number of the observation points in the data set
is N , we must find an integer i such that

$$T_{i-1} < N \leq T_i \quad .$$

We order the nodes d_1, $1 \geq 1$, in s^i in such a way
that their centres be geometrically relatively uniform
in the set $D \cap M_1$ for every N .

What concerns the dipole moments $(\alpha_4^{(1)}, \alpha_5^{(1)}, \alpha_6^{(1)})$ we
simply use a random procedure in the rectangle

$$0 \leq \phi < 360^0 \quad , \quad 150^0 \leq \theta \leq 180^0$$

for finding the longitude and the latitude of the pole in the
spherical coordinates (see (8) but for $\alpha_4, \alpha_5, \alpha_6$). The
magnitude of the dipole is put 1 or for the geomagnetic
case is given a value I_0 which is guessed by the data.

To symmetrize the above procedure even more, we may
consider the points of the reversed (with respect to the axes
z) simplex:

$$T_1^0 = (-a/2, -\gamma b^2 - R^2, R/3)$$

$$T_2^0 = (a/2, -\gamma b^2 - R^2, R/3)$$

$$T_3^0 = (0, -R/\gamma 2, R/3)$$

$$T_4^O = (0,0,2R/3)$$

and the zero point $\quad S_0^O = (0,0,0)$.

Finally, let us remark that the above scheme may be applied with minor changes to the local and global inverse gravimetric and magnetometric problems.

5. Selecting the data

The solution of the inverse problem is even more complicated when a considerable error is present in the data. This is typical for the paleomagnetic data Parkinson (1983).

Below we describe one way in which we selected the paleomagnetic data. No obstacles are seen the same algorithm to be applied in principle to other sorts of data especially when there is a reasonable hypothesis on the structure and the magnitude of the sources.

Let us suppose that the observation points of the data x_j lie on the Earth's surface S (which we think to be a sphere). The error in the data vector F_j (which in the case of paleomagnetism is a unit vector) is given by an angle E_j: the true value of F_j is with 95% in the surface circle defined by

$$\{ x: (x,F_j) \geq \cos E_j \}$$

where $(,)$ is the scalar product in $\mathbb{R}^3$.

The values of E_j which may be found in degrees in the paleomagnetic catalogues are quite considerable, sometimes $15°, 20°$ even $25°$.

The problem arises how to select some data from the given data set so that it would provide consistent and independent information.

In a very simple and approximate way this may be done by the help of a table of the "influence circles" for every point on the Earth's surface S.

This table is produced as follows:

We fix a dipole

$$\tilde{\alpha} = (0,0,0,\alpha_4,\alpha_5,\alpha_6)$$

which is central and the values of the moment vector $(\alpha_4,\alpha_5,\alpha_6)$ are chosen in accordance with our general hypothesis about the most probable its state. Let us notice that the variations of these parameters do not affect considerably the "influence circles" map and we may use this map to get an orientation on the significance of the data.

After that we compute the vector $F(x,\tilde{\alpha})$ for a sufficient number of points $x \in S$, the points with

$$\text{latitude} = i.5^\circ \text{ , for } i = 0,1,2,\ldots,72 \text{ and}$$

(10)

$$\text{longitude} = j.5^\circ \text{ , for } j = 0,\pm1,\pm2,\ldots,\pm18 \text{ .}$$

Let us recall that in spherical coordinates ϕ = latitude, θ = 90° − longitude.

Now for every spherical circle around x and defined by the angle γ,

$$C(x,\gamma) = \{ y \in S: (y/\|y\|,x/\|x\|) \geq \cos\gamma \}$$

we consider the subset of the unit sphere in R^3

$$G(x,\gamma) = \{ F(y,\tilde{\alpha} : y \in C(x,\gamma) \}$$

which contains the vector $F(x,\tilde{\alpha})$.

There is an angle γ_1 such that the spherical circle around the vector $F(x,\tilde{\alpha})$, given by

$$\{ z: \|z\| = 1, (z,F(x,\tilde{\alpha}) \geq \cos\gamma_1 \} ,$$

contains the set $G(x,\gamma)$. We shall choose the angle γ_1 which is the minimal with this property. Let us notice that when γ increases from 0° upward then γ_1 also increases monotonely from 0° upward.

Finally, we can write the table of the influence circles . It contains the latitudes and the longitudes of the grid (10) together with a one-to-one correspondence between the values of γ,

$$\gamma = 0^\circ,1^\circ,\ldots,50^\circ$$

and the corresponding values of γ_1,

$$\gamma_1 = a_0, a_1, \ldots, a_{50} \; .$$

To draw the influence circles map we proceede as follows. For every point x_j we take a point from the grid with longitude and latitude as close as possible to x_j, and after that we look at the value of γ_1 which is the closest to E_j. Through the correspondence $\gamma_1 \dashrightarrow \gamma$ we find the corresponding value of $\gamma = \gamma_j$. This is the radius of the influence circle. This spherical circle may be drawn on the map of the Earth. Some circles intersect more, others less. So looking at this "influence circles" map we may choose from the initial data set $(x_1, F_1), (x_2, F_2), \ldots, (x_N, F_N)$ a subset of data where the overlapping of the circles is as minimal as possible.

To our opinion this selection of data is the most natural.

REFERENCES

Alfven, H., 1981: Cosmic Plasma, D. Reidel Publishing Company, Dordrecht, Holland.
Kounchev, O., 1990: The elliptical current loop model of Earth's magnetic and paleomagnetic field sources, to appear in Model Optimization in Exploration Geophysics. Proceedings of a Seminar held at the Free University in Berlin, 1989.
Parkinson, W.D., 1983: Introduction to Geomagnetism, Scottish Academic Press, Edinburgh and London.
Smythe, W.R., 1950: Static and dynamic electricity, McGraw-Hill, New York.
Tamm, I.E., 1966: Foundations of electricity theory, Nauka, Moscow (in Russian).

The Applicability of Two-dimensional Inversion Filters in Magnetic Prospecting for Buried Antiquities

G. N. Tsokas, C. B. Papazachos

Geophysical Laboratory, Dept. of Geology, University of Thessaloniki, 540 06 Thessaloniki, Greece

ABSTRACT

Inversion filters can be computed on the basis that a particular signal can be considered as the convolutant of two distinct and analytically determined functions. Therefore, if such filters are convolved with the measured field, they provide the distribution of the function which was not inverted. If the non-inverted function has been chosen in a manner to define the surface projection of the buried target, a plan view of the subsurface conditions has been provided.

The vertical sided prism is justified as the model which resembles the majority of the expected structures in an Archaeological site. Hence, the effect which that produces forms the signal upon which manipulations are done. Several filters are presented which were computed on that basis.

Applications on synthetic and actual data are presented in order to clarify the merits and disadvantages of the technique.

INTRODUCTION

The aim of "archaeological geophysics" is to locate and identify buried ancient relics. Up to date, several techniques have been tested with moderate to high success on this matter.

However, because problems of exproprietation of lands are very commonly met by Archaeologists, more information is needed than a simple pinpointing of localities that might conceale an antiquity. The survey is considered to be succesfully completed if it has as final product a map which more of less gives the plan view of the concealed structures, i.e., the result that would have been drawn if an excavation had taken place. A map in that form aims to help the Archaeologists to decide upon continuation, postpone, alterate or even cancel the progression of workings of civil engineering projects if they are connected with an archaeological site. Of course, the utility of a geophysical product is not exhausted in such cases of "resque excavation" only. The exploitation of such results in designing and better planning any excavation is obvious.

The present study comprises an effort to achieve the presentation of the geophysical maps over buried antiquities in an image form. That image should give an outline of the dimensions of buried structures, should present their centers in exact locations (within the range of accuracy), and of course it should be as reliable as possible.

The problem is attempted to be solved by means of inverse filtration. The convolutional model of the total magnetic field anomalies, which is necessary for this case, was proposed by Bhattacharrya and Navolio (1975) and Bhattacharrya and Chan (1977).

Karousova and Karous (1989) proposed the exploitation of inverse filtration in the geophysical search for antiquities. They constructed the appropriate filters for the vertical cylinder and infinite prism using the equations of Logacev and

Zacharov (1973). Such filters, if convolved with the total field anomaly profiles, produced pronounced and informative results.

Tsokas et al. (1989) attempted to modify the procedure described by karousova and Karous (1989). They used filters which were produced employing the anomaly caused by a rectangular prism model and convolved them with every profile of the geophysical map. In such a way, the most commonly met model was used in a pseudo-two-dimensional procedure. Additionally they used as filter's truncation length the lag where the autocorrelation of the shape function vanishes to zero. Where, shape function is the convolutant to be inverted and thus result into a filter.

The present study is aimed towards a fully two-dimensional procedure. Such a procedure, should be relatively simple and easy applicable, in order to meet the requirements of all kind of excavation projects.

THE CONVOLUTIONAL MODEL

McGrath and Hood (1973) proposed an algorithm to produce the magnetic effect over any prism shaped body. These geometrical models were generated from the thin-plane model by numerical integration. According to that algorithm the following equation was considered (Grant and West, 1965, McGrath and Hood, 1973) which gives the magnetic effect over a thin-plate:

$$\Delta T\ (x,y) = J\ s\ b\ c\ [f(x,y+Y)-f(x,y-Y)], \tag{1}$$

$$\text{where,}\ \ f(x,y+Y)=T_1 -T_2 -T_3 \cdot (T_4 +T_5)-T_6 \cdot T_7 \tag{2}$$

Figure (1) shows a dipping thin-plane which strikes along Y axis of an orthogonal coordinate system. The symbols used to annotate several quantities have the following explanation:

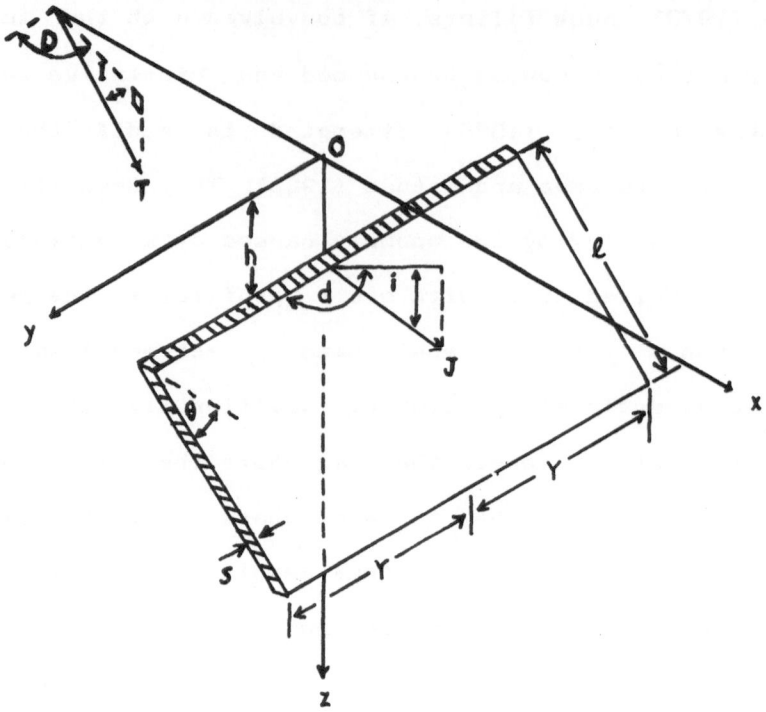

Fig.1. View of the thin-plane model (McGrath and Hood 1973).

J=intensity of Magnetization,

T=Earth's total magnetic field intensity,

s=thickness of the plate,

b=$(\sin^2 i+\cos^2 i\cdot\sin^2 d)^{1/2}=\sin i/\sin\gamma$,

c=$\sin^2 I+\cos^2 I\cdot\sin^2 D)^{1/2}=\sin I/\sin\beta$,

I=angle of inclination of the earth's magnetic field,

 i=angle of inclination of the magnetization (-180 to 180 degrees),

d=angle between the horizontal projection of J and the positive Y-axis,

D=angle between the horizontal projection of T and the positive Y-axis,

Y=half-strike length of the plate,

$c^2=(x-1\cos\theta)^2+(h+1\sin\theta)^2$,

θ=angle of dip of plate,

l=width of the plate,

h=depth of the plate,

$E^2 = x^2 + h^2$,

$B = x \cdot \sin\theta + h \cdot \cos\theta$

$A = h \cdot \sin\theta - x \cdot \cos\theta$,

β=angle of inclination of the component of T in the xz plane,

$= \tan^{-1} (\tan I / \sin D)$,

γ=angle of inclination of the component of J in the xz plane,

$= \tan^{-1} (\tan i / \sin d)$.

The quantities at the right hand side of equation (2) are defined as follows:

$$T_1 = \frac{y+Y}{\left(c^2 + (y+Y)^2\right)^{1/2}} \cdot \left[\frac{(x - l \cdot \cos\theta)\cos\alpha - (h + l\sin\theta)\sin\alpha}{c^2}\right]$$

$$T_2 = \frac{y+Y}{\left(E^2 + (y+Y)^2\right)^{1/2}} \left[\frac{x\cos\alpha - h\sin\alpha}{E^2}\right],$$

$$T_3 = \frac{1}{B^2 + (y+Y)^2} \cdot \left[\frac{A+l}{\left[c^2 + (y+Y)^2\right]^{1/2}} - \frac{A}{\left[E^2 + (y+Y)^2\right]^{1/2}}\right] \cdot (y+Y),$$

$$T_4 = \left(\cos\alpha \cdot \cos\theta - \cos\beta \cdot \cos\gamma + \cos\beta \cdot \cot D \cdot \cos\gamma \cdot \cot d\right) \cdot (y+Y),$$

$$T_5 = \left(\cos\gamma \cdot \cot d \cdot \sin(\theta-\beta) + \cos\beta \cdot \cot D \cdot \sin(\theta-\gamma)\right) \cdot B,$$

$$T_6 = \cos\gamma \cdot \cot d \cdot \cos(\theta-\beta) + \cos\beta \cdot \cot D \cdot \cos(\theta-\gamma),$$

$$T_7 = \frac{1}{\left[c^2 + (y+Y)^2\right]^{1/2}} - \frac{1}{\left[E^2 + (y+Y)^2\right]^{1/2}}$$

α=β+γ-θ .

Let us now consider a model which resembles (or suits to) the majority of structures occurred in archaeological search. Of course, this can be nothing else than a rectangular vertical sided prism. This choise is well justified since the most commonly met features can be represented as a series of such "blocks". For instance, the concealed features are usually building foundation relics, roads, tombs, wall relics, ditches, e.t.c.

Consequently, following the initial algorithm, we consider the vertical sided rectangular prism of figure (2) as been represented by a series of thin plates. On the validity of the superposition principle the magnetic effect which the prism produces can be calculated as the sum of the effects of the individual thin plates. This is given by the following equation:

$$\Delta T(x,y) = J \; s \; b \; c \sum_{i=1}^{N} [f_i(x,y+Y) - f_i(x,y-Y)], \qquad (3)$$

where, N is the number of thin-plates which represent the prism. The expression of total field anomaly is given in equation (3) as a simple summation instead of the elliptical integration used by McGrath and Hood (1973). This is a compromise for the shake of the simplicity of the whole procedure. However, if the number of thin plates used to generate the prism is big enough, the difference is negligible.

Fig.2.
View of a vertical sided prism
composed of a number of thin plates.

Equation (3) can be written as the product of two functions, say D and R. If we define

$$D = J \qquad\qquad\qquad (4)$$

then, this factor can be reasonably termed as "amplitude function" because it modulates the strength of the anomaly. The function

$$R(x,y) = s \; b \; c \sum_{i=1}^{N} [f_i(x,y+Y)-f_i(x,y-Y)] \qquad (5)$$

can be termed "shape function" since it controls the shape of the anomaly. The shape function must become effectively zero very rapidly as the distance from the body center increases. I.e.

$$R(x,y) \longrightarrow 0 \quad \text{as} \quad x \text{ or } y \longrightarrow \pm \infty \qquad (6)$$

Hence, equation (3) can be presented as

$$\Delta T(x,y) = D \cdot R(x,y) \qquad (7)$$

In order to proceed in a convolutional model the following assumptions must hold:

a) Magnetization of the thin plates is presumably of induced type or, instead, it should be considered along a Known stable direction.

b) Any structure of irregular shape can be simulated by an ensemble of elementary bodies of simple geometric shape. These pieces should be similar in shape and buried at equal depths.

The posetion of the former assumptions is not at expence of the generality. Induced magnetization can be considered as basically responsible for the anomalies over buried relics except pronounced situations (like buried Kilns), where, there is no need for any sort of filtration. On the other hand, the second assumption is self evident in "archaeological geophysics".

Let consider an ensemble of bodies of attitude similar to the one given in figure (2), which are placed at the same depth at points (x ,y) given by

$$x_1 = 1 \Delta_x \qquad 1 = L_1, \ldots \ldots L_2$$

$$y_a = m\Delta_x \qquad m = M_1, \ldots \ldots , M_2$$

127

where, Δ_x and Δ_y are the digitization intervals' along x and y axis respectively. For the shake of simplicity we can consider that

$$\Delta_x = \Delta_y .$$

The bodies are considered equally magnetized. On the validity of superposition principle the total field anomaly which they produce can be written as:

$$\Delta T (x_i, y_j) = \sum_{\ell=L_1}^{L_2} \sum_{m=M_1}^{M_2} D_{\ell,m} \cdot R(x_i - x_\ell, y_j - y_m) \quad (8)$$

Equation (8) is an expression of digital two dimensional convolution which becomes more obvious if we rewrite it as:

$$T_{ij} = \sum_{\ell=M_1}^{L_2} \sum_{m=M_1}^{M_2} D_{\ell m} \cdot R_{i-\ell, j-m} \quad (9)$$

where $\Delta T(x_i, y_j) = T_{ij}$ and $R(x_i - x_1, y_j - y_m) = R_{i-1, j-m}$. The same equation can be now presented in a simpler form as

$$T = D * R \quad , \quad (10)$$

The spatial distribution of the amplitude function D can be obtained from equation (10) by convolution with the inverse filter R^{-1}. Hence,

$$D = T * R^{-1} \quad , \quad (11)$$

given that,

$$R * R^{-1} = \theta$$

where, θ, is a unit diagonal matrix.

The minimization of the sum of the square errors $E = R * R^{-1} - \theta$ produces the optimum filter coefficients $\overline{R}_{ij}^{-1}$. Reasonably, $\overline{R}_{ij}^{-1}$ denotes the truncated filter series because inversion of R_{ij} results in an infinitely extending in both directions series (z transform theory).

Such a finite in both directions filter which is optimum in a least square sense, is given by

$$\sum_{j=L_1}^{L_2} \sum_{i=M_1}^{M_2} \overline{R}_{i,j}^{-1} \cdot A_{k-i,\ell-j} = R_{-k,-\ell}$$

(12)

where,

$$A_{k,1} = \sum_{j=L_1}^{L_2} \sum_{i=M_1}^{M_2} R_{i+k,j+1} \; R_{i,j}$$

is the autocorrelation of the shape function (Kanasewitch 1975).

INVERSION FILTERS

According to the analysis of the previous paragraph, the inversion filters which are produced by equation (12) if convolved with the total magnetic field anomaly should produce two dimensional pulses. Presumably, the subsurface situation which results into the particular magnetic field distribution is composed by structures which can be built up by combinations of the used model. The amplitude of such pulses should be equal to the magnitude of magnetization of the model prism used. Of course, we are not using infinitely long filters and buried bodies are not actually of perfect geometrical form, and burial depth varies. Hence, it is impossible to achieve such a scope in a strictly mathematical manner.

Nevertheless, employing equation (12), several filters were produced. Four of them have been tabulated in the following tables (I) to (IV) and they were annotated as F2, F2A, F3A, F4 respectively. The parameters of the model used to produce the respective filter are also given at the tables. The above filters were selected for demonstration purposes. The difference between the filter F2 and F2A is that the later consists of 7X7 positions against 5X5 positions of the first. The filter F3A was derived using exactly the same model as in the cases F2 and F2A but the prism is located deeper in the F3A model. In all three first cases the model was a with side of 1m. The last filter, F4, was derived using a vertical sided parallilepiped.

The burial depth associated with each model is not the actual depth but the distance between the magnetometer's sensor to the top surface of the model.

TABLE (I)

FILTER F2

PARAMETERS OF THE USED MODEL

Inclination of the Earth's total field:54°

Declination of the Earth's total field with respect to y-axis:90°

Burial depth (depth to the top of the prism):1m

Depth extend:1m

Thickness along N-S direction (X axis):1m

Half strike lenght (y axis):0.5m

Dip of the prism:90°

Number of thin plates used to generate the prism:11

PRODUCED FILTER

0.002	0.006	0.01	0.006	0.002
-0.01	-0.005	0.05	0.005	-0.01
0.06	-0.29	0.64	-0.29	0.06
-0.03	0.16	-0.47	0.16	-0.03
0.007	-0.06	0.2	-0.06	0.007

x ↑   y ⟶   N ↑

TABLE (II)

FILTER F2A

Inclination of the Earth's total field:54°

Declination of the Earth's total field with respect to y-axis:90°

Burial depth (depth to the top of the prism):1m

Depth extend:1m

Thickness along N-S direction (X axis):1m

Half strike lenght (y axis):0.5m

Dip of the prism:90°

Number of thin plates used to generate the prism:11

PRODUCED FILTER

0	0.001	0.006	0.011	0.006	0.001	0
-0.001	-0.001	-0.003	0.002	-0.003	-0.001	-0.001
-0.0002	-0.0025	0.006	0.025	0.006	-0.0025	-0.0002
-0.01	0.041	-0.17	0.38	-0.17	0.041	-0.01
0.0003	-0.014	0.089	-0.298	0.089	-0.014	0.0003
0.0001	0.0024	-0.043	0.166	-0.043	0.0024	0.0001
-0.0016	-0.0005	0.0156	-0.0709	0.0156	-0.0005	-0.0016

x $y \longrightarrow$ N

TABLE (III)

FILTER F3A

Inclination of the Earth's total field:54°

Declination of the Earth's total field with respect to y-axis:90°

Burial depth (depth to the top of the prism):2m

Depth extend:1m

Thickness along N-S direction (X axis):1m

Half strike lenght (y axis):0.5m

Dip of the prism:90°

Number of thin plates used to generate the prism:11

PRODUCED FILTER

0.01	-0.03	0.03	0.09	0.03	-0.03	0.01
-0.05	0.09	-0.12	-0.05	-0.20	0.09	-0.05
0.19	-0.54	1.02	-1.02	1.02	-0.54	0.19
-0.06	2.12	-4.74	6.79	-4.74	2.12	-0.60
0.51	-1.82	4.23	-6.47	4.23	-1.82	0.51
-0.29	1.04	-2.49	3.88	-2.49	1.04	-0.29
0.11	-0.42	0.98	-1.49	0.28	-0.42	0.11

X

N

y ⟶

TABLE (IV)

FILTER F2A

Inclination of the Earth's total field:54°

Declination of the Earth's total field with respect to y-axis:90°

Burial depth (depth to the top of the prism):1m

Depth extend:1m

Thickness along N-S direction (X axis):2m

Half strike lenght (y axis):0.4m

Dip of the prism:90°

Number of thin plates used to generate the prism:13

PRODUCED FILTER

-0.001	0.002	0.011	0.003	0.011	0.002	-0.001
0.003	-0.011	0.021	-0.037	0.021	-0.011	0.003
0.009	-0.026	0.032	0.024	0.032	-0.026	0.009
0.019	-0.040	0.015	0.072	0.015	-0.040	0.019
0.026	-0.034	-0.021	0.022	-0.021	-0.034	0.026
0.023	-0.018	-0.016	0.022	-0.016	-0.018	0.023
-0.003	0.032	-0.060	0.074	-0.060	0.032	-0.003

X ↑   y ⟶   N ↑

133

The system of equations (12) was solved using the Gauss-Jordan method in order to confront to the general simplicity for the whole procedure. However, if filters of relatively large length in both directions are attempted to be computed, equations (12) become singular. Thus, if one wishes to produce by his own large filters, is better to use another method to overcome this problem.

Figure (3) and (4) shows the 2 dimensional convolution of filters F2 and F2A with the respective shape functions which were inverted to produce these filters. Both figures provide a test on the whole procedure. Furthermore, they give a check on the performance of the filters. As it was expected, the filter F2A (7X7 positions) which is longer in both directions than the F2 (5X5 positions) functions better. The result is much more pronounced for F2A, less noisy, and more sharp. The last factor gives to the filter an approach to the desired response. In other words, the response of F3A comes closer to the surface projection of the disturbing body.

APPLICATION ON SYNTHETIC DATA

In order to check the performance and applicability of the presented filters, two artificially constructed data sets were employed. In both cases, we used models which simulate structures which often occur in the archaeological search. The total magnetic field effect was calculated using the algorithm given by Bhattacharrya (1964).

Figure (5a) shows the response after convolution of filter F2 with the anomaly produced by a thin rectangular slab buried at 1m depth. The slab is considered positively magnetized with respect to the envinronment possessing a susceptibity contrast of 0.0005

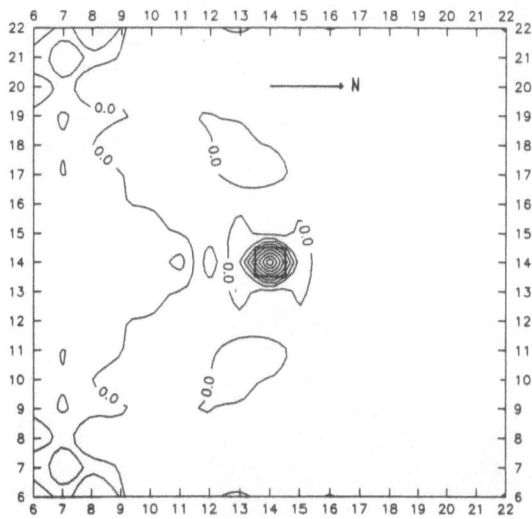

Fig.3. The results of convolution of filter F2 with the respective shape function

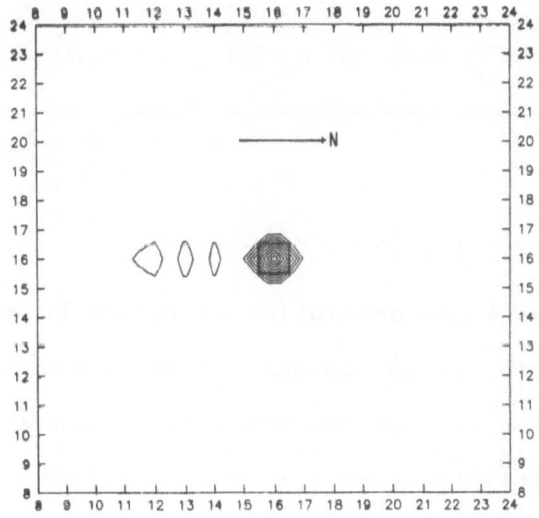

Fig.4. The results of convolution of filter F2A with the respective shape function.

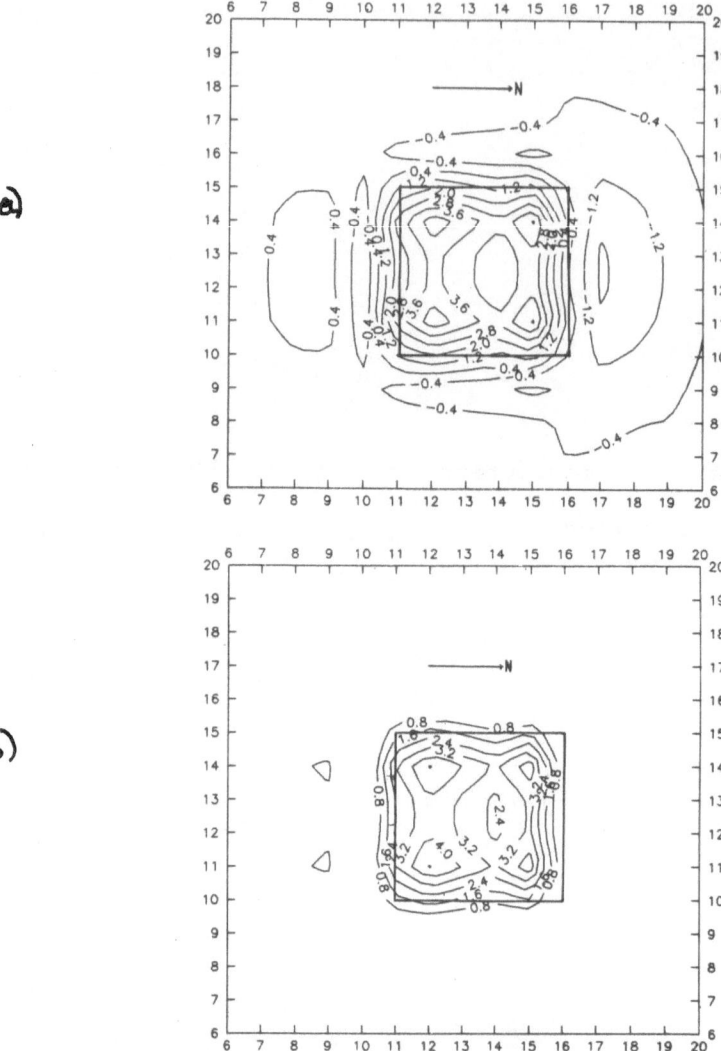

Fig.5. a) The result of convolution of filter F2 with a produced total magnetic field anomaly using a horizontal thin slab model. The plan view of the model is shown with solid line. The susceptibility contrast was consider to be +0.0005 (C.G.S.). b) The same map as in a) after cut-off of the lower contour levels.

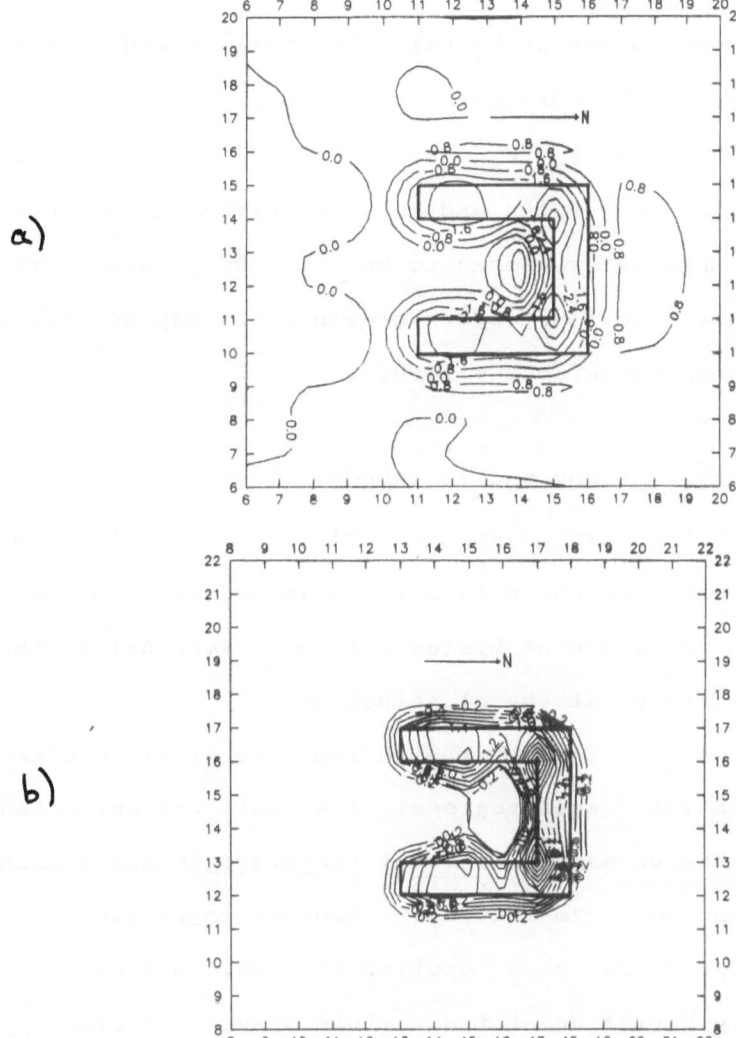

Fig.6. a) The result of convolution of filter F2 with a calculated total magnetic field anomaly using the model whose plan view is shown by solid line. The susceptibility contrast was considered to be +0.0005 (C.G.S.). b) The same map as in a) after cut-off of the higher contour levels.

(C.G.S.). It has a depth extent of 1m and a crossection of 5X5m. The result can be further enhanced if we cut off the lower contours up to some level (figure 5b).

The result of convolution of F2 with the total field anomaly produced by a model which represents the ruins of the foundation of a building are shown in figure (6a). The ruins supposed to be composed of walls of 5m length and 1m wide buried at 1m depth. The susceptibility contrast supposed to be -0.0005 (C.G.S.). The cut-off of the higher contour levels provides the map of figure (6b) which is a rather satisfactory result.

APPLICATION ON THE SITE OF EUROPOS (N.GREECE)

Figure (7) shows the total magnetic field map over a location at the acropolis of the ancient city Europos in N.Greece. Europos was an ancient commercial center byside of the river Axios. The city was mainly flourished at the classical era.

The map of figure (7) presents the residual total field after the subtraction of a 1st degree regional. The data was collected using a proton-precession magnetometer at traverses 1m apart each from the other, stepwise at 1m intervals. Another magnetometer of the same type served as base station in order to reduce the values for the daily variation of the earth's magnetic field.

Figure (8) shows the result after convolution of the total field values with the filter F2A. The positive values of the distribution of amplitude function (fig.8.) have been plotted on figure (9a) while the negative ones on figure (9b). The map of positive values shows very well the location and spatial extent of structures which possess high susceptibility relativelly to the envinronment. In other words, the location of features like "destruction phases", brick walls, ditches, e.t.c. On the

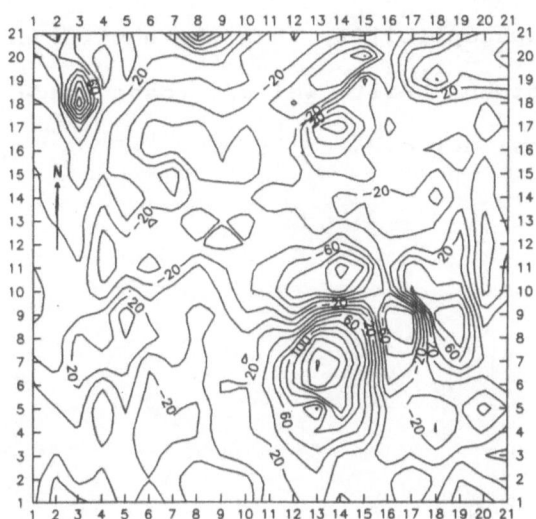

Fig.7. Residual total field map of a subregion of the archaeological site at the acropolis of Europos (N. Greece). Contours are in Gammas.

Fig.8. The result of convolution of the filter F2A with the data of figure (7).

contrary, the map of negative values pinpoints structures which possess lower susceptibility than the hosting soil. The later are features like stone built walls, roads made by hewn stones, foundation ruins, e.t.c.

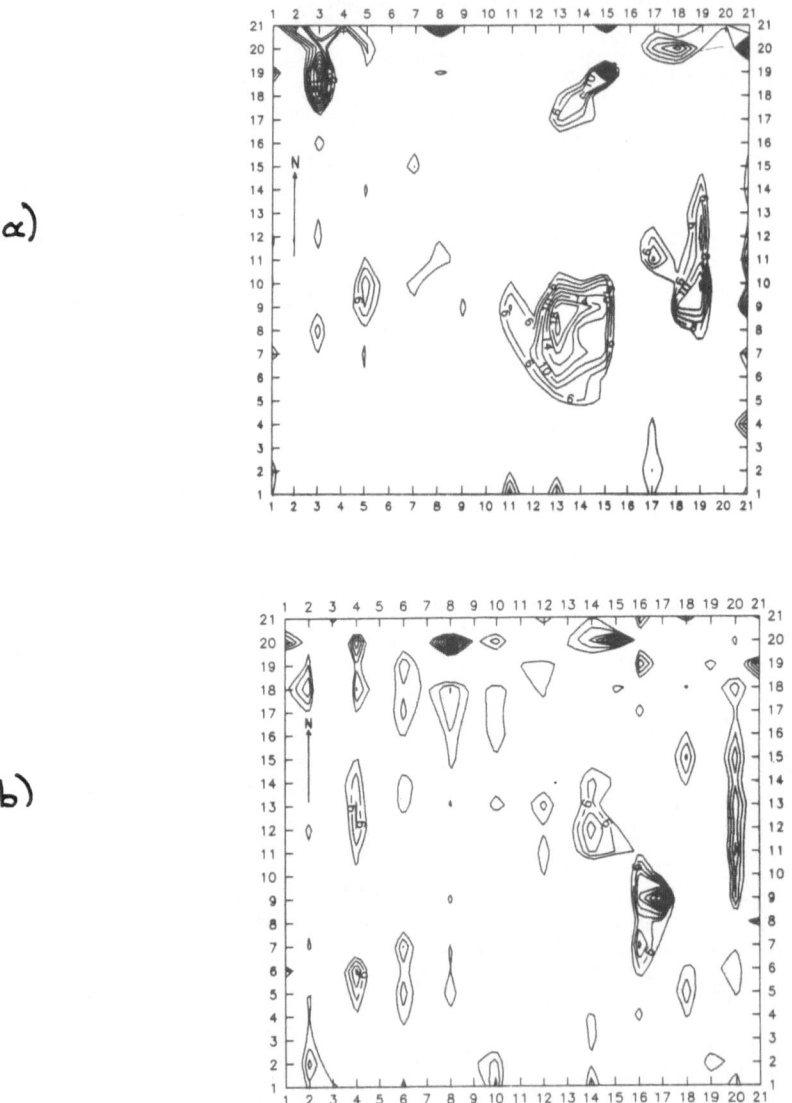

Fig.9. The positive contour levels of figure (8) in a) and the negative ones in b).

DISCUSSION

With respect to the presented examples, either synthetic or actual, it is obvious that the proposed procedure reaches its best performance if the positive and negative values are plotted separately. This is a compromise in order to avoid confusion which might be created if one attempts to interpret the initial amplitude function distribution maps. Alternatively, one may hatch the areas of positive values on the unique map, or present the result as dot-density image. In any case of presentation, the common feature is that the inversion filters function in a manner similar to reduction to the pole. Furthermore, the limits of positive or negative anomalies give a fair approach to the horizontal dimensions of the buried targets.

A crucial point of the whole procedure is the selection of burial depth for the models used to produce the filters. At the presented example of Europos, the burial depth was known to be at about 0.5m below ground surface. The magnetometers sensor was fixed at 0.5m above surface. Hence the overall burial depth was 1m. An attempt to apply the filter F3A which was constructed for 2m burial depth produced noisy results as shown in figure (10). Hence, an estimate of depth level where the antiquities are expected is necessary. This is easily accomplished if a trial pit exists at the site under study. Alternatively, one can use the Historical-Archaeological or any other sort of information about the site.

The overall procedure which was presented in these pages was formulated in a simplified manner as already stated. The simplicity results in increased applicability of the inversion filters without loss of desired information. However, it would be better for anyone who wishes to produce his own filters, to have

Fig.10. The result of convolution of the filter F3A with the data
of figure (7).

them ready and tested on one or two subgrids of the area to be
studied. Next, the filters can be applied on large scale at the
same area.

Another shortcoming of the simplicity rule is that solution
of equation (12) can be achieved by any other method than the
Gauss-Jordan used by the authors, resulting thus in more accurate
and larger filters.

The question on the appropriate filter length has been posed
by Tsokas et al. (1989) who constructed 1-dimensional filters in
an attempt to modify the original procedure of Karousova and
Karous (1989). The suitable length is that where the
autocorrelogram of the shape function (input wavelet) vanishes to
zero. In that case compatibility with reflection seismics is
achieved (Yilmaz 1986).

However, such a consideration in the present case will result in large filters which are inconvenient for the involved calculations.

The filtration by convolution of the total magnetic field data with the proposed filters is easily applied. This is an advantage of the technique because it can be carried out at the fields by a portable computer yielding thus the results rapidly "in situ".

Inversion filters can be evaluated on the basis of models other than the vertical sided rectangular prism. For instance, vertical or horizontal cylinders simulate some of target structures in the "archaeological geophysics" like kilns, pits, ditches, e.t.c. However, the ensemple of vertical sided prisms can effectively built the majority of models. This aspect is self evident because we are dealing with ruins of past human activity. In addition, it is confirmed by the experience of anyone who is involved with the matter. However, one can follow the described procedure and produce filters for cylidrical structures using an appropriate formula from the literature.

CONCLUSIONS

The proposed procedure of construction of inverse filters and convolution of them with the total magnetic field data seems to function satisfactorily. The results are maps of the spatial distribution of the amplitude function. The anomalies on those maps are located directly over the disturbing structures. The spatial extent of the structures is approximately given by the extent of the anomalies after cutting-off of some levels. The levels which must be removed should have the opposite sign of the susceptibility contrast of the target structure with the envinronment.

143

The results can be considered as an approximate image of the plan view of the relics. This fact contributes into rapid location and identification of excavational targets.

REFERENCES

Bhattacharyya,B.K. Magnetic anomalies due to prism shaped bodies with arbitrary polarization, Geophysics, 29, 517-531, 1964.

Bhattacharyya,B.K. and Navolio,M.D. Digital convolution for computing gravity and magnetic anomalies due to arbitrary bodies. Geophysics 40, 981-992, 1975.

Bhattacharrya,B.K. and Chan,K.C. Computation of gravity and Magnetic anomalies due to inhomogeneus distribution of magnetization and density in a localized region. Geophysics, 42, 602-609, 1977.

Grant,F.S. and West,G.F. Interpretation theory in applied Geophysics. McGraw-Hill Book Co., New York, 1965.

Kanasewitch, E.R. Time sequence Analysis in Geophysics. The University of Alberta press, Alberta, 1975.

Karousova and Karous, M. Deconvolution of $\Delta.T.$ profile curves (personal communication), 1989.

Logacev, A.A. and Zacharov, V.P. Magnetic prospecting, Moscow, 1973.

McGrath,P.H. and Hood,P.J. An automatic least-squares multimodel method for magnetic interpretation. Geophysics, 38, 349-358, 1973.

Tsokas,G.N., Papazachos,C.B. and Loucoyiannakis, M.Z. Inversion filters for the transformation of Geophysical data from Archaeological sites based on the vertical sided finite prism model. Paper submitted to Geoexploration, 1989.

Yilmaz,O. Seismic data processing. Elsevier, Amsterdam 1986.

3. Geoelectrics and Electromagnetics

Resolving Resistive Layers using Joint Inversion of LOTEM and MT Data

A. Hördt[1], K.-M. Strack[1], K. Vozoff[2], P. A. Wolfgram[1],

[1] University of Cologne, Albertus-Magnus Platz, 5000 Köln 41, Germany
[2] Macquarie University, Sydney 2109, Australia

Abstract
Hydrocarbon exploration with electromagnetic methods is mainly done in areas where seismics do not give satisfactory results, for example in areas with volcanic cover. Most commonly a horizontally layered earth is used for interpretation, because in many sedimentary basins 1-D models are a reasonable description of the true earth, or they give starting models for 2-D or 3-D interpretation. Moreover, for transient EM the numerical modeling for 2-D or 3-D structures is extremely time consuming. Even though the progress in 3-D modeling during the last few years has been enormous, the improvement of 1-D interpretation is still important.

Introduction
Among the electromagnetic methods, Magnetotellurics (MT) is most frequently used in oil exploration. However, one of the major weaknesses of MT is the poor resolution of thin resistive layers. We propose to use the joint inversion scheme described by Vozoff and Jupp (1975) to overcome this problem. They applied the algorithm to MT and DC resistivitiy soundings and found significant improvements in joint interpretation. Similar results were obtained by Gomez- Trevino and Edwards (1983) for controlled-source EM with Schlumberger soundings and by Raiche at al. (1985) for coincident loop TEM with Schlumberger soundings.

Using an example with synthetic data, we show that the resolution of resistive layers can be significantly improved using joint inversion of MT with the electric fields of the LOTEM method. In an example with field data the joint inversion result compares best of all inversions used with well log information.

The LOTEM method
The LOTEM method is a transient EM system which uses a grounded horizontal electric dipole (HED) as transmitter, which is kept fixed for many different receiver stations (Fig. 1). The transmitter-receiver distance is comparable to the exploration depth, which is the reason for calling the method Long Offset TEM.

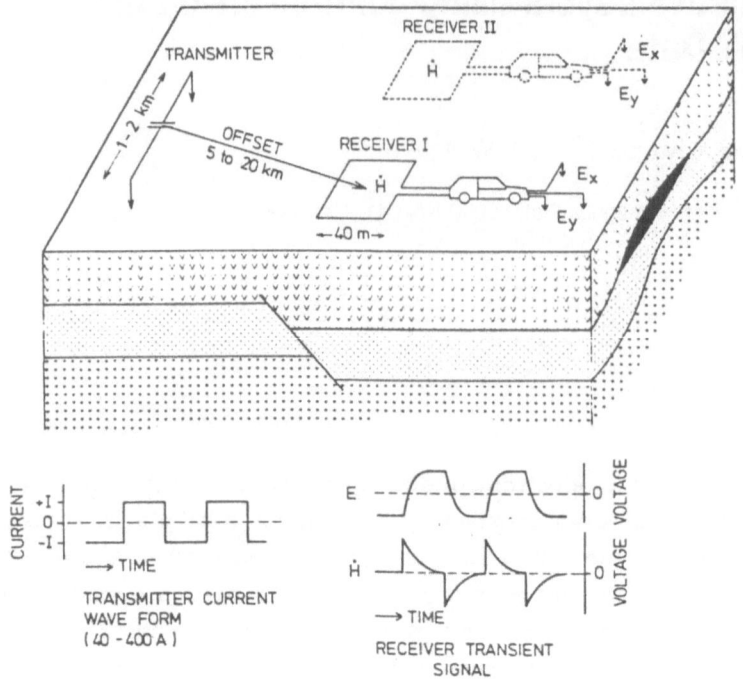

Fig. 1: Survey configuration for the LOTEM method (Strack et al., 1989).

The two horizontal electric field components and the time-derivative of the vertical magnetic field are measured as voltages and interpreted in time domain. It is important to point out that through the electrodes of the HED an additional vertical current system is injected directly into the ground. It can be shown from electromagnetic theory that MT, which uses pure inductive sources, involves only horizontal currents provided that the earth is horizontally layered. The electric fields of the LOTEM method depend also on vertical currents and thus can help to resolve resistive layers (Verma and Mallik, 1979).

The inversion method

Our implementation of the joint inversion is based on a modified Marquardt algorithm which was described by Jupp and Vozoff (1975). Assume there are n measured data points, which are sorted into the data vector $y = (y_1....y_n)$. We use logarithmic data as well as logarithmic model parameters. This constrains the model parameters to positive numbers and improves the conditioning of the problem (Jupp and Vozoff, 1975, and Raiche et al., 1986).

Each of the data points has a corresponding relative error σ_i, $i=1,...n$. The parameter vector $\underline{p} = (p_1....p_m)$ describes the earth model. For a horizontally layered earth it consists of the logarithms of the layer resistivities and layer thicknesses.

The goal of the inversion is then to minimize:

$$\chi^2 = \frac{1}{n} \sum_{i=1}^{n} (f_i - y_i)^2 w_i^2 \tag{1}$$

where $\underline{f}(\underline{p}) = (f_1(\underline{p}),...,f_n(\underline{p}))$ is the model function for the current parameter vector $\underline{p}$, and $w_i = 1/\sigma_i$ is the weight of the data point i. The minimized value χ is the root mean square deviation between measured and calculated data, where every data point is weighted with its relative error σ. The weighting of the data by comparing the residual of each data point with its expected error was proposed by Jackson (1972).

To derive the iterative inversion scheme, the model function $\underline{f}$ is linearized for a given starting model $\underline{p_0}$ by expanding it into a Taylor series up to first order:

$$\underline{f}(\underline{p}) \approx \underline{f}(\underline{p}_0) + J \Delta \underline{p}, \tag{2}$$

where $\Delta \underline{p} := \underline{p}_1 - \underline{p}_0$ is the parameter change vector and the Jacobian J is an n×m matrix defined by:

$$J_{ij} = \frac{\partial f_i}{\partial p_j} \Big|_{\underline{p} = \underline{p}_0}$$

Solving

$$\frac{\partial \chi^2}{\partial p_j} = 0 \quad \text{for } j=1,...,m \tag{3}$$

yields the linear equation system:

$$J^T W^2 J \; \Delta \underline{p} = J^T W^2 \underline{\varepsilon}, \tag{4}$$

where $\underline{\varepsilon} = \underline{y} - \underline{f}(\underline{p}_0)$ may be called the discrepancy vector (Lines and Treitel 1984), and W is an n×n diagonal matrix containing the weights. This equation system has to be solved for the parameter correction step $\Delta \underline{p}$. The starting model $\underline{p}_0$ is then replaced by $\underline{p}_1 = \underline{p}_0 + \Delta \underline{p}$, and the process iterates until a given convergence criterium or a maximum number of iterations is reached.

A difficulty with the solution of (eq. 4) occurs when the problem is ill-posed, which means that the matrix $J^T W^2 J$ is singular or nearly singular. The way to overcome this problem is the regularization of $J^T W^2 J$ by adding a constant diagonal damping matrix (Lines and Treitel, 1984):

$$(J^T W^2 J + \mu^2 I_m) \Delta \underline{p} = J^T W^2 \underline{\varepsilon}, \tag{5}$$

where I_m is the m×m identity matrix and μ is known as the Marquardt parameter.

Instead of inverting eq. (5) directly, we use the Singular Value Decomposition (SVD) described by Golub and Reinsch (1970), which is applied to the weighted Jacobian $J_w = W J$:

$$J_w = U S V^T, \qquad (6)$$

where U is an $n \times m$ matrix which contains the data space eigenvectors and satisfies $U U^T = I_n$, V is an $m \times m$ matrix which contains the parameter space square-roots of the eigenvalues of $J_w^T J_w$ which are called the singular values of J_w.

Using this decomposition the solution of equation (5) is:

$$\Delta \underline{p} = V \ \text{diag}\left(\frac{1}{S_i}\right) \ \text{diag}\left(\frac{S_i^2}{S_i^2 + \mu^2}\right) \ U^T W \ \underline{\varepsilon} \qquad (7)$$

The reason for this eigenanalysis is that it is a useful tool to examine the inversion results in detail. The columns of the V-matrix are the parameter space eigenvectors and they are linear combinations of the physical parameters (thicknesses and resistivities). Thus, even though one often cannot resolve all parameters indepently, using the SVD one can at least tell which combination of the parameters is resolved.

Moreover, the second diagonal matrix in eq. (7), tells us how important a corresponding eigenvector is for the result. If a singular value S_i is small compared to the Marquardt parameter μ, the influence of the corresponding eigenvector will be damped out. If S_i is large, the eigenvector is important. Thus, the eigenanalysis helps to classify the parameters into important, unimportant and irrelevant parameters (Jupp and Vozoff, 1975).

The $S_i^2/(S_i^2 + \mu^2)$ are called damping factors. In our implementation we use modified damping factors with an exponent of 4 instead of 2, which may be called a second order Marquardt algorithm. In this case the Marquardt factor acts more like a threshold and is used to control the correction step during each iteration. It is convenient to start with 10 per cent of the largest eigenvalue, which means that at the beginning of the inversion only important parameters are changed. During the following iterations it will be decreased until a lower threshold, usually one per cent, is reached (Jupp and Vozoff, 1975).

The idea of the joint inversion is to fit two data sets of different measurements with one earth model. Using the algorithm described above, this is achieved by combining the two data sets into one vector, as well as the standard deviations, the model functions and the Jacobians, while keeping one parameter vector to fit the data:

$$\underline{y} = \begin{pmatrix} \underline{y}_1 \ (\text{LOTEM}) \\ \underline{y}_2 \ (\text{MT}) \end{pmatrix}, \quad \underline{\sigma} = \begin{pmatrix} \underline{\sigma}_1 \ (\text{LOTEM}) \\ \underline{\sigma}_2 \ (\text{MT}) \end{pmatrix}, \quad J = \begin{pmatrix} J_1 (\text{LOTEM}) \\ J_2 (\text{MT}) \end{pmatrix}, \quad \underline{f} = \begin{pmatrix} \underline{f}_1 (\text{LOTEM}) \\ \underline{f}_2 (\text{MT}) \end{pmatrix}$$

When jointly inverting two different data sets with one earth model, it is important to make sure that both data sets are equally weighted. In our implementation this is guaranteed by the use of the relative errors, which makes the Jacobian scale free (Jupp and Vozoff, 1975).

Example with synthetic data

Two artificial data sets were generated for a three-layer model, where a resistive layer is embedded between two more conductive layers, which is known as a K-type model (Fig. 2). Even though the resistivity contrast is quite large, the MT data show only little structure, which is due to the well-known fact that the resolution of MT is poor for thin resistive layers. It seems to be the same for the LOTEM Ex curve, but here the shape of the curve cannot be related to the conductivity structure in a simple way. The late-time limit of the curve, which is equal to the DC limit, depends on the conductivity structure of the entire depth range and thus contains information also about the second layer (Kaufmann and Keller 1983).

For these two data sets three inversions were run: the two separate inversions and one joint inversion. The starting model chosen for these inversions was a three-layer model, in which all layers had the same resistivity ($50\,\Omega\,m$), to leave the inversion as free as possible.

The results of the three inversions are compared in Fig. 3. As expected, the MT inversion recognizes the K-type structure, but the second layer is thicker and less conductive than in the true model. The result of the LOTEM-Ex inversion is very different from the true model. Instead of a K-type model an A-type model (increasing resistivities) was fit to the data, only the first layer resistivity is the same as in the true model. The result of the joint inversion cannot be distinguished from the true model on this plot.

We repeated these inversion trials with a number of different starting models. The results were very similar except when the true model was entered as starting model.

To examine this effect in more detail, we will now analyse the V-matrices for these three inversions (Fig. 4). As explained above, the columns of the V-Matrix show which linear combination of the parameters is resolved independently, whereas the damping factors show how important that parameter combination is for the result. Note that this applies to the final model of the inversion and not to the model which was used to generate the synthetic data. It is only valid for the region around the final model where the linearity assumption is valid.

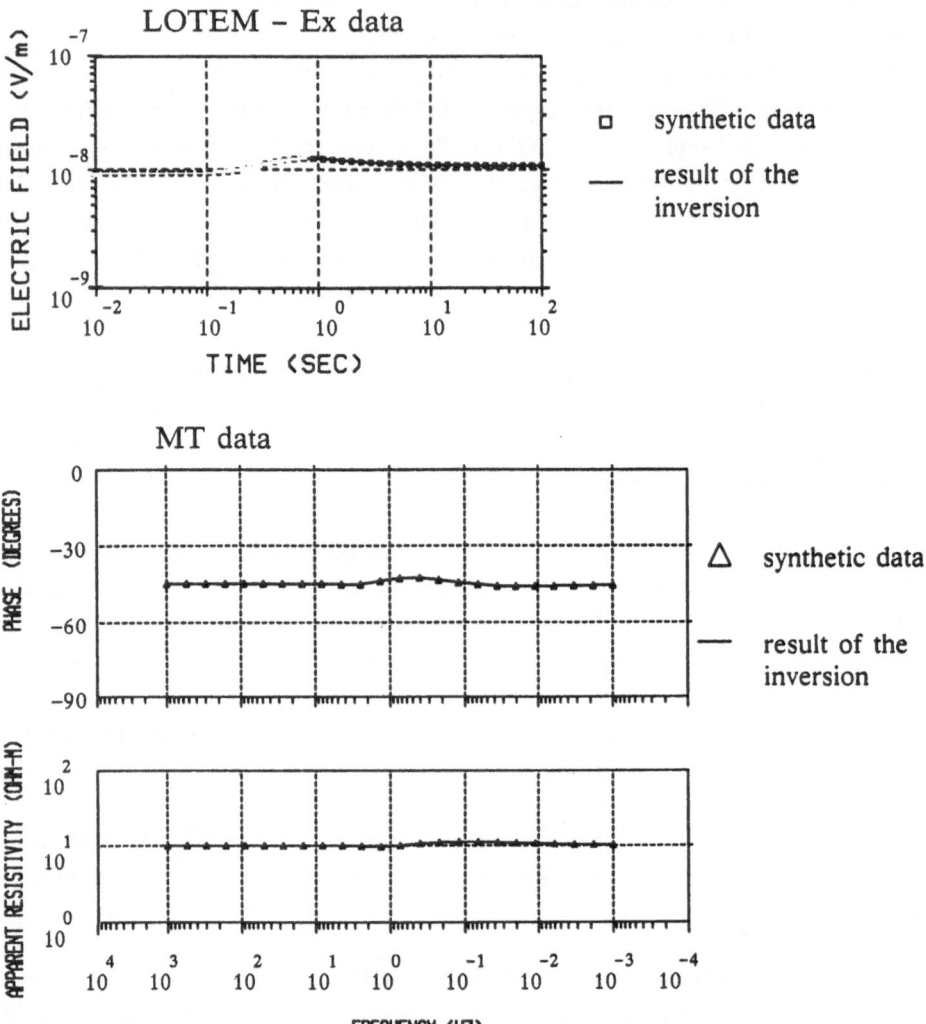

Earth model:

$\rho_1 = 10\ \Omega\,m$	$h_1 = 2000\ m$
$\rho_2 = 500\ \Omega\,m$	$h_2 = 500\ m$
$\rho_3 = 10\ \Omega\,m$	

Fig. 2: Synthetic data sets for a 3 – layer K – Type model

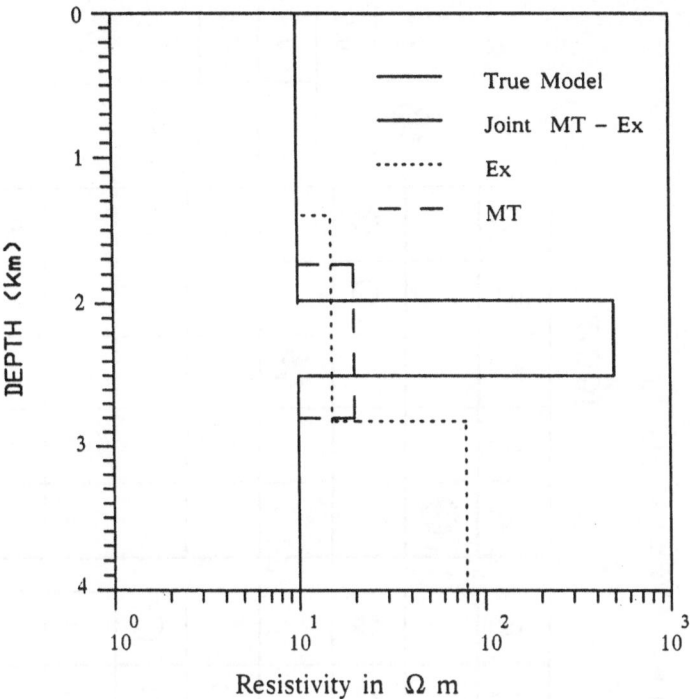

Starting model:

$$\rho_1 \ = \ \rho_2 \ = \ \rho_3 = 50 \ \Omega m$$

$$h_1 = \ 1800 \ m \quad h_2 = 1000 \ m$$

Fig. 3: Comparison of the inversion results for the K – Type model

For the LOTEM-Ex inversion the resistivity of the first layer is resolved independently, the second eigenparameter is a linear combination of the first
two layer parameters, and in the third column the resistivity of the third
layer is dominating. This means that the increase in resistivity (see ρ_3 in
Fig. 3) is well resolved. P4 represents mainly the second layer thickness and
P5 is irrelevant since the damping factor is 0.

It may be surprising that the LOTEM-Ex inversion finds a model which is
completely different from the true model, and that it does not resolve any
structure under the resistive layer. The explanation is that the electric field
at late times is dominated by the DC limit, which is mainly dependent on
the resistivity increase. The information about the conductive basement is
contained in the inductive part of the signal, but it is too small to be important.

153

V – MATRICES K – TYPE

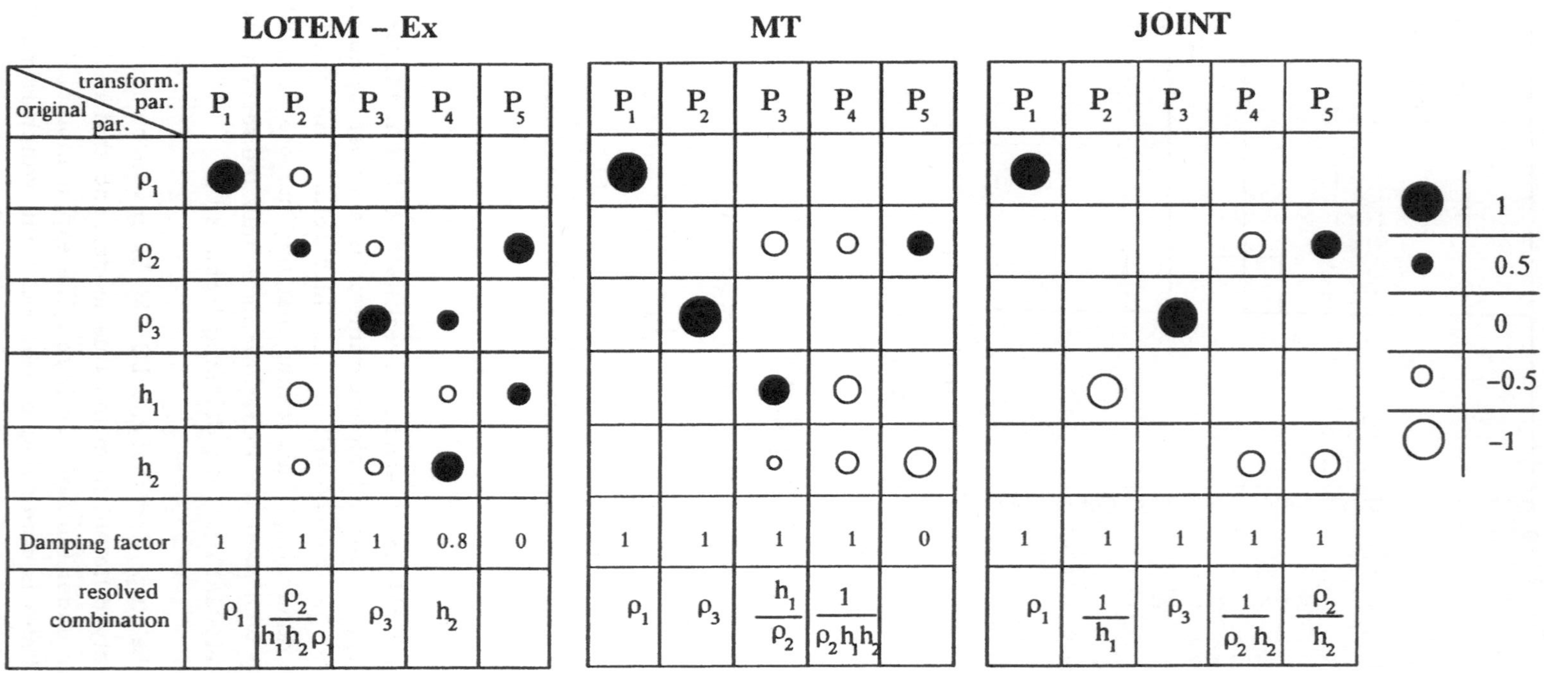

Legend:

1	● (large filled)
0.5	• (small filled)
0	
-0.5	○ (small open)
-1	◯ (large open)

LOTEM – Ex

transform. par. \ original par.	P_1	P_2	P_3	P_4	P_5
ρ_1	●	○			
ρ_2		•	○		●
ρ_3			●	•	
h_1		○		○	•
h_2		○		●	•
Damping factor	1	1	1	0.8	0
resolved combination	ρ_1	$\dfrac{\rho_2}{h_1 h_2 \rho_1}$	ρ_3	h_2	

MT

	P_1	P_2	P_3	P_4	P_5
ρ_1	●				
ρ_2		●	○	○	•
ρ_3			●		
h_1			•	○	◯
h_2			○	○	
Damping factor	1	1	1	1	0
resolved combination	ρ_1	ρ_3	$\dfrac{h_1}{\rho_2}$	$\dfrac{1}{\rho_2 h_1 h_2}$	

JOINT

	P_1	P_2	P_3	P_4	P_5
ρ_1	●				
ρ_2				○	•
ρ_3			●		
h_1		◯			
h_2				◯	◯
Damping factor	1	1	1	1	1
resolved combination	ρ_1	$\dfrac{1}{h_1}$	ρ_3	$\dfrac{1}{\rho_2 h_2}$	$\dfrac{\rho_2}{h_2}$

Fig. 4: V – Matrices for the inversions of the K – type model

The V-matrix of the MT-inversion is somewhat simpler. P1 and P2 clearly correspond to ρ_1 and ρ_3. P3 and P4 are linear combinations of h_1, h_2 and ρ_2. P5 would be required to resolve all 5 parameters, but it is irrelevant since the corresponding damping factor is 0.

In the joint inversion, P1, P2 and P3 correspond to ρ_1, ρ_3 and h_2. P4 is equivalent to the resistivity-thickness product and P5 is equivalent to the conductivity-thickness product of the second layer. Now P5 is no longer damped out and all 5 parameters are resolved. The effect of the joint inversion here is that the resolution properties of the two methods are combined. The LOTEM-Ex data contain information about the resistivity increase, the MT data resolve the structure underneath the resistor.

A case history with field data

The field data were measured in the Münsterland area in West Germany. The MT data (Fig. 5) consists of two parts: the high frequency data were measured with a controlled source method (CSAMT), (Fuerch, 1984). Only one transmitter was used, so that no strike direction can be determined. The low frequency data were measured by the University of Münster between 1981 and 1982 (Buechter, 1983). The data used here were rotated into the direction of the CSAMT data. This was a valid procedure, because the MT data do not show any significant 2-D effect in the overlapping frequency range.

The MT station is close to the borehole Münsterland 1, which was drilled in 1962/63 down to 5960 m. The laterolog data were reduced to a blocked resistivity depth section by Buechter (1983). This section was used to derive starting models for the inversions and will be compared with the results.

LOTEM data consisting of the y-component of the electric field, (Fig. 5) were recorded about 1 km from the MT-station. In contrast to artificial data, the early-time limit of the electric field is zero for field data. This effect is due to the response of the receiver system, and it is overcome by convolving the forward model with the measured system response during the inversion. We would like to mention that the data quality is not very good, neither for the MT, nor for the LOTEM.

The results of the three inversions are compared with the laterolog information in Fig. 6. The LOTEM-Ey-inversion agrees closely with the log in the first 3 km, but the LOTEM-data do not resolve the deeper sections. The result of the MT-inversion is quite different from the laterolog section. In particular the second, more conductive layer is much thicker than in the LOTEM-Ey inversion. In the joint inversion the different resolution properties are combined and the result now agrees closely with the well log information over the whole depth range of the borehole.

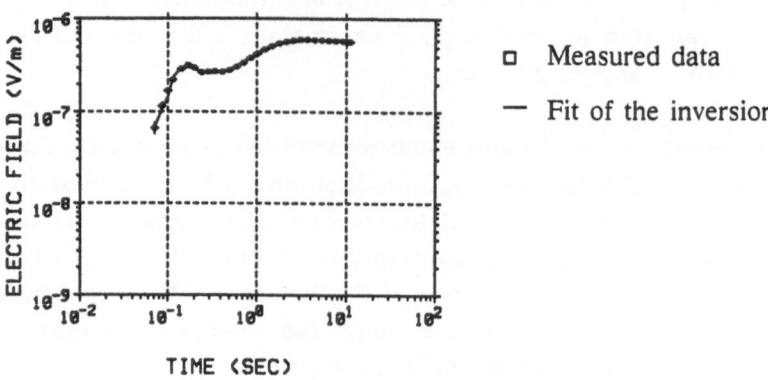

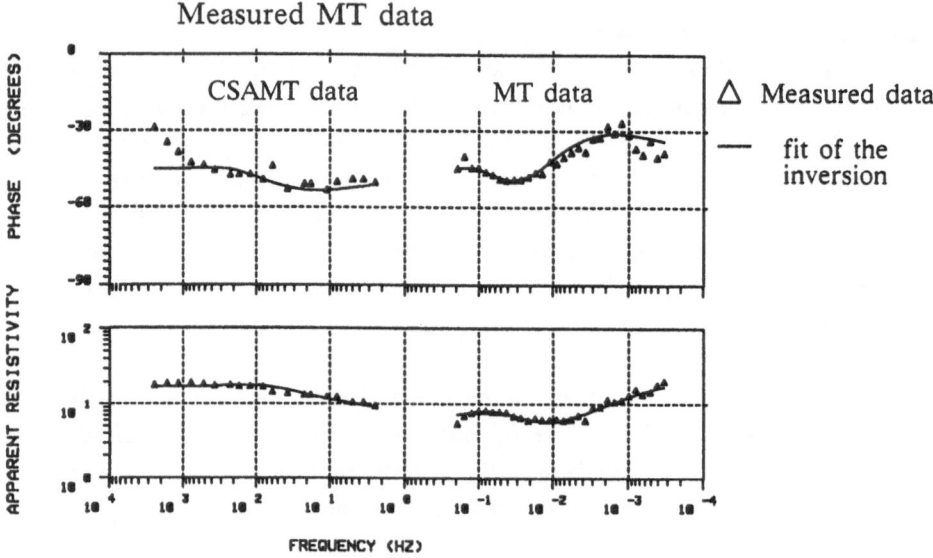

Fig. 5: Measured LOTEM data (top) and MT data (bottom)
from Münsterland/West Germany.

The improvement of the inversion result compared to the well log information
is quite obvious. It should be regarded as one example, where joint inversion
gives a satisfactory result even though the data quality is bad.

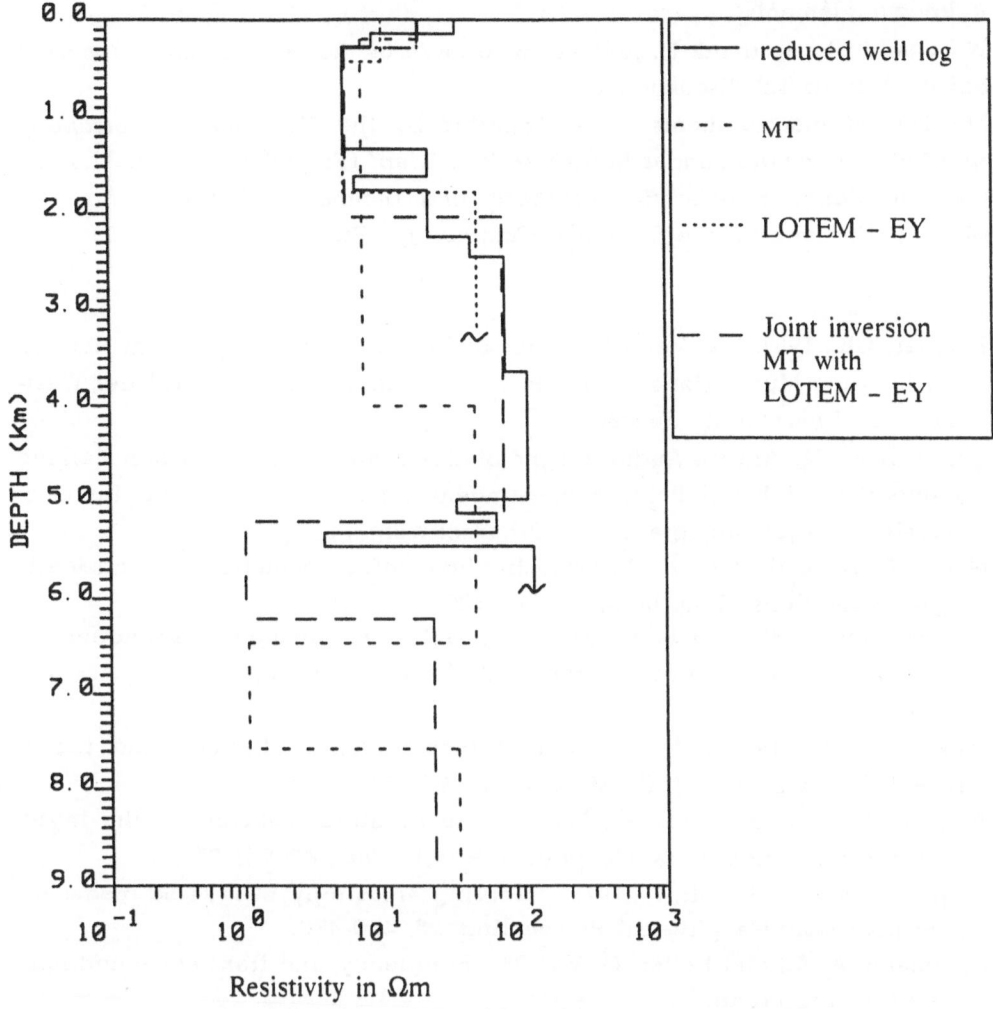

Fig. 6: Comparison of the inversion results of the Münsterland data

Conclusions

Joint inversion of MT data together with the electric field of the LOTEM method can vastly improve the interpretation in the case of thin resistive layers. The synthetic data example, where a K-type model could be resolved only by joint inversion, was very typical for these kinds of models. The inversion scheme, a second order Marquardt algorithm with singular value decomposition, can provide useful insight into the resolution properties of different methods.

The field data example showed that joint inversion can improve the interpretation even if the data quality is not very good. The LOTEM-Ey data provided information about the shallower structures, while the MT data contained information about the deeper sections, and through their combination the whole depth section could be resolved.

Acknowlegdements

We would like to thank H. Jödicke, who gave us the MT data and supported our work in useful discussions.

The LOTEM-measurements were supported by the *"Minister für Forschung und Technologie des Landes Nordrhein Westfalen"* (Project No. 03E6360A), the German *"Bundesminister für Forschung und Technologie"* (Project No. TH/0159/85-DE) and by the *European Community* (Project No. 0326550B).

References

Büchter, Ch., 1983, Die Verteilung der elektrischen Leitfähigkeit im Bereich der Bohrung Münsterland 1: Diplom thesis, Institut für Geophysik der Westfälischen Universität Münster.

Fürch, N., 1984, Aktive Audio-Magnetotellurik an der Bohrung Münsterland 1 und Vergleich mit Bohrlochmessungen: Diplom thesis, Institut für Geophysik und Meteorologie der TU Braunschweig.

Golub, G. H., and Reinsch, C., 1970, Singular value decomposition and least-squares solutions: Num. Math. **14**, 403-420.

Gomez-Trevino, E., and Edwards, R. N., 1983, Electromagnetic soundings in the sedimentary basin of Southern Ontario-A case history: Geophysics, **48**, 311-326.

Jackson, D. D., 1988, Interpretation of inaccurate, insufficient, and inconsistent data, Geophys. J. R. astr. Soc., **28**, 97-109.

Jupp, D. L. B., and Vozoff, K., 1975, Stable iterative methods for the inversion of Geophysical Data: Geophys. J. R. astr. Soc., **42**, 957-976.

Jupp, D. L. B., and Vozoff, K., 1977, Resolving anisotropy in layered media by joint inversion: Geophysical Prospecting, **25**, 460-470.

Kaufmann, A. A., and Keller, G. V., 1983, Frequency and transient soundings: Elsevier, Amsterdam.

Lines, L. R., and Treitel, S., 1984, Tutorial: A review of least-squares Inversion and its application to geophysical problems: Geophysical Prospecting, **32**, 159-186.

Raiche, A. P., Jupp, D. L. B., and Vozoff, K., 1985, The joint use of coincident loop transient electromagnetic and Schlumberger sounding to resolve resistive layers: Geophysics, **50**, 1618-1627.

Strack K.-M., Hanstein, T. H., and Eilenz, H. N., 1989, LOTEM data processing for areas with high cultural noise levels: Phys. Earth Plan. Interior, **53**, 261-269.

Verma, R. K., and Mallik, K., 1979, Detectability of intermediate conductive and resistive layers by time domain electromagnetic sounding: Geophysics **44**, 1862-1878.

Vozoff, K., and Jupp, D. L. B., 1975, Joint inversion of geophysical Data: Geoph J. R. astr. Soc., **42**, 977-991.

An Analysis of the Broad Spectrum Apparent Complex Resistivity Method

H. Holstein, J. W. Wilkinson

Department of Computer Science, University College of Wales, Aberystwyth, Dyfed SY23 3BZ, United Kingdom

Abstract

The measurement of broad spectrum apparent complex resistivity, obtained from a grounded pair of dipoles energised over a wide range of transmitter frequencies, is analysed by model calculations and shown to be a powerful geophysical method of delineating subsurface structures. For a horizontally layered earth model, synthetic complex resistivity data from a single dipole-dipole station is shown to lead to a better resolution of model parameters than can be obtained in the corresponding DC resistivity case. Used from a single station, broad spectrum complex resistivity provides a method of frequency depth sounding. The depth sounding data are analysed in the linearised regime of parameter estimation, where the number of well-separated independent parameter combinations that can be determined from the data leads to a quantification of the content of resistivity data.

The effect of dipole separation on parameter resolution is also investigated. Simultaneous inversion of data from multiple stations of different dipole separations leads to a further refinement in interpretation. This manner of parameter estimation combines both frequency and spatial depth sounding measurement modes. There can be a substantial data overlap between these two modes. However, provided that measurement stations are not placed too densely, the data from complex resistivity measurements will yield better parameter resolution than the corresponding DC resistivity measurement from the same stations.

1 Introduction

Apparent resistivity is a bulk response of material to electrical energisation, conceptually similar to standard DC apparent resistivity, except that the energisation is caused by an oscillating rather than a direct current source. The complex resistivity model calculations in this article are computed for a pair of grounded dipoles, shown schematically in figure 1. For a received voltage amplitude of V volts and phase lead ϕ radians over the energising current of amplitude I amps, the apparent complex resistivity $\rho_{app}(\omega)$ at a given circular frequency of ω radians per second is given by

$$\rho_{app}(\omega) = GVe^{i\phi}/I \tag{1}$$

in units of Ohm.metres (Ω.m), where both V and ϕ are frequency dependent and G is the geometric factor. With reference to figure 1, the geometric factor for the grounded dipole pair is given by

$$G = (2\pi)/(1/r_{AC} - 1/r_{AD} - 1/r_{BC} + 1/r_{BD}), \tag{2}$$

in units of metres.

The measurement of complex resistivity as a function of energising frequency is of geophysical interest, since the shape of the complex resistivity spectrum is formed in response to the electrical properties of the subsurface. This offers the possibility of inverting the frequency data to estimate model parameters. At low frequencies (0.1 Hz to 100 Hz) the standard frequency domain IP regime is obtained, while at frequencies above 1 kHz the method remains largely unexplored, despite prior recognition of its potential (Sumner, 1976).

The observed dependence of complex resistivity on energisation frequency may be ascribed to several physical mechanisms. The material may itself have frequency dependent electrical properties. The dependence at low frequencies forms the basis of IP measurement, as described by Pelton *et al.* (1978). Secondly, signal attenuation increases with increasing circular frequency ω and layer depth h, as governed by the negative exponential of the dimensionless quantity

$$(\mu\sigma\omega h^2)^{\frac{1}{2}} \tag{3}$$

where σ and μ are the conductivity and electromagnetic permeability that are characteristic of the material. This mechanism is loosely described by the "skin depth" phenomenon. It gives an intuitive understanding why a broad resistivity spectrum responds to decreasing depth with increasing frequency. A third mechanism is the mutual inductive coupling between the transmitting and receiving dipoles via the ground. At low frequencies the effect is proportional to the dimensionless quantity

$$\mu\sigma\omega d^2 \tag{4}$$

where d is the centre-to-centre separation of the dipoles. As the dependence on μ in (3) and (4) shows, both the latter mechanisms have their origins in electromagnetic induction.

Pelton *et al.* (1978) regarded the inductive response as an unwanted disturbance. They confined observations to the IP regime, even at frequencies as high as 100 kHz, by limiting

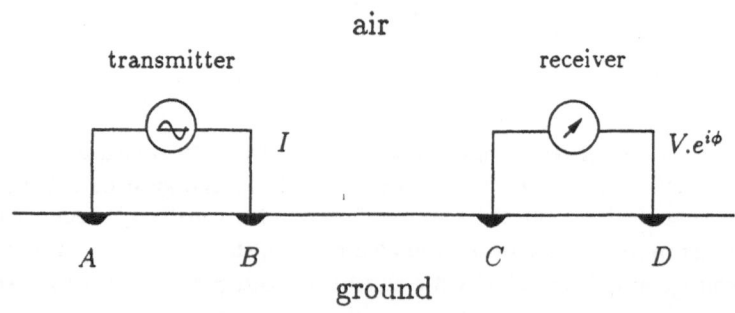

Figure 1: Schematic diagram of a complex resistivity measurement system consisting of a pair of grounded dipoles with electrodes A, B and C,D. The dipole separation is given by the distance $r_{AB}/2 + r_{BC} + r_{CD}/2$, and the dipole lengths are r_{AB} and r_{CD}.

the dipole separation d to a maximum of 1 metre. Our approach is complementary, in that we are concerned with the interpretation of apparent resistivity under conditions which include predominantly inductive effects, by using dipole separations d typically between 100 metres and 1000 metres and an operating frequency range of 0.1 Hz to 100 kHz. IP effects, if present, manifest themselves by a small slope in the apparent resistivity amplitude and by a small apparent resistivity phase at low frequencies. At intermediate frequencies, IP and inductive effects may be comparable, and inductive effects generally dominate at frequencies ω and dipole separations d which make the dimensionless quantity (4) of order unity.

Field equipment for broad spectrum resistivity measurement has become commercially available in recent years. In this article, we aim to give a theoretical justification for the effectiveness of the method.

In order to emphasise that spectral variations in complex resistivity do not necessarily have their origin in frequency dependent material properties of the ground, we make the simplification of taking all earth model parameters to be frequency independent. Thus we omit IP effects entirely. They can be incorporated into the model calculations by a frequency dependent Cole-Cole factor in each material conductivity.

In section 2 we outline the manner in which the apparent complex resistivity may be calculated for a horizontally layered earth model. Section 3 gives a method of assessing the content of the synthetic resistivity data, by determining the extent to which it is capable of resolving model parameters during data inversion. This approach is applied to the model calculations of section 4, which illustrate the effect of varying the earth model on the resistivity spectrum, and to the calculations of section 5, which illustrate the effect of varying the dipole separation for a fixed earth model. In each case, the complex resistivity data is found to possess a higher information content for parameter estimation than the corresponding DC results. The significance of the results is summarised in the conclusions.

2   Model equations for the computation of complex resistivity

Derivations of the electromagnetic field due to a horizontal electric dipole over a horizontally layered earth are to be found, for example, in Sunde (1949) and Wait (1982). Here we collect together the main results for the calculation of apparent complex resistivity as measured by a pair of grounded dipoles.

With reference to figure 1, let the transmitter dipole be represented by a straight wire between grounded electrodes A and B placed along the x-axis of a horizontal x-y coordinate system. Assuming that the current in the transmitter dipole is given everywhere along its length by $Ie^{i\omega t}$, let the electric field at the point (x, y) due to a length $d\xi$ of the transmitting dipole at $x = \xi$ be given by $d\mathbf{E}(x - \xi, y, \omega) e^{i\omega t}$. Then the field at (x, y) due to the whole transmitter is the sum of contributions from elementary dipoles along the length of the transmitter:

$$\mathbf{E}(x, y, \omega) e^{i\omega t} = \int_T \left(\frac{d\mathbf{E}}{d\xi} \right)_{(x-\xi, y, \omega)} d\xi \, e^{i\omega t} \tag{5}$$

161

where the integration is carried out along the x-directed transmitter path T from A to B. This field gives rise to a voltage $(Ve^{i\phi})\,e^{i\omega t}$ between receiver electrodes C, D given by:

$$V(\omega)\,e^{i\phi(\omega)} = \int_R \mathbf{E}(x, y, \omega)\,.\,d\mathbf{s} \tag{6}$$

where the line integral is taken along the path R, with vector length element $d\mathbf{s}$, connecting the receiver electrodes C and D. Only the horizontal projection of the electric field enters this equation.

For a horizontally layered earth, Maxwell's equations of electromagnetism indicate that the horizontal projection of the integrand in (5) is expressible in terms of two scalar functions P and Q:

$$\left(\frac{d\mathbf{E}}{d\xi}\right)_{(x-\xi, y, \omega)} = I\left(\nabla\frac{\partial Q\,(r, \omega)}{\partial x} - \hat{\mathbf{x}}\,P\,(r, \omega)\right) \tag{7}$$

where

$$r = \sqrt{\left((x - \xi)^2 + y^2\right)} \tag{8}$$

and $\hat{\mathbf{x}}$ is a unit vector in the x-direction. Q represents a scalar potential and $\hat{\mathbf{x}}\,P$ a vector potential. The apparent complex resistivity ρ_{app} is then given by (1), (6), (5) and (7):

$$\rho_{app}(\omega) = GVe^{i\phi}/I =$$

$$G\left\{Q(r_{AC}, \omega) - Q(r_{AD}, \omega) - Q(r_{BC}, \omega) + Q(r_{BD}, \omega) - \int_R\int_T P(r_{\xi s}, \omega)\,d\xi\,\hat{\mathbf{x}}\,.\,d\mathbf{s}\right\}, \tag{9}$$

where $r_{\xi s}$ is the distance between typical pair of points on the transmitter and the receiver.

The functions P and Q are solutions of the Helmholtz form of Maxwell's equation, at the air-ground interface, and are expressible as Hankel transforms over the zero order Bessel function J_0:

$$P(r, \omega) = \frac{i\mu_0\omega}{2\pi}\int_0^\infty F\,(\lambda,\,\omega,\,\mathbf{p})\,J_0\,(\lambda r)\,d\lambda, \tag{10}$$

$$Q(r, \omega) = \frac{1}{2\pi\sigma_1}\int_0^\infty G\,(\lambda,\,\omega,\,\mathbf{p})\,J_0\,(\lambda r)\,d\lambda/\lambda. \tag{11}$$

Q is normalised with respect to the top layer conductivity σ_1. The dimensionless complex functions F and G depend on the model parameters, denoted by a vector $\mathbf{p}$ with n components

$$p_1,\,p_2,\,\dots p_n. \tag{12}$$

When, as considered here, the permeability of each layer is the same as the free space permeability μ_0, and the only model parameters are the layer conductivities and the layer heights, then the functions F and G are computable from recurrence relations given in Wait (1982). In formula (9) for the apparent complex resistivity, model dependencies are inherited through the functions P and Q. The geometric details of the dipole configuration enter through arguments of Q, the paths of integration in the double integral, and the geometric factor G.

The function Q, being required only between end points of the dipoles, represents a potential field contribution, which persists in the DC limit of zero frequency. The term involving P represents mutual inductive coupling between the transmitter and receiver in the presence of the ground, and grows proportionally to (4) as ω increases from zero.

162

All integrals in (9), (10) and (11) are evaluated numerically. The integrand (11) appears to be troublesome at $\lambda = 0$. However, the combination of such integrands as required in (9) is finite at $\lambda = 0$. Indeed, subtracting a term $G(0, \omega, \mathbf{p})$ from the function G in the integrand of (11) leads to a well defined integral, while the sum of the offsets required in (9) satisfies

$$G(0, \omega, \mathbf{p}) \int_0^\infty (J_0(\lambda r_{AC}) - J_0(\lambda r_{AD}) - J_0(\lambda r_{BC}) + J_0(\lambda r_{BD})) \, d\lambda/\lambda$$

$$= G(0, \omega, \mathbf{p}) \ln \left(\frac{r_{AD}\, r_{BC}}{r_{AC}\, r_{BD}} \right), \qquad (13)$$

a result that can be deduced from Abramowitz and Stegun (1965).

When the dipole separation d is much less than a free space wavelength at the frequency ω, that is

$$\mu_0 \varepsilon_0 \omega^2 d^2 \ll 1, \qquad (14)$$

where ε_0 is the free space permittivity, a good approximation for evaluating F and G is the *quasi-static* approximation, in which ε_0 is set to zero. This corresponds to neglecting displacement currents in the air layer. Under these circumstances, the integrals for P and Q may be rapidly evaluated using the adaptive filtering algorithm of Anderson (1979).

Departures from the quasi-static condition (14) arise at large dipole separations and high frequencies, requiring calculations of the apparent complex resistivity to proceed without the quasi-static approximation. The integrand G in (11) is then found to undergo a rapid change at $\lambda^2 = \mu_0 \varepsilon_0 \omega^2$, making the filter method of evaluating the Hankel transform inappropriate. Instead, we integrate the non-smooth part of the integrand by reverting to numerical quadrature of subintegrals taken over the successive zeros of the Bessel function. This situation arises in section 4, where we have $d = 1000$ m, $\omega = 2\pi.10^5$ radians/s, giving $\mu_0 \varepsilon_0 \omega^2 d^2 \approx 4$. In general however, regimes for which the quasi-static approximations is valid will be adequate for model parameter estimation.

3   A content measure for complex resistivity data

For a given earth model characterised by a parameter vector $\mathbf{p}$ of components (12), expression (9) is a forward modelling function for computing the apparent complex resistivity $\rho_{app}(\omega)$ at each of a number of selected frequencies ω_j, $j = 1, \dots, m$. In principle, each such datum $(\rho_{app}(\omega_j), \omega_j)$ furnishes one equation for the determination of the components of $\mathbf{p}$. Since the synthetic resistivity spectrum is smooth, such data for $j = 1, \dots, m$ will not be independent. Thus it is of interest to estimate the number of well-separated independent relations furnished by the data, since this will lead to an equivalent number of independent relations among the components of $\mathbf{p}$. We define the data content to be measured by this number. For the given model, no inversion process will be able to improve on this number of independent parameter relations.

We follow other authors (see for example Hohmann and Raiche (1988), Hördt *et al.* (1991), Press *et al.* (1986)) in establishing the degree of data independence via a singular value decomposition of the system Jacobian. The number of its normalised eigenvalues which are greater than some given threshold $\varepsilon > 0$ will be used to define our measure of data content. The procedure is justified below. To simplify the discussion, we shall use a forward modelling function $\mathcal{F}$ that computes only the magnitude of the apparent complex resistivity. Discarding the phase entails some loss of information, but this is offset by the difficulty of obtaining accurate phase measurements in practice, and by the fact that model amplitude and phase are correlated.

Consider the data inversion problem, linearised about the parameter point $\mathbf{p} = \mathbf{p}_0$. A change of $\delta\mathbf{p}$ in the parameter vector leads to a data vector change whose jth component is

$$\mathcal{F}(\omega_j, \mathbf{p}_0 + \delta\mathbf{p}) - \mathcal{F}(\omega_j, \mathbf{p}_0). \tag{15}$$

To first order, the vector change to the data is given by

$$\mathbf{J}_0 \, \delta\mathbf{p}, \tag{16}$$

where $\mathbf{J}_0$ is the $m \times n$ Jacobian matrix of the modelling function $\mathcal{F}$, whose element in row j and column k is the derivative

$$\frac{\partial \mathcal{F}(\omega_j, \mathbf{p})}{\partial p_k} \tag{17}$$

evaluated at $\mathbf{p} = \mathbf{p}_0$. Let $\mathbf{J}_0$ have ordered singular eigenvalues

$$s_1 \geq s_2 \geq \ldots \geq s_n \geq 0 \tag{18}$$

with associated right singular eigenvectors

$$\mathbf{v}_1, \mathbf{v}_2, \cdots, \mathbf{v}_n \tag{19}$$

which form an orthonormal basis for the parameter space $\mathcal{R}^n$. The inversion problem at $\mathbf{p} = \mathbf{p}_0$ investigates which parameter perturbations $\delta\mathbf{p}$ satisfy the given synthetic data to within some specified error δ:

$$\|\mathbf{J}_0 \, \delta\mathbf{p}\|_2 \leq \delta. \tag{20}$$

In the 2-norm, the region in parameter space satisfying the above is the error ellipsoid, given in terms of the singular eigenvalues and vectors by

$$s_1^2 (\delta\mathbf{p}.\mathbf{v}_1)^2 + s_2^2 (\delta\mathbf{p}.\mathbf{v}_2)^2 + \cdots + s_n^2 (\delta\mathbf{p}.\mathbf{v}_n)^2 \leq \delta^2. \tag{21}$$

If, given ε, r is the first index for which

$$s_{r+1}/s_1 \leq \varepsilon < s_r/s_1, \tag{22}$$

then the directions

$$\mathbf{v}_{r+1}, \mathbf{v}_{r+2}, \cdots, \mathbf{v}_n \tag{23}$$

admit parameter changes of order $O(1/\varepsilon)$ times parameter changes in the direction $\mathbf{v}_1$, while being confined to the same error ellipsoid. We consider the model incapable of resolving parameter components in the directions (23). Conversely, we consider the model capable of resolving parameter components in directions

$$\mathbf{v}_1, \mathbf{v}_2, \cdots, \mathbf{v}_r \tag{24}$$

with decreasing sensitivity factors from 1 to s_r/s_1. The ratios of the singular eigenvalues s_i/s_1 are the normalised singular eigenvalues of the model's Jacobian matrix. Thus we shall carry out a singular value decomposition of the model's Jacobian matrix to see the distribution of normalised singular eigenvalues that exceed some given threshold value ε. The number r above the threshold ε gives the dimension of the subspace, associated with basis (24), in which we consider the parameter vector to be well-determined.

The practical evaluation of the Jacobian matrix was carried out in terms of a scaled parameter $\mathbf{p}$, whose components are the logarithms of their physical values. The derivative (17) was obtained by differentiating a least squares cubic Chebyshev fit to five interpolation points of $\mathcal{F}$, placed symmetrically around $\mathbf{p}_0$. In test cases, the cubic approximation was found to give a good representation of skewness introduced by the logarithmic parameters.

The interpolation and differentiation routines, as well as the singular value decomposition routine, were taken from a mathematical software library.

In the next section we have adopted a threshold value of

$$\varepsilon = 0.01. \tag{25}$$

This means that we reject parameter components whose standard error is 100 times worse than the standard error of the best determined parameter component. Since we give all the normalised singular eigenvalues, the number of parameter components rejected for any other threshold may easily determined.

4 Complex resistivity calculations for a family of earth models

The apparent complex resistivity measured at a single dipole-dipole station in the frequency range 0.1 Hz to 100 kHz was computed for a family of six earth models, corresponding to a buried conducting layer of various thicknesses. In this manner we demonstrate the effect of model variation on the observed synthetic resistivity spectra. The results are shown in figure 2. After describing the model and the computed spectra, we analyse the data content of the model spectra and contrast the apparent complex resistivity and apparent DC resistivity cases.

The earth models consist of three horizontal layers, and are distinguished by the thickness of the middle layer. Using the notation h_i, ρ_i to denote respectively the height and resistivity (reciprocal conductivity) of layers $i = 1, 2, 3$, the middle layer thicknesses were chosen as

$$h_2 = 0 \text{ m (homogeneous half-space)}, \ 1 \text{ m}, \ 2 \text{ m}, \ 10 \text{ m}, \ \infty \text{ (two-layer case)}, \tag{26}$$

and the remaining parameters, common to all six models, are given in Table 1.

layer i	height h_i(m)	resistivity $\rho_i(\Omega.m)$
1	200	350
2	as in (26)	0.35
3	∞	350

Table 1: Horizontally layered earth models

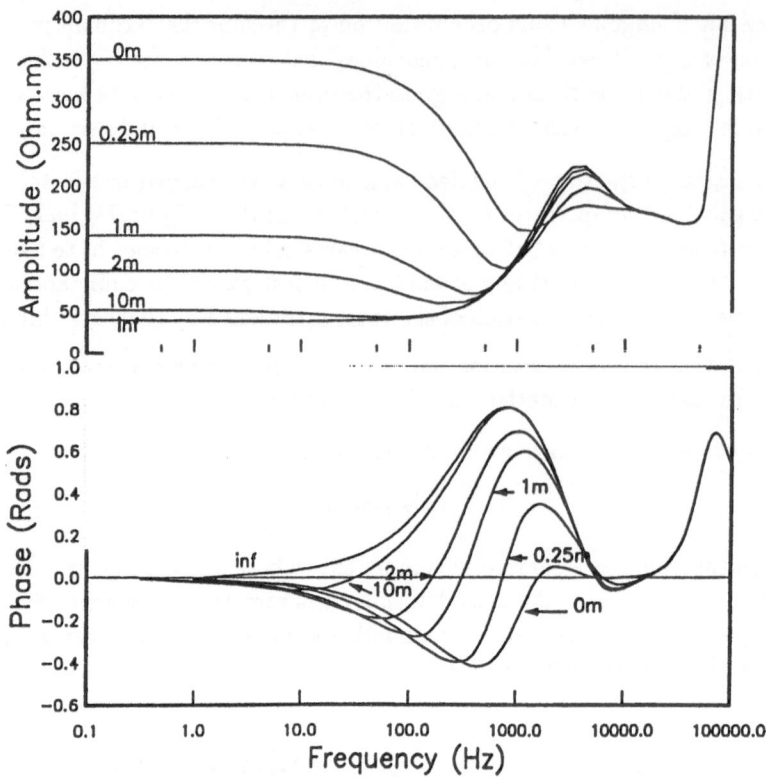

Figure 2: Computed complex resistivities for six three-layer models. The number on each curve gives the middle layer thickness. Above: Complex resistivity amplitude. Below: Complex resistivity phase.

As dipole configuration, an in-line dipole pair was used, with dipole lengths r_{AB}, r_{CD} of 200 m and a centre-to-centre separation of 1000 m (see figure 1).

Figure 2 shows the computed apparent resistivity magnitude and phase spectra for the six models. Up to 100 Hz, the curves are essentially flat, and cannot reveal any structure of the model. In this regime standard resistivity measurements are made. In the range 100 Hz to 10 kHz, however, each curve exhibits dramatic changes of slope, over- and under-shoot, in response to the underlying model parameters. Beyond 10 kHz the curves respond only to the top layer, as expected from skin depth considerations. The final amplitude rise indicates that air displacement currents have become important. The frequency range in which the curves coalesce cannot reveal further substructure, and is therefore not of interest in the determination of subsurface parameters. To determine the extent to which these curves can reveal the model parameters, we have sampled the amplitude data at the nine frequencies given in Table 2. The frequencies were chosen at rational fractions of the logarithmic decade, sufficiently dense to reproduce the synthetic curves by interpolation. With respect to these frequencies, the Jacobians for the six models were calculated. The results of the singular value decomposition gave rise to the normalised singular eigenvalues, summarised in Table 3. The underlined values correspond to ratios of singular eigenvalues that exceed the threshold ε, chosen at 0.01. Inspection of the singular eigenvectors (not presented here) reveals that the parameters h_2 and ρ_2 are not determined

Frequencies (Hz)				
0.1	31.6	100	1000	10000
		215	2150	
		464	4640	

Table 2: Sample frequencies of resistivity data

h_2	s_1/s_1	s_2/s_1	s_3/s_1	s_4/s_1	s_5/s_1
0	1.000	0.4879	0.0004		
0.25	1.000	0.5958	0.3610	0.0545	0.0008
1	1.000	0.8461	0.4182	0.0167	0.0005
2	1.000	0.9009	0.2844	0.0083	0.0011
10	1.000	0.8882	0.0575	0.0034	0.0012
∞	1.000	0.8818	0.0060		

Table 3: Normalised singular eigenvalues for six models characterised by h_2.

separately, but only through their conductance ratio h_2/ρ_2. Such geo-electrical equivalence arises in models with a conducting layer, which no resistivity measurements can resolve. The equivalence accounts for the presence of one small singular eigenvalue ratio for each model. Apart from this inherent lack of parameter resolution, however, Table 3 shows that the remaining parameter combinations are completely resolved in the first three and the last models. There is a progressive worsening of parameter resolution as the conducting layer increases in thickness, leading to one further small eigenvalue ratio in models 4 and 5. The poorer resolution is expected due to the difficulty of resolving resistive structures below a conducting layer.

The power of the complex resistivity method is now demonstrated : at most two parameter combinations have not been determined from the observations of a single dipole-dipole station. In the case of DC resistivity, only one parameter component could be determined from the measurement at a single station. Equivalent information could only be obtained from the simultaneous interpretation of data from a number of suitably placed DC measuring stations, at least as many as there are independent parameter combinations to be resolved.

The ability of single station complex resistivity measurements to yield depth sounding information equivalent to that gained from an array of DC dipoles, justifies our use of the description "frequency depth sounding" for the broad spectrum complex resistivity method.

5 Complex resistivity calculations for different dipole separations

In this section we investigate the ability of broad spectrum complex resistivity method to provide information about an earth model with a buried conducting layer, whose detection is made difficult by the presence of a conducting overburden. We assess the information content of complex resistivity data as measured from single stations of different dipole separations, and from the simultaneous inversion of data from several dipole separations.

layer	height (m)	resistivity (Ω.m)
1	5	10
2	300	200
3	20	2
4	∞	200

Table 4: Horizontally layered earth model with overburden

We also give a comparison with the corresponding DC results. The horizontally layered earth model is given in Table 4. It presents the non-trivial problem of detecting a buried conducting layer at a depth of 305m in the presence of a conducting overburden.

According to an empirical rule, the semi centre-to-centre separation of a dipole-dipole measuring system is of the same order as its the maximum depth penetration. Accordingly, we have chosen dipole separations of 400 m, 600 m and 1000 m with which to investigate the model. At the first separation, little evidence of the conducting layer is expected. At the second, some influence of the conducting layer might be detectable, while the largest separation should respond to the conducting layer. As well as investigating the sensitivities of the individual dipole pairs to the model structure, we investigate the sensitivity of the three data sets taken together.

Figure 3 shows the computed apparent complex resistivity amplitude spectra for the three dipole separations. The frequency range is 0.1 Hz to 10 kHz. Beyond this range the three curves coalesce. The curves do not have the variety of features exhibited in figure 3, due to the presence of the overburden. Nevertheless, the manner in which the curves progress to the point of coalescing is in response to the layering parameters.

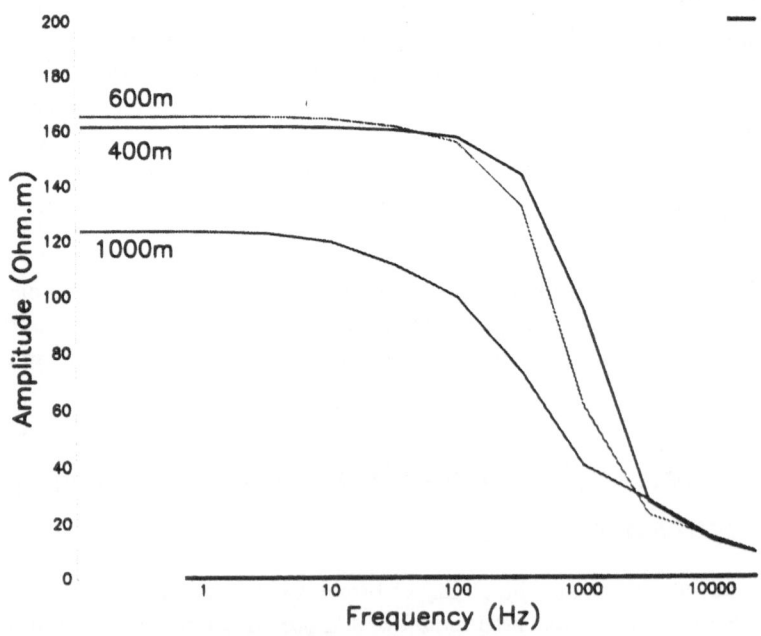

Figure 3: Computed amplitude of apparent complex resistivity for the earth model in Table 4, for three dipole measuring systems. Each curve is annotated with the centre-to-centre dipole separation.

To determine the extent to which these curves can reveal the model parameters, we have again sampled the spectra at the frequencies shown in Table 2, and have subjected the three Jacobians, one for each dipole separation, to the singular value analysis. The normalised singular eigenvalues are given in Table 5.

d	s_1/s_1	s_2/s_1	s_3/s_1	s_4/s_1	s_5/s_1	s_6/s_1	s_7/s_1
400	1.000	0.0858	0.0656	0.0073	0.0030	0.0004	0.0001
600	1.000	0.1889	0.1250	0.0165	0.0025	0.0005	0.0003
1000	1.000	0.2927	0.0955	0.0500	0.0017	0.0008	0.0007

Table 5: Normalised singular eigenvalues for an earth model investigated at different dipole separations d metres.

The underlined values correspond to ratios of singular values that exceed the threshold ε, chosen at 0.01. Thus, three parameter combinations are resolved for the case $d = 400$ m, and four parameter combinations are resolved in the cases $d = 600$ m and $d = 1000$ m, out of a possible seven parameter combinations. Again we find the loss of one parameter combination due to the geo-electrical equivalence of all models with the same value of the conductance ratio h_3/ρ_3 in the conductive layer 3. Corresponding DC measurements can only isolate one parameter combination per station.

The data from the three stations at different dipole separations is used to best advantage when interpreted simultaneously. The analysis for the simultaneous inversion extends straightforwardly - the Jacobian is formed from the three data sets, under the assumption of equal weighting. The analysis can also be carried out for the three DC data points, taken at 0.1 Hz for each separation. The normalised singular eigenvalues for these two cases are presented in Table 6.

	s_1/s_1	s_2/s_1	s_3/s_1	s_4/s_1	s_5/s_1	s_6/s_1	s_7/s_1
complex resistivity	1.000	0.4571	0.1402	0.0380	0.0045	0.0024	0.0008
DC resistivity	1.000	0.4121	0.0535				

Table 6: Normalised singular eigenvalues for the joint data from an earth model investigated at three of different dipole separations.

The simultaneous inversion of the complex resistivity data gives an improved resolution of the parameters, shown by an increase in nearly all singular value ratios. However, still only four values exceed the threshold of 0.01. By comparison, the DC data has an increased resolution from one to three parameter combinations - a consequence of the individual measurements acting together in the manner of spatial depth sounding.

These results point to an equivalence between frequency and spatial depth sounding methods. A single station measuring broad spectrum resistivity with a dipole separation d yields the same information as a suitable number of DC stations employing dipole separations of d and less. Consequently, simultaneous interpretation of data obtained from a sufficiently dense spacing of measuring stations yields the same information for broad spectrum and DC resistivity methods. At sparser spacings deployed in surveys, however, the broad spectrum resistivity method offers a better parameter resolution than the corresponding DC method, or conversely, offers the possibility of obtaining the same information from fewer stations.

We conclude this section by exhibiting the sensitivities of selected parameter combinations to a goodness of fit to the model. We calculate the error contours directly from the differences (15) rather than from the linearisation (16), and can therefore exhibit the non-quadratic nature of the parameter estimation problem. The figures of error contours highlight how the dipole separation affects the determination of particular parameter combinations, and indicate the manner in which simultaneous inversion of data sets from different dipole separations can improve parameter estimates. Throughout the figures presented, the position of $\mathbf{p_0}$ is indicated by the symbol "$+$".

To quantify the error measure when the model vector parameter $\mathbf{p_0}$ is changed to $\mathbf{p}$, we define an error function $\epsilon(\mathbf{p})$ by

$$\epsilon(\mathbf{p}) = \mathbf{max}(j = j_1 \ldots j_5) \log_{10} \sum_j \left(\mathcal{F}(\omega_j, \mathbf{p}) - \mathcal{F}(\omega_j, \mathbf{p_0}) \right)^2, \tag{27}$$

which represents a logarithm of the sum of squares of the five largest deviations between the spectra for $\mathbf{p}$ and $\mathbf{p_0}$, sampled over a fixed set of frequencies. For the examples used here, we have taken two frequencies per decade, at equal logarithmic intervals.

Figures 4 to 7 illustrate error contours for the estimation of parameters h_1 and h_2. At 400 m separation, the contours are open with a vertical bias (figure 4), indicating that the depth of the conducting layer is not resolved. On the other hand, there is a rapid increase in error in the horizontal direction, indicating that the thickness h_1 of the overburden is well resolved. This is as expected for a dipole whose depth penetration is does not reach the conducting layer.

The situation gradually changes as the dipole separation is increased, through figures 4 to 6. Figure 6 shows a predominately horizontal alignment of contours, indicating that the depth of the conducting layer is well resolved, but not the overburden. Again, this is expected for a dipole whose separation is over twice the depth of the conducting layer, and large compared with the thickness of the overburden.

When the error function (27) is applied to all three sampled data sets simultaneously, the result is a closed set of contours shown in figure 7. Not only are parameters h_1 and h_2 thereby completely resolved, but the contours have crowded towards the target point, indicating steeper gradients and hence better error sensitivity.

Figures 8 to 11 give error contours for the depth of the conducting layer versus its resistivity. The 400 m separation gives little indication of the true minimum, and indeed shows an additional false minimum. The larger separations (figures 9 and 10) give improved but still ambiguous valley-shaped regions for the minimum. The slightly different orientations of the valleys allows the minimum to be somewhat refined in the composite error contours of figure 11. The depth of the conducting layer is well resolved, and a lower bound on the resistivity can be determined. The contours far from the minimum are almost identical to those in figure 10, indicating that the shorter separations affected estimates only in the region of the minimum.

The final figure 12 shows error contours for the 1000 m case of $\log h_3$ against $\log \rho_3$, with a slope of -1, directly demonstrating the geo-electrical equivalence for all models with the same conductance ratio h_3/ρ_3, already mentioned above. In this case the contour plots for the smaller dipole separations were similar, except more widely spaced, so that a composite error plot would be identical to the 1000 m case, with the narrower separations giving no extra information.

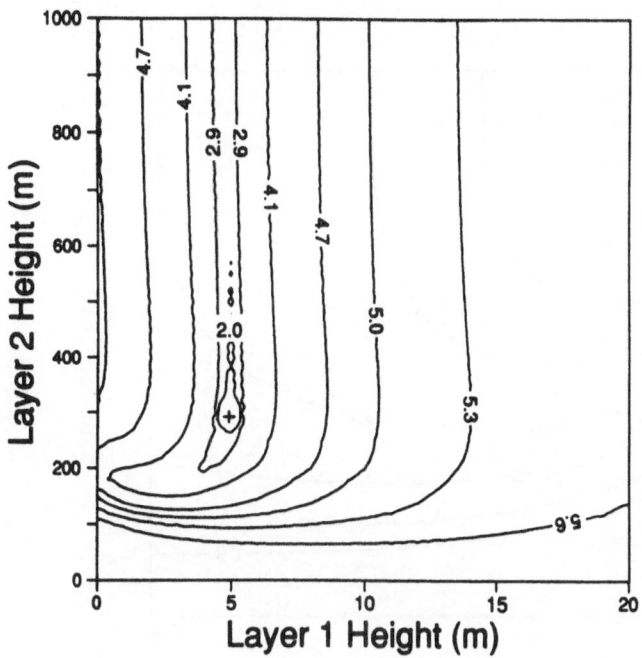

Figure 4: Error contours for h_1 and h_2 : dipoles of 400 m separation.

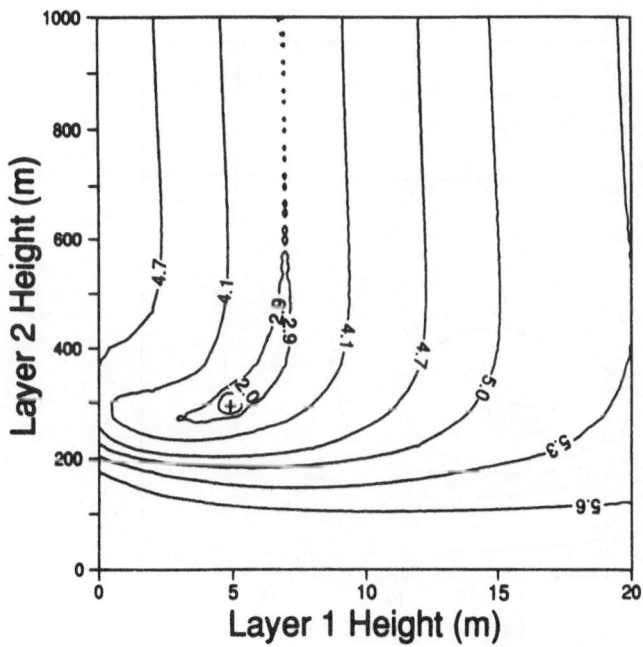

Figure 5: Error contours for h_1 and h_2 : dipoles of 600 m separation.

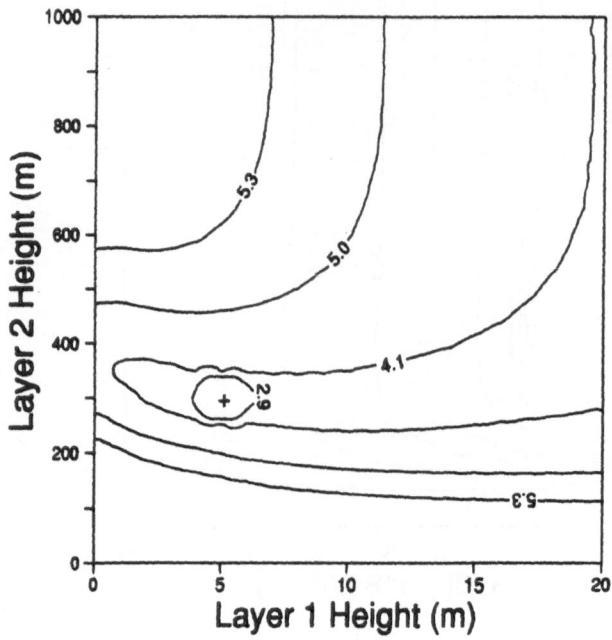

Figure 6: Error contours for h_1 and h_2 : dipoles of 1000 m separation.

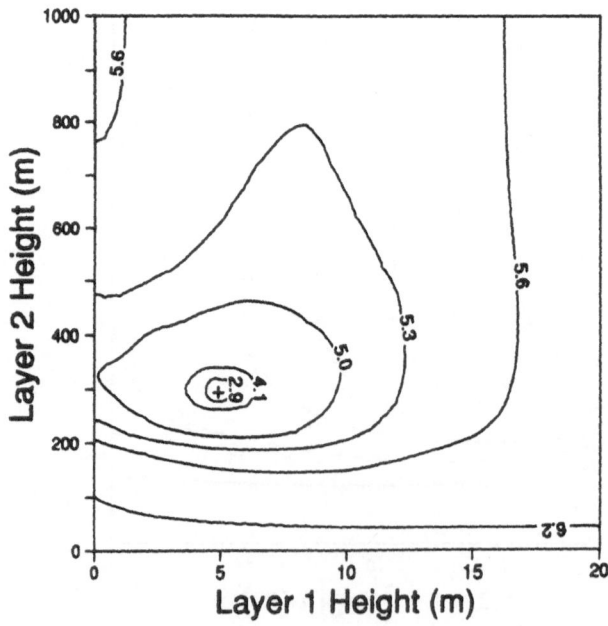

Figure 7: Composite error contours for h_1 and h_2 : dipoles of 400 m, 600 m and 1000 m separation.

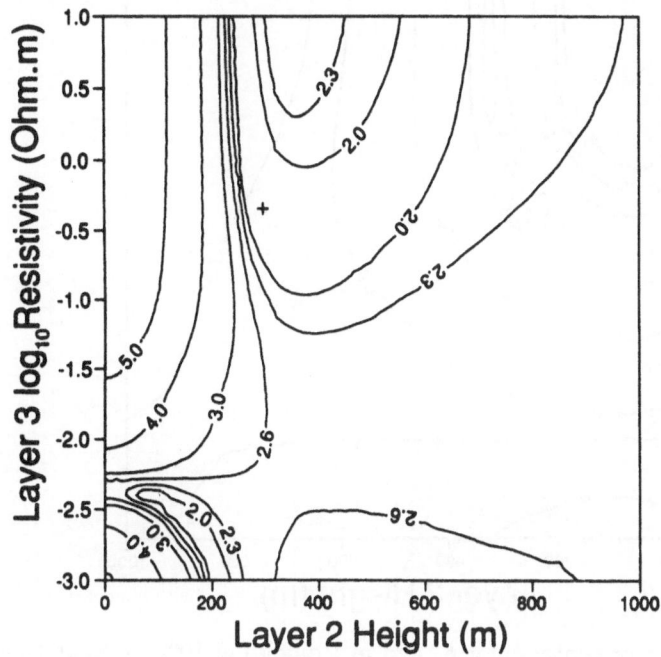

Figure 8: Error contours for h_2 and ρ_3 : dipoles of 400 m separation.

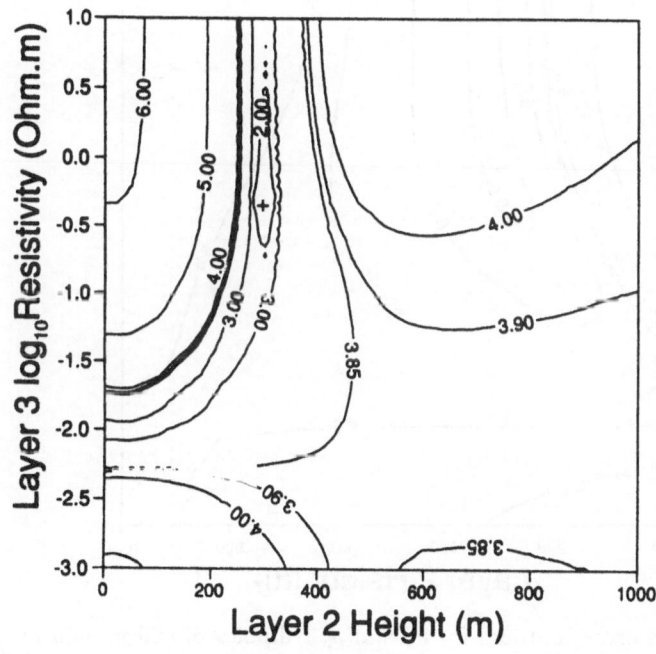

Figure 9: Error contours for h_2 and ρ_3 : dipoles of 600 m separation.

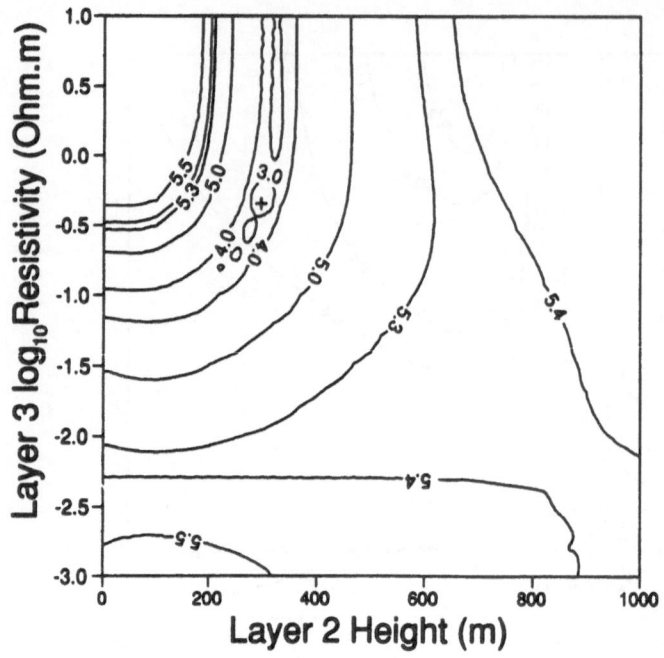

Figure 10: Error contours for h_2 and ρ_3 : dipoles of 1000 m separation.

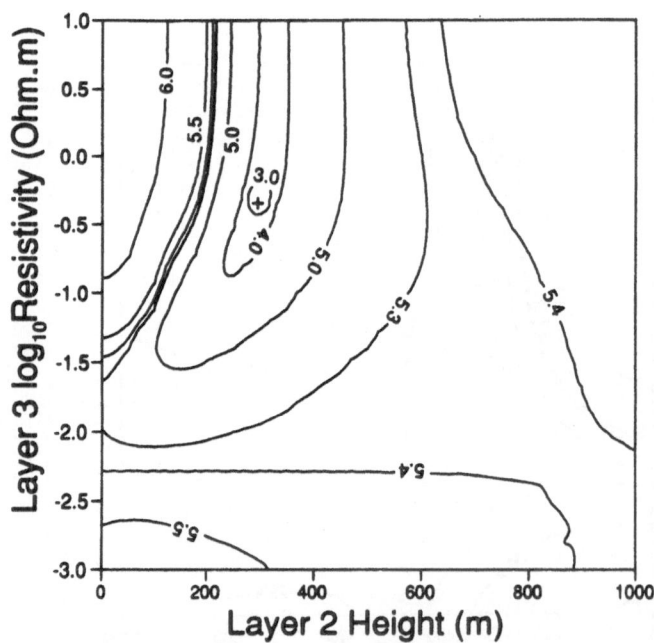

Figure 11: Composite error contours for h_2 and ρ_3 : dipoles of 400 m, 600 m and 1000 m separation.

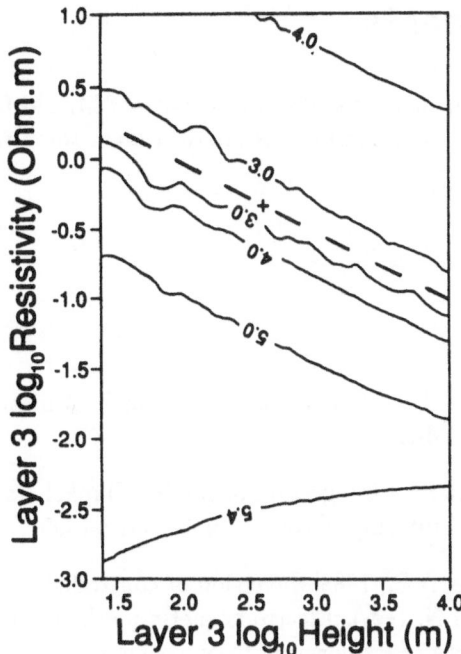

Figure 12: Error contours for $\log_{10} h_3$ and $\log_{10} \rho_3$: dipoles of 1000 m separation.

The figures illustrate a variety of parameter sensitivities, in particular the effect of interpreting data from multiple dipole separations, as in spatial depth sounding. The figures must be viewed with some caution, however, since they only give a two dimensional projection of the parameter space. They complement the *local* description of the n-dimensional error ellipsoid.

6   Conclusions

We have analysed the broad spectrum complex resistivity method of ground exploration through synthetic model calculation over horizontally layered earths, and have shown that single station measurement of the apparent complex resistivity, obtained for a broad range of energising frequencies, is equivalent to DC resistivity measurements made from a number of stations of different separations. This allows us to regard broad spectrum resistivity measurement as a form of frequency depth sounding. When the measuring stations are not too densely spaced, the deployment of broad spectrum resistivity will yield more information for model parameter estimation than the corresponding DC resistivity measurements. These conclusions are drawn on the basis of a linearised local process of parameter fitting. We have also illustrated some global aspects of the problem by exhibiting error contours for selected cases. In particular, we have illustrated how different dipole separations can affect the sensitivity of parameter estimation.

While the interpretation of broad spectrum complex resistivity is in general more difficult to carry out than for conventional resistivity methods, this study suggests that there are benefits to be gained from this method of geo-exploration.

7 Acknowledgements

This study was undertaken during a Raw Materials Contract MA1M-0007-E supported by the Commission of European Communities. Numerical algorithms were taken from the NAG library.

8 References

Abramowitz, M., and Stegun, I.A., (eds.) 1965. Handbook of mathematical functions. Section **11.1.20**. Dover Edition.

Anderson, W.L., 1979. Numerical integration of related Hankel transforms of orders 0 and 1 by adaptive digital filtering. Geophysics **44**, 1287-1305.

Hohmann, G.W., and Raiche, A.P., 1988. Inversion of controlled-source electromagnetic data. In: Nabighian, M.N. (ed.), Electromagnetic methods in applied geophysics. Volume 1, Theory. Society of Exploration Geophysicists.

Hördt, A., Gorenflo K.-M., Vozoff, K., and Wolfgram, P.A., (1991). Resolving resistive layers using joint inversion of LOTEM and MT data. In: *these proceedings*.

Pelton, W.H., Ward, S.H., Hallof, P.G., Sill, W.R., and Nelson, P.H., 1978. Mineral discrimination and removal of inductive coupling with multifrequency IP. Geophysics **43**, 588-609.

Press, W.H., Flannery, B.P., Teukolsky, S.A., Vetterling, W.T., 1986. Numerical recipes. Cambridge University Press.

Sumner, J.S., 1976. Principles of induced polarisation for geophysical exploration. Elsevier.

Sunde, E.D., 1949. Earth conduction effects in transmission systems. D. van Nostrand Co. Inc.

Wait, J.R., 1982. Geo-electromagnetism. Academic Press.

4. Seismics

An Efficient Algorithm for Impedance Reconstruction by the Modified Gelfand-Levitan Inverse Method

A. K. M. Sarwar[1], K. W. Holladay[2]

[1] Dept. of Geology & Geophysics, University of New Orleans, New Orleans, LA 70148, USA
[2] Dept. of Mathematics, University of New Orleans, New Orleans, LA 70148, USA

ABSTRACT

We present a fast approximate algorithm for reconstructing one-dimensional acoustical impedance profiles from reflection seismograms. Carroll and Santosa (1981) treated an equation different from the Schrodinger wave equation and derived the modified Gelfand-Levitan (MGL) integral equation. Their time domain interpretation (Carroll and Santosa, 1982) of the MGL equation was more natural than the usual Gelfand-Levitan equation in the context of exploration seismology. Santosa (1982) gave a numerical solution of the MGL based on Gaussian elimination, but that approach requires a large amount of computational time and memory.

We refine the MGL equation by a few changes of variables. The refined form of the equation yields a Toeplitz system which can be solved by Levinson recursion. The discrete version of the transformed impedance equation reconstructs the acoustical impedance profile of a Goupillaud type layered medium on an even grid. Our matrix solution is analogous to the Levinson recursion solution for the Wiener shaping filter. Since the systems developed in this paper can be solved more efficiently than general symmetric positive definite systems, our approach is a more efficient method for reconstructing impedance profiles.

1. INTRODUCTION

This paper presents an efficient algorithm using the modified Gelfand-Levitan (MGL) integral equation for reconstruction of the acoustical impedance log of a layered medium. This study considers only

the one-dimensional problem in which propagation is constrained along
the vertical axis and lithology varies vertically in terms of the
acoustical impedance as a function of one-way travel time. Since there
is no suitable mathematical apparatus for the exact solution of more
practical three-dimensional problems, the one-dimensional model is
often used as a conceptual model of the earth in exploration
seismology.

A compressional wave is assumed to be generated at the surface of
the earth. Geophones placed at the source position record reflections
from the heterogeneous half space of density ρ and compressional wave
velocity α. This zero-offset section cannot be obtained directly from
an experiment. However a common-depth-point section simulated from a
seismic acquisition can be considered an approximate zero-offset
section. The forward problem is to determine the response of the
receiver from the given density and wave velocity. The inverse problem
is to determine the acoustical impedance from the given response of the
receiver at the surface of the earth.

The Gelfand-Levitan integral equation was developed in 1951 for
solving quantum mechanical inverse spectral problems by studying the
Schrodinger wave equation. Burridge (1980) presented a time domain
derivation of the Gelfand-Levitan integral equation following a
spectral approach similar to Alekseev (1962). Later Carroll and
Santosa (1982) presented a time-domain interpretation of the MGL
equation and Santosa (1982) presented its numerical solution for the
one-dimensional propagation of shear waves. Santosa's numerical
solution was based on a primitive rule which needs large computational
time and memory.

The Gelfand-Levitan method of Burridge and the modified Gelfand-
Levitan method of Carroll and Santosa are different. Excellent papers
exist on Gelfand-Levitan and related methods in exploration seismology
including those of Claerbout (1976), Berryman and Greene (1980) and
Bube and Burridge (1983). We also mention the works in the general area
of exact inversion by Goupillaud (1961), Sondhi and Gopinath (1971),
Newton (1981), Santosa and Schwetlick (1982), Carrion and Patton (1983),
Carrion and Kuo (1984), Gray and Symes (1984), Gray (1984) and Carrion
and Foster (1985). A large number of papers by the group led by Carroll
at the University of Illinois at Urbana contributed significantly to
exact inverse scattering. Since the method of Carroll and Santosa is
in a more natural setting in the context of exploration seismology, we

present an efficient algorithm by refining Carroll and Santosa's modified Gelfand-Levitan method.

In section 2, the modified Gelfand-Levitan method is summarized and compared to the Gelfand-Levitan method of Burridge. We refine the MGL equation by a few changes of variables and solve the resulting Toeplitz system in section 3. A brief conclusion is given in section 4.

2. THE MODIFIED GELFAND-LEVITAN METHOD OF INVERSION

2.1 Preliminary Equations

The governing equations for the vertical propagation of compressional waves in a layered acoustic medium are

$$p_\tau(\tau,t) = - z(\tau) \, w_t(\tau,t) \tag{2.1.1}$$

and

$$p_t(\tau,t) = - z(\tau) \, w_\tau(\tau,t), \tag{2.1.2}$$

where $p(\tau,t)$ is hydrostatic pressure, $w(\tau,t)$ is particle displacement, $z(\tau)$ is acoustical impedance and t is time. The subscript denotes partial derivative. The one-way travel time, τ, is defined for a velocity function $\alpha(z)$ by

$$\tau(z) = \int_0^z \frac{dz'}{\alpha(z')} . \tag{2.1.3}$$

Decoupling the hyperbolic wave equations (2.1.2), we obtain

$$w_{\tau\tau}(\tau,t) - q(\tau)w_\tau(\tau,t) - w_{tt}(\tau,t) = 0, \tag{2.1.4}$$

where the potential $q(\tau)$ is given by

$$q(\tau) = - \dot{z}(\tau)/z(\tau). \tag{2.1.5}$$

Here $\dot{z}(\tau)$ is the derivative of $z(\tau)$. The Fourier transform of equation (2.1.4) gives

$$\hat{w}_{\tau\tau}(\tau,\omega) - q(\tau)\hat{w}_\tau(\tau,\omega) + \omega^2\hat{w}(\tau,\omega) = 0. \tag{2.1.6}$$

Equations (2.1.5) and (2.1.6) are treated by Caroll and Santosa. The solution of (2.1.6) in terms of the modified Gelfand-Levitan integral equation is directly related to the inverse problem of impedance reconstruction.

We can derive the alternative equation treated by Gelfand and Levitan (1951) and Burridge (1980) by defining a new variable

$$\Phi(\tau,t) = \eta(\tau)w(\tau,t), \tag{2.1.7}$$

where $\eta(\tau)$ is the square root of $z(\tau)$. Substitution of (2.1.7) into (2.1.4) leads to

$$\Phi_{tt}(\tau,t) - \Phi_{\tau\tau}(\tau,t) + v(\tau)\Phi(\tau,t) = 0, \qquad (2.1.8)$$

where the potential is given by

$$v(\tau) = \ddot{\eta}(\tau)/\eta(\tau). \qquad (2.1.9)$$

Fourier transformation of (2.1.8) gives

$$\hat{\Phi}_{\tau\tau}(\tau,\omega) - v(\tau)\Phi(\tau,\omega) + \omega^2\Phi(\tau,\omega) = 0. \qquad (2.1.10)$$

Equation (2.1.10) is the well known Schrodinger wave equation studied by Gelfand and Levitan for solving quantum mechanical inverse spectral problems. Burridge (1980) treated equations (2.1.8) and (2.1.9).

2.2 Inverse Solution of Carroll and Santosa

The modified Gelfand-Levitan method of inversion summarized here follows the presentations of Carroll and Santosa (1981) and Santosa (1982). Santosa's paper concerns shear waves whereas this paper presents the inversion of compressional waves.

The solution to equation (2.1.4) can be obtained in terms of the Jost solutions $\exp(\pm i\omega\tau)$, as τ goes to ∞, which represent the general incoming and outgoing waves. Using the Paley-Wiener theorem for entire functions (Chadan and Sabatier, 1977) the Povzner-Levitan representation can be expressed as

$$\hat{\Phi}(\tau,\omega) = \cos \omega\tau + \int_0^\tau k(\tau,s)\omega\sin\omega s \; ds. \qquad (2.2.1)$$

The solution of equation (2.1.4) in terms of the outgoing Jost solution and its analytic properties lead to

$$w(\tau,t) = -\int_0^\infty \Phi(\tau,\omega)[\sin(\omega t)/\omega]\hat{v}(\omega) \; d\omega, \qquad (2.2.2)$$

where $\hat{v}(\omega)$ is the spectral density given by

$$\hat{v}(\omega) = -(2\omega/\pi)\int_0^\infty w(0,t)\sin\omega t \; dt. \qquad (2.2.3)$$

By combining (2.2.1) and (2.2.2) and using completeness, Carroll and Santosa (1981) obtained the modified Gelfand-Levitan integral equation:

$$K(T,t) + \tilde{h}(T,t) = \int_0^T h(t',t)K(T,t')dt', \qquad t \le T \qquad (2.2.4)$$

in which the kernel h is the reflection data, $\tilde{h}$ is the integral of h, K(T,t) is an unknown function and T is the record length of a one-way travel time section. Equation (2.2.4) contrasts with the linear Gelfand-Levitan equation of Burridge (eq. 2.27, 1980) obtained from equations (2.1.8) and (2.1.9)

$$f(x,T) + \int_0^x K(x,t)f(t,T) \, dt + K(x,T) = 0, \qquad 0 < T < x \qquad (2.2.5)$$

where f is the data and K is the unknown function. A numerical solution of the GL equation was presented by Bube and Burridge (1983) which is essentially similar to that of Claerbout (1976).

Equation (2.2.4) of Carroll and Santosa and the Gelfand-Levitan equation (2.2.5) of Burridge seem to be almost the same, both being Fredholm integral equations of the second kind. However, they differ significantly in certain details. In contrast to the GL equation, the kernel h(t,t') of equation (2.2.4) is differentiated with respect to t'. The unknown function K in (2.2.4) has odd parity; this follows from the fact that the kernel in equation (3.1) below has odd parity. The unknown function K in (2.2.5) has even parity (Burridge, 1980). In the GL method, A is C^2; while in the new method of Carroll and Santosa, Z is only C^1. We shall call this the modified Gelfand-Levitan (MGL) method of Carroll and Santosa. The data of the MGL method can be directly interpreted in terms of an exploration seismic experiment (Sarwar, 1988, figure (1) while the data of the GL equation can not be interpreted directly. Therefore, the MGL equation is a more natural setting for exploration seismology.

3. TOEPLITZ SYSTEM FOR MGL INTEGRAL EQUATION

By refining the MGL equation, we have been able to recast the numerical solution as a Levinson recursion for a Toeplitz matrix. This is more suitable than Gaussian elimination (Santosa, 1982). Our matrix is analogous to the Levinson recursion solution for the Wiener shaping filter.

Carroll and Santosa's time domain equations (eq 3.9 and eq 2.22, Santosa, 1982) for the impedance reconstruction are given by

$$K_T(t) + \tfrac{1}{2}[\tilde{h}(T+t) - \tilde{h}(T-t)] = \tfrac{1}{2}\int_0^T [h(t+t') - h(|t-t'|)]K_T(t') \qquad (3.1)$$

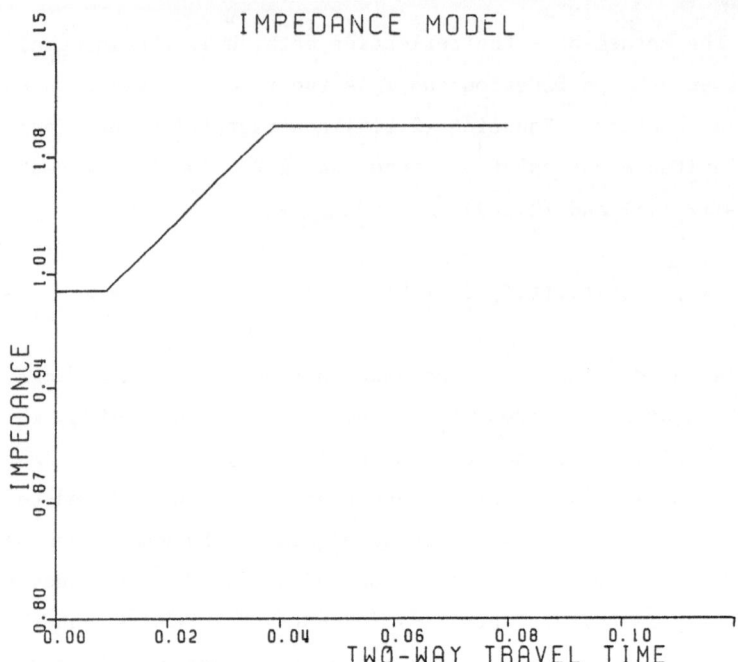

Figure 1.   A three-layered impedance model.   The second layer has a
gradational impedance.

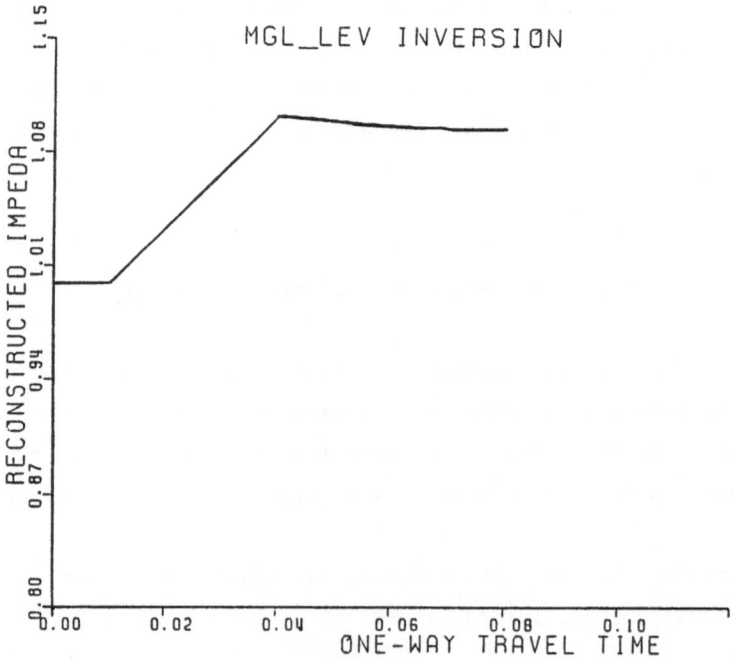

Figure 2.   Synthetic data due to the model, given in Figure 1, have been
inverted by the approximate algorithm and result is shown above.
The reconstruction is satisfactory.

and

$$Z(T) = 1 / (1 - K_T(T))^2.\qquad(3.2)$$

Breaking up the difference in the integral and then substituting x for $-t'$ in the first integral, we obtain

$$\text{LHS} = \tfrac{1}{2} \int_0^{-T} h(t-x)K_T(-x)\ dx - \tfrac{1}{2} \int_0^T h(|t-t'|)K_T(t').\qquad(3.3)$$

In the first integral we can replace $K_T(-x)$ by $-K_T(x)$ and reverse the limits of integration. Since x is negative and t is positive, $(t - x)$ is positive and $|t - x| = t - x$. Therefore, if we rename x as t' in the first integral, the integrals can be written in the form

$$\text{LHS} = -\tfrac{1}{2} \int_{-T}^0 h(|t-t'|)K_T(t')\ dt' - \tfrac{1}{2} \int_0^T h(|t-t'|)K_T(t')\ dt',$$

and the integrals can be combined to give

$$K_T(t) + \tfrac{1}{2}[\tilde{h}(T+t) - \tilde{h}(T-t)] = -\tfrac{1}{2} \int_{-T}^T h(|t-t'|)K_T(t')\ dt'.$$

For the next transformation, we need to rewrite $K_T(t)$ as an integral using the delta function. This gives

$$-\int_{-T}^T [\delta(t-t') + \tfrac{1}{2}h(|t-t'|)]K^T(t')\ dt' = \tfrac{1}{2}[\tilde{h}(T+t) - \tilde{h}(T-t)].$$

In the integral, we replace t by $(T - x)$ to get

$$-\int_{-T}^T [\delta(T-x-t') + \tfrac{1}{2}h(|T-x-t'|)]K_T(t')\ dt' = \tfrac{1}{2}[\tilde{h}(2T-x) - \tilde{h}(x)].$$

Making the substitution $T - t' = y$ in the integral yields

$$\int_{2T}^0 [\delta(y-x) + \tfrac{1}{2}h(|y-x|)]K_T(T-y)\ dy = \text{RHS}.$$

We now reverse the limits of integration and evaluate the delta function part of the integral to get

$$-K_T(T-x) - \tfrac{1}{2} \int_0^{2T} h(|y-x|)K_T(T-y)\ dy = \text{RHS}.$$

Finally, we multiply by -1, rename y as t', rename x as t and reverse the expression in the absolute value. This gives the desired form of the MGL

$$K_T(T-t) + \tfrac{1}{2} \int_0^{2T} h(|t-t'|)K_T(T-t') \, dt' = \tfrac{1}{2}[\tilde{h}(t) - \tilde{h}(2T-t)] \qquad (3.4)$$

We discretize this equation by considering a discrete sequence of times equally spaced at some interval delta. The number of times is determined by the number of observations of reflection data. We will denote the i th reflection data value by h(i). Equation (3.4) determines the $K_T(T)$ value needed in (3.2) by integrating over a period of length 2T. to obtain the value of the descretized impedance function at a given time i, we will set up a system of equations for 2i + 1 values of the discretized K_T function. With this in mind, we fix an even number N. One half N corresponds to T. We will be seeking an N+1 vector K of values. K[i] corresponds to the value of $K_T(t)$ at time (N/2 - i) where i ranges from 0 to N. The values of $K_T(t)$ for negative and zero times are defined using the oddness of $K_T(t)$. We will show that the solution of the system of equations defined below necessarily satisfies the consistency condition

$$K[N - i] = -K[i] \qquad (3.5)$$

expressing the fact that K must be odd.

The coefficient matrix of the system of linear equations arises from the left hand side of (3.4). The integral is replaced by a sum and we get an N+1 by N+1 coefficient matrix, A, with entries

$$A[i,j] = \delta[i,j] + h(|i - j|) \qquad (3.6)$$

where $\delta[i,j]$ is the i,j th entry of the identity matrix. As for K, the indices i and j range from 0 to N. The right hand side of the system is computed using an integral of h. Since h is only known on a discrete set of values, it is natural as necessary to compute the integral using some discretized integration method such as the trapezoid rule. However it is computed, we need an integral $\tilde{h}(t)$ of h(t). This integral is extended by convention to give 0 for $\tilde{h}(0)$. We then define an N+1 column vector H as

$$H[i] = \tfrac{1}{2}(\tilde{h}(i) - \tilde{h}(N - i)). \qquad (3.7)$$

As above, the index i ranges from 0 to N. This gives the system of equations

$$AK = H. \qquad (3.8)$$

Notice that A is a symmetric Toeplitz matrix and the H satisfies

$$H[N - i] = -H[i]. \tag{3.9}$$

We can formulate these facts in terms of matrix equations. Let E be the matrix with 1's down the non-main diagonal and 0's elsewhere. This N+1 by N+1 matrix is called the exchange matrix because multiplying a matrix on the left by E reverses the order of the rows. The first and last rows exchange, the second and N th rows exchange, etc. Multiplying a matrix on the right by E performs a similar exchange on the columns. Toeplitz matrices are persymmetric; that is, they are symmetric about the non-main diagonal. The persymmetry property can be expressed as the following matrix equation

$$A = EA^T E.$$

Since A is also symmetric, this simplifies to

$$A = EAE. \tag{3.10}$$

Equation (3.9) for H can be also restated using E

$$EH = - H. \tag{3.11}$$

We can now give a simple proof that the solution, K, of equation (3.8) must be an odd function as required by equation (3.5). First substitute (3.10) into (3.8)

$$H = AK = EAEK.$$

Next multiply by E and use (3.11) as well as the fact that EE is the identity.

$$-H = EH = EEAEK = AEK.$$

This can be rewritten as

$$A(-EK) = H.$$

Since A is nonsingular, the solution to (3.8) is unique and we have obtained

$$-EK = K,$$

which can be rewritten as

$$EK = - K. \tag{3.12}$$

Equation (3.12) is a matrix restatement of (3.5) similar to the restatement of (3.9) as (3.11). An immediate corollary of this proof is the fact

$$0 = K[N/2] = K_T(0).$$

For N=2, the system assumes the following form. The 3 vector K
that is being solved for has components

$$K[0] = K_T(1), \qquad K[1] = K_T(0), \qquad K[2] = K_T(-1).$$

The 3 vector H has components

$$H[0] = \tfrac{1}{2}[h(0) - h(2)], \qquad H[1] = 0, \qquad H[2] = -H[0].$$

The 3 x 3 matrix A has entries

$$\begin{bmatrix} 1+\tfrac{1}{2}h(0) & \tfrac{1}{2}h(1) & \tfrac{1}{2}h(2) \\[2ex] \tfrac{1}{2}h(1) & 1+\tfrac{1}{2}h(0) & \tfrac{1}{2}h(1) \\[2ex] \tfrac{1}{2}h(2) & \tfrac{1}{2}h(1) & 1+\tfrac{1}{2}h(0) \end{bmatrix}$$

The system (3.8) is a symmetric Toeplitz system and can be solved
by Levinson recursion. The Gaussian elimination applied by Santosa
(1982) for solving N simultaneous equation requires memory $O(N^2)$ and
time $O(N^3)$. The Levinson recursion for Toeplitz matrices requires
memory $O(N)$ and time $O(N^2)$. A derivation of Trench's improvement of
Levinson's algorithm and code for implementing it can be found in Golub
and Van Loan (1983) section 5.7. The algorithm presented there involves
a variant that uses Durbin's solution to an associated Yule-Walker
system. Kung and Hu (1983) proposed a highly concurrent algorithm which
requires time $O(N)$ to solve a Toeplitz system of N equations on a
vector computer with N processors. Recently "fast" methods involving
the Fast Fourier Transform have been proposed. Bunch (1985) discusses
two of these methods. The Bitmead-Anderson algorithm takes $O(N \log N)$
time and $O(N)$ memory. Bunch indicates that this algorithm will begin to
be faster than the Trench algorithm when N is in the low thousands.

The numerical stability of the Levinson algorithm for solving
Toeplitz systems has been studied by Cybenko (1980). Bunch (1985)
compares the stability of several methods. Most methods are stable for
positive definite symmetric matrices. Since most methods preserve
Toeplitz structure by not pivotting, even for indefinite symmetric
matrices, the keys to good behavior are the condition numbers of the
leading submatrices. For our case, the leading submatrices are the
coefficient matrices used to compute the previous values of the
impedance. There is also an excellent paper by Gray and Symes (1985) on
the stability of the one-dimensional seismic inverse problem.

4. CONCLUSIONS

We have presented a numerical approach for reconstructing acoustical impedance profiles from the impulse response. This approach involves solving a symmetric, positive definite, Toeplitz system of equations. Since such systems can be solved more efficiently than general symmetric, positive definite systems, our approach is a more efficient method for reconstructing impedance profiles.

REFERENCES

Alekseev, A., 1962. Some inverse problems of the theory of wave propagation, parts I & II: Izv. Akad. Nauk. SSSR, Ser. Geofiz., 11, 1515-1531.

Berryman, J.G., and Greene, R.R., 1980. Discrete inverse methods for elastic waves in layered media, Geophysics, 45, 213-233.

Bube, K.P., and Burridge, R., 1983. The one-dimensional inverse problem of reflection seismology: SIAM Rev., 25, 497-559.

Burridge, R., 1980. The Gelfand-Levitan, the Marchenko, and the Gopinath-Sondhi integral equations of inverse scattering theory, regarded in the context of inverse impulse response problems, Wave Motion, 2, 305-232.

Bunch, J.R., 1985. Stability of methods for Toeplitz systems: SIAM J. Sci. Stat. Comput., 6, 349-364.

Carrion, P.M., and Foster, D.J., 1985. Inversion of seismic data using the precritical reflection and refraction data, Geophysics, 50, 759-765.

Carrion, P.M., and Kuo, J. T., 1984. A method for computation of velocity profiles by inversion of large-offset records, Geophysics, 49, 1249-1258.

Carrion, P.M., and Patton, W., 1983. "Criteria for resolution and recon-struction of the acoustic impedance, JGR, 88, 10349-10358.

Carroll, R., and Santosa, F., 1981. Scattering techniques for a one-dimensional inverse problem in geophysics: Math. Meth. in the Appl. Sci. 3, 145-171.

Carroll, R., and Santosa, F., 1982. Stability for the one-dimensional inverse problem via the Gelfand-Levitan equation: Applicable Analysis, 13, 271-277.

Claerbout, J.F., 1976. Fundamentals of Geophysical Data Processing: With Applications to Petroleum Prospecting. New York, McGraw-Hill, Inc.

Cybenko, G., 1980. The numerical stability of the Levinson-Durbin algorithm for Toeplitz systems of equations: SIAM J. Sci. Stat. Comput., 1, 303-319.

Driessel, D.R., and Symes, W. W., 1983. Stable triangular decomposition of symmetric kernels, Inverse Scattering: Theory and Application, ed. Bednar, J. B., Redner, R., Robinson, E., and Weglein, A., SIAM.

Gelfand, I. M., and Levitan, B.M., 1951. On the determination of a differential equation by its spectral function: Izv. Akad. Nauk. SSSR, Ser. Mat., 15, 309-360. (Am Math. Soc. Trans., 1955, Series 2.1, 253-304.)

Goupillaud, P.L., 1961. An approach to inverse filtering of near-surface layer effects from seismic records, Geophysics, 26, 754-760.

Gray, S.H., 1984. The relationship between 'direct, discrete', and iterative, continuous' one-dimensional inverse method, Geophysics, 49. 54-59.

Gray, S.H., and Symes, W.W., 1984. Stability considerations for one-dimensional inverse problems: Geophys. J.R. astr. Soc., to appear.

Kung, S.Y., and Hu, Y.H., 1983. A highly concurrent algorithm and pipe-lined architecture for solving Toeplitz systems: IEEE Trans. Acous. Speech Sig. Proc., ASSP-31, 66-76.

Newton, R.G., 1981. Inversion of reflection data for layered media: A review of exact methods: Geophys. J.R. astr. Soc., 65, 191-215.

Santosa, F., 1982. Numerical scheme for the inversion of acoustical impedance profile based on the Gelfand-Levitan method, Geophys. J.R. astr. Soc., 70, 229-243.

Santosa, F., and Schwetlick, H., 1982. The inversion of acoustical impedance profile by methods of characteristics, Wave Motion, 4, 99-110.

Sarwar, A.K.M., 1988. Some closed-form solutions of the impedance inverse problem, J. Acoust. Soc. Am., 83, in press.

Sondhi, M.M. and Resnick, J.R. 1983. The inverse problem for the vocal tract: numerical methods, acoustical experiments, and speech synthesis, J. Acoust. Soc. Am., 73, 985-1002.

Sondhi, M.M. and Gopinath, B., 1971. Determination of vocal-tract shape from the impedance response at the lips, J. Acoust. Soc. Am., 49, 1867-1873.

Imaging of the Earth by Iterative Reconstruction Methods

R. Guney, E. F. Benson, A. K. M. Sarwar

University of New Orleans, New Orleans, LA 70148, USA

Abstract

Crossborehole imaging (tomography) is an inversion technique that recreates the velocity field between two wells using observed traveltimes. The standard procedure is based on the decomposition of the crosshole area into a number of square-shaped cells. Two iterative techniques, the Algebraic Reconstruction Technique (ART) and the Simultaneous Iterative Reconstruction Technique (SIRT), are the basis for reconstructing images from observed traveltimes. These two techniques are compared to show the advantages and disadvantages for image reconstruction using synthetic data.

The initial velocity estimates required for these techniques was found to play a very important role in image resolution and in the speed of convergence to the solution. Here, three different initial velocity estimates are used including averaged, layered and estimated velocities. The layered initial velocity estimate was found to work best for both algorithms. The effects of different initial models on reconstruction are discussed for various models. In order to simulate real data, noise with zero mean and various standard deviations have been added to the traveltimes .

The results of this study indicate that the ART method converges more rapidly than the SIRT method, but the SIRT method is more stable in the presence of noise.

1   Introduction

The goal of seismic cross-sectional imaging is to determine the physical parameters of the subsurface from seismic data. The tomographic inversion method has been introduced to seismology by Backus and Gilbert(1970). This method, based on linearized inversion also employed by Wiggins (1972) and Aki et al.(1976), has been used in various problems such as earthquake data analysis and seismic statics. More recent work in seismology includes that of Tarantola and Valette (1982), and Mora(1987) who applied a generalized inversion algorithm to solve nonlinear and linear problems.

Although the mathematical principles of imaging are the same for medical and non-medical applications, due to the geometrical difficulties such as restrictions on the collection of projections from different angles (full coverage), reconstruction quality is different. Because of the ability to have a wide range of angular coverage, reconstruction in medical imaging is better than the techniques used to image the earth's interior.

Several different reconstruction algorithms have been proposed by Herman (1980), and Gordon et al.,(1970). These reconstruction algorithms can be categorized into two groups: the direct method of image reconstruction which solves Radon's integral equation using various mathematical procedures, and the indirect iterative method which attempts to iteratively reconstruct the digitally represented velocity field. Image reconstruction from large and complex data sets cannot be considered to be a simple linear algebra problem (Gordon et al. 1970) because:

I The number of unknowns in most cases is greater than the number of equations; consequently there are an infinite number of possible solutions.

II The observed data are subject to measurement error which further complicates the choice of elements in the image.

III The large number of equations (projections) and unknowns makes application of mathematical techniques for solving linear the algebra problem very difficult.

Besides linear algebra problems, noisy traveltime data, ray bending, and incomplete ray coverage may often cause the appearance of artifacts and image distortions in the reconstructed image.

This paper focuses on the problem of seismic velocity estimation from traveltimes in a two-dimensional crosshole geometry where seismic sources are located in one borehole while the receivers are located in another. Two reconstruction techniques (ART and SIRT) are compared using synthetic data. In order to simulate real data, noise at various levels are added and tested for both algorithms.

2 Model Image and Projection Representation

2.1 Model Setup

The parameters for surface seismic modeling include velocity contrasts and traveltimes, while velocities and depth-offset are used for crosshole modeling. The source function estimated for the surface seismic study is another important parameter needed to create the reflection model, while the crosshole survey does not need it. The area to be imaged with a conventional reflection survey does not need to be discretized, but this is necessary in a tomographic survey (Figure 1).

The cell size of the model is constant throughout the reconstruction steps. This plays a very important role on reconstruction quality. Increasing cell size means increasing the area over which we obtain velocity values, thus defining the ray coverage. Conversely if cell size remains constant, increasing the number of cells also increases ray coverage.

2.2 Ray Tracing

Modeling the sound waves traveling through a velocity medium by ray tracing requires solving the ray equation for a given background velocity, reflection and refraction boundaries and shot/receiver geometry. Crossborehole imaging is based on the approximation

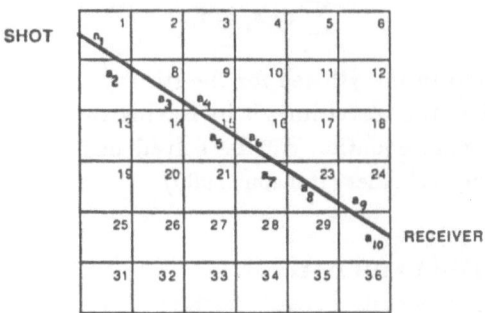

Figure 1: Ray segmentation: The area of interest is discretized into square shaped cells. The representative ray (bold line) between shot and receiver samples ten grid elements (cell) with length a_i in each. The grid element has slowness s_{ij}.

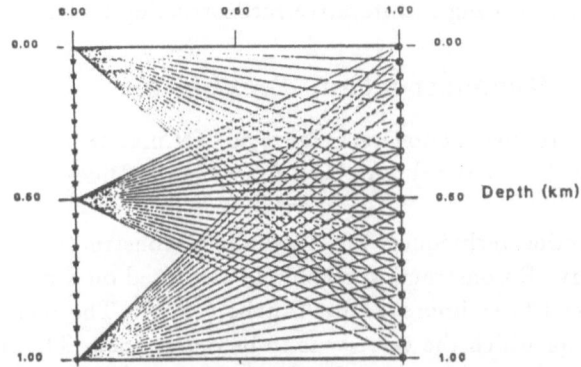

Figure 2: Shot-receiver geometry representing crosborehole seismic survey. Straight rays are traced from three sources placed in one borehole through a constant velocity model to twenty-one receivers in the other borehole.

that seismic energy traveling from source to receiver propagates along straight paths. The image field is scanned with direct rays which means that the rays are not reflected or refracted.

In this study, straight rays are traced through the cells to calculate the traveltimes needed for reconstruction. Straight ray approximation is valid if the velocities do not vary substantially across the area to be imaged. The models used for this study are considered to have slowly varying velocities. The shot/receiver geometry and associated ray coverage are shown in Figure 2. In these examples the ray segment lengths are held constant.

2.3   Traveltime Computing

The traveltimes for reconstruction are obtained by ray tracing. The medium being imaged is discretized into a grid of square shaped cells (pixels) . The traveltime t_i for each ray connecting source and each receiver through the area is calculated by multiplying the slowness, s_{ij}, by the ray length a_{ij} in each cell that a ray passes through. This is written as

$$\sum_{i=1}^{l} a_{ij} s_j = t_i \qquad (1)$$

where a_{ij} is the ray segment in the jth cell for the ith ray

Another way to calculate the traveltimes is by solving the 2-D acoustic wave equation (Claerbout, 1976). The wave equation can be solved using explicit second-order and fourth-order finite difference schemes (Benson, 1990).

3 Image Reconstruction

The first step in imaging the subsurface in the ray tomography procedure outlined by McMechan (1983) is to approximate the seismic energy traveling in straight paths between sources and receivers. The traveltimes are calculated using the length of the ray paths and the velocities between the source and receiver.

The task of creating the image involves the reconstruction of the velocities using the observed traveltimes. A recursive scheme developed for inverse processes reconstructs the image from traveltimes using two iterative reconstruction techniques.

3.1 Iterative Reconstruction Techniques

The iterative reconstruction techniques for estimation of velocity fields from the projection (observed traveltime) have been described by Dines et al.,(1979), and Gordon et al.,(1970).

Two reconstruction techniques, the Algebraic Reconstruction Technique and the Simultaneous Iterative Reconstruction technique are based on a method first proposed by Kacmarz (1937) and later improved by Tanebe (1971). The region of interest encompasses a 2-D area for which the velocity is to be determined. The traveltimes associated with each ray for a medium are given by the line integral,

$$\int \frac{da}{v(x,y)} = t_i \qquad (2)$$

where $v(x,y)$ is the velocity of medium and da is the ray segment length of the ith ray passing through the corresponding cell. After discretizing the area between source and receiver boreholes the line integral becomes a finite sum,

$$\sum_{j}^{N} s_j a_{i,j} = t_i \qquad (3)$$

where s_j is the constant slowness in jth cell, t_i is the traveltime for the ith ray and a_{ij} is the ray segment length for ith ray and jth cell. Let M be the total number of rays and N the total number of cells. For complete projection there are M equations (rays),with N unknowns, which are written in linear form as,

$$
\begin{aligned}
a_{11}s_1 + a_{12}s_2 + \cdots + a_{1N}s_N &= & t_1 \\
a_{21}s_1 + a_{22}s_2 + \cdots + a_{2N}s_N &= & t_2 \\
&\vdots& \\
a_{M1}s_1 + a_{M2}s_2 + \cdots + a_{MN}s_N &= & t_M
\end{aligned}
\qquad (4)
$$

or, in matrix form

$$[a][s] = [t].\tag{5}$$

The equation given above relates the slowness s, of the object to the traveltimes, t, that are observed from the rays propagating through it. When the properties of the medium are given, one can show how energy propagates through the medium.

The problems of finding an unknown vector s fall into two groups depending on the nature of the linear system. When there are more equations (projections) than unknowns ($(M > N)$, then $[a][s]$ is overdetermined . Usually an overdetermined system has no exact solution. Minimization techniques can be employed. If there are fewer equations than unknowns ($M < N$), the system is underdetermined and has an infinite number of solutions.

If M and N are small, conventional matrix theory could be used to invert the system of equations. However, in its applications of image reconstruction (seismic, radio astronomy, medical) the approximate size of the matrix is roughly known. N may be as large as 65,000 for 256 × 256 images and in most cases M may have the same magnitude. Increasing the number of cells per area results in better resolution. For such a large matrix, direct matrix solution is impractical. Another consideration is the presence of noise for large matrices. In the presence of noise even for a system with $M < N$ and small N, direct solution is also computationally impractical. ART and SIRT are techniques used to solve large matrices by iterative methods.

The techniques developed here are based on the traveltime equation

$$\sum_{j=1}^{N} a_i s_{ij} = t_i^c,\tag{6}$$

beginning with an initial slowness estimate where a_i is the ray segment length, s_{ij} is the slowness, and t_i^c represents the calculated traveltime for the $(p-1)$th iteration. Subtracting t_i^c from t_i gives the residual traveltime,

$$r_i^{(p)} = t_i^o - t_i^c.\tag{7}$$

The ultimate aim is to make residual traveltime as close to zero as possible. The slowness correction vector at pth iterations $\vec{\Delta s}$, and is written as

$$\Delta s_{ij}^{(p)} = \frac{r_i^{(p)}}{\sum a_{ij}^2},\tag{8}$$

where i is the ray, j is the cell number. In the reconstruction procedure each cell has a slowness correction vector if the ith ray passes through the cell.

3.1.1 Algebraic Reconstruction Technique

Several algorithms which share the same basic operating procedure are based on the Algebraic Reconstruction Technique (ART). A model of the image is iteratively adjusted to reduce the error between an observed traveltime and a corresponding updated traveltime (after the first iteration) across the model. The mathematical basis for the technique is the Kaczmarz method described in previous section. In the same section we saw that a field represented by a digital image which is made up of many boxes is assigned a value which represents the mean value of the velocity field. It has been shown that any projection across the original field can be given by the linear equations (4). The residual traveltime is defined as the time difference between the observed traveltime t_i^o, and the calculated traveltimes, t_i^c. In other words $r^{(p)}$ is the error between the pth observed projection and the calculated projection over the pth estimate of the image. The error vector or residual vector is given by

$$r_i^{(p)} = t_i^o - t_i^c. \tag{9}$$

To update the slowness field, the slowness correction vector is calculated by dividing the residual traveltime by the corresponding sum of the squared ray segments and scaled with the corresponding ray segment such that

$$\Delta s_{ij}^{(p)} = \frac{r_i^{(p)}}{\sum_j a_{ij}^2} a_{ij}. \tag{10}$$

The updated slowness vector is calculated by adding the slowness correction vector components $\Delta s_{ij}^{(p)}$ to the corresponding initial slowness vector components s_{ij}^0,

$$s_{ij}^{(p+1)} = s_{ij}^0 + \Delta s_{ij}^{(p)} \tag{11}$$

where i indicates ray and j indicates cell number. The ART algorithm described here considers only one ray at a time when the image vector is adjusted. In other words updating of the slowness vector is done by using a ray-by-ray process. Updating the slowness in every iteration step reduces the amount of computer memory required to produce an acceptable reconstruction. However, ART is quite sensitive to noise. The advantage of the technique is that it converges faster than any other existing algorithm but it has a very high instability rate. To a large extent it is this last observation which led us to an improvement of the algorithm. ART at some point gives a good estimate of the velocity, but it becomes unstable since it tries to satisfy inconsistent real data. To resolve this we can accept a solution with a certain tolerance rather than seek an exact result to satisfy the observed traveltime. In this case we can avoid the problem of inconsistency.

3.1.2 Simultaneous Iterative Reconstruction Technique

The Simultaneous Iterative Reconstruction Technique (SIRT) was first proposed by Gilbert(1972) as an alternative to ART. This technique, unlike ART, makes adjustments to the image based upon the information contained within the entire set of projection data. In other words, the updating process is done iteratively to account for all ray segments.

The SIRT algorithm is based on the following equations,

$$s_{ij}^{(p+1)} = s_{ij}^{(p)} + \frac{1}{M} \sum_j \Delta s_{ij}^{(p)}, \tag{12}$$

where the slowness $\Delta s_{ij}^{(p)}$ is given by

$$\Delta s_{ij}^{(p)} = a_{ij} \frac{r_i^{(p)}}{\sum_j a_{ij}^2}, \tag{13}$$

and the residual traveltime is

$$r_i^{(p)} = t_i^0 - t_i^c. \tag{14}$$

M is the total number of rays passing through the individual cell, and each cell has a different M.

196

SIRT converges more slowly than classical ART but has an advantages regarding stability. The stopping criteria for both methods are given in the next subsection.

3.1.3   Stopping Criteria

Iterative Reconstruction requires a stopping criterion since one cannot continue the calculation to infinity without knowing the convergence to the desired solution. It is necessary to know at what iteration to terminate the process. Various criteria for stopping have been devised by Gordon et al. (1970). Three measures for the convergence of the slowness vector s_i are used. One is the discrepancy $D^{(p)}$ between the measured traveltime t_i and calculated traveltime t_i^c given by

$$D^{(p)} = \sqrt{\frac{1}{M} \sum_{i=1}^{M} (t_i^o - t_i^c)}, \tag{15}$$

where M is the total number of rays. A second is variance $V^{(p)}$ calculated from differences between the updated slowness vector $s_j^{(p)}$ and the mean vector $\bar{s}$ given by,

$$V^{(p)} = \frac{1}{N} \sum_{j=1}^{N} (s_j^{(p)} - \bar{s}). \tag{16}$$

Finally the entropy E is used as

$$E^{(p)} = \frac{-1}{ln(N)} \sum (\frac{s_j^{(p)}}{\bar{s}}) ln(\frac{s_j^{(p)}}{\bar{s}}), \tag{17}$$

where N is the total number of cells.

If the set of equations given by (4) is complete and consistent, the discrepancy, $D^{(p)}$, will tend to zero while the variance $V^{(p)}$ will tend to a minimum and the entropy, E, to a maximum with an increase of iteration (p). The purpose of minimizing variance or maximizing entropy is to find the smoothest of all possible images.

In real applications the starting model is usually smooth so that the variance actually increases and the entropy decreases in order to satisfy the data. $D^{(p)}$, $V^{(p)}$, $E^{(p)}$ values show convergence of the solution to the true image. In fact, these measures may continue to converge while the solution diverges unacceptably from the true image (Gordon,1970).

As mentioned above, the divergence of the Algebraic Reconstruction algorithms with inconsistent data is nevertheless acceptable; therefore, the computation should be terminated before D begins to diverge. Herman (1980) used the stopping criteria $|V^{(p)} - V^{(p+1)}| < V^{(p)}/100$.

4   Computer Simulation

Two iterative techniques have been tested for their performance on several velocity model reconstructions. Controlled computerized modeling (simulation) enables us to understand various aspects of the algorithms described in the previous section. The initial velocity estimates play an important role in the reconstruction of images. Here three different initial velocity estimates are used.

I The average velocity estimate is calculated by dividing the sum of the total lengths of all the rays by the sum of all the observed traveltimes.

II The layered velocity estimate is calculated by dividing the total lengths of the parallel rays by the corresponding traveltimes for each source/receiver pair.

III The estimate velocity is taken from known physical data such as well logs or by making an initial guess as to what the velocity is.

4.1   Reconstruction without Noise

In this section, effect of various initial models on final reconstructions are presented. The model used was circular with a velocity of 4.2 km/sec and a background velocity of 3.9 km/sec (Figure 3) . Three initial velocity model (Figure 3a-c) are used in the ART and SIRT algorithm. Two initial model calculations (layered, averaged) are based on the observed traveltimes. The other initial velocity model is directly obtained from geologic data.

All reconstructions are carried out on a 64 by 64 sampling grid with 4096 rays (projections) and 4096 picture elements. Figure. 4 ,Figure 6, and Figure 8 show the reconstructions comparing ART with SIRT methods using the three initial velocity estimates

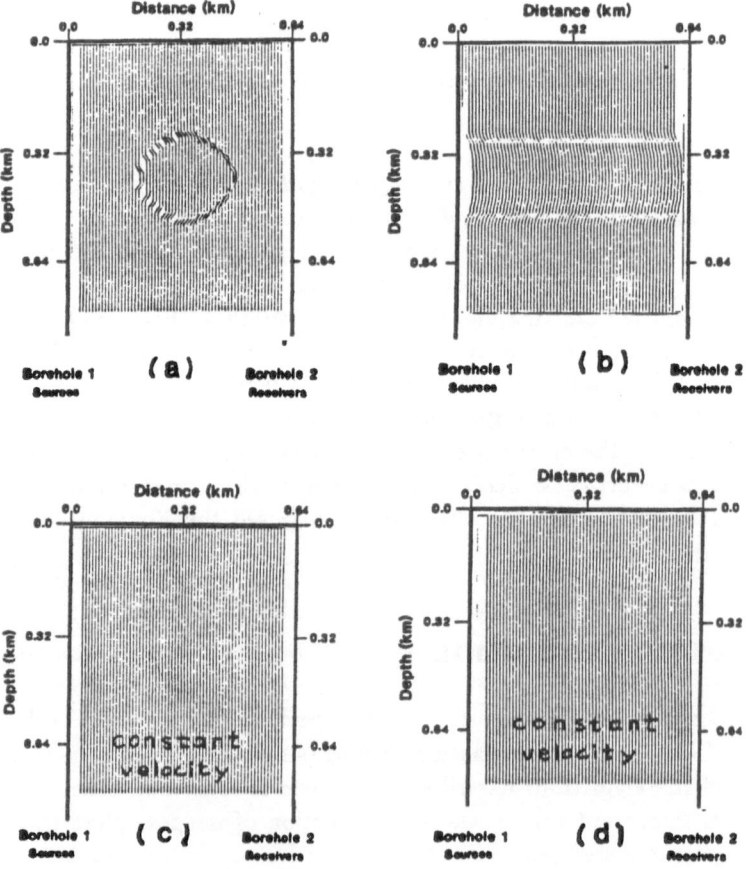

Figure 3: (a) Circular velocity model represented by 64 by 64 matrix. (b) layered, (c) average, and (d) estimated initial models.

for 5th, 10th and 20 th iterations, respectively. Examining line plots the reconstruction velocities for the three velocity estimates shows that reconstruction for the layered estimate is better than the reconstruction for the average and estimated initial velocities (Figures 5- 9). ART converged at iteration 10, but the image was poor at this iteration for SIRT (Figure 7).

Figure 4 a,b,c shows the reconstructions of the circular model using the layered ,average and estimated initial velocity models. Figure 4 d,e,f shows the reconstructions using the SIRT. As can be seen ART still converges to the solution faster. The reconstructions using the average velocity estimate, Figure 6, and the estimated initial model, Figure 8 also show that ART converges to solution faster. Comparisons between the initial velocity estimates show that the layered velocity estimate, Figure 8, reconstructed the model better and faster than the other two initial velocity estimates

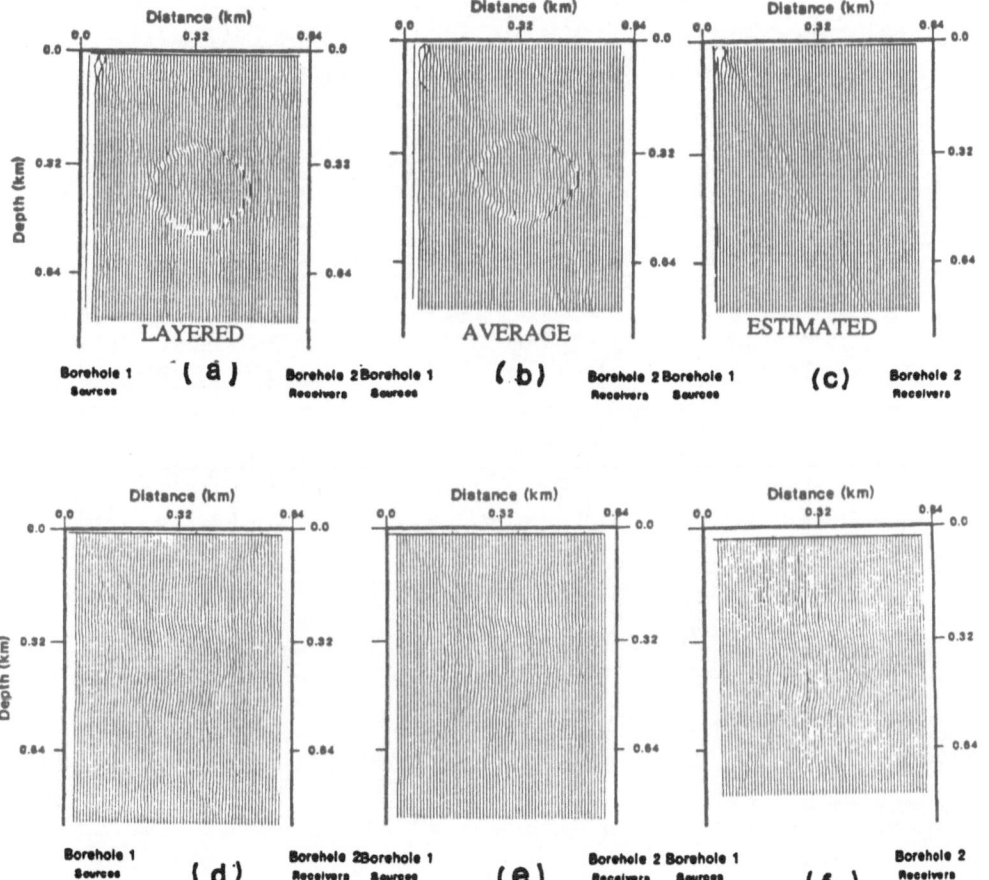

Figure 4: Reconstructions of a circular velocity model with three different initial models: (a) layered, (b) average, (c) estimated at the 5 th iteration with ART and reconstructions for (d) layered, (e) estimated and (f) averaged for the SIRT at the 5 th iteration, respectively.

Figure 5, 7, and 9 show the line plots for the three initial velocity estimates at iteration 5, 10, and 20, respectively. The best results are achieved at the 20th iteration for all the initial velocity estimates but the layered initial velocity is the best of the three. This is indicated for both ART and SIRT.

These examples were free of noise but to simulate real data, noise should be introduced.

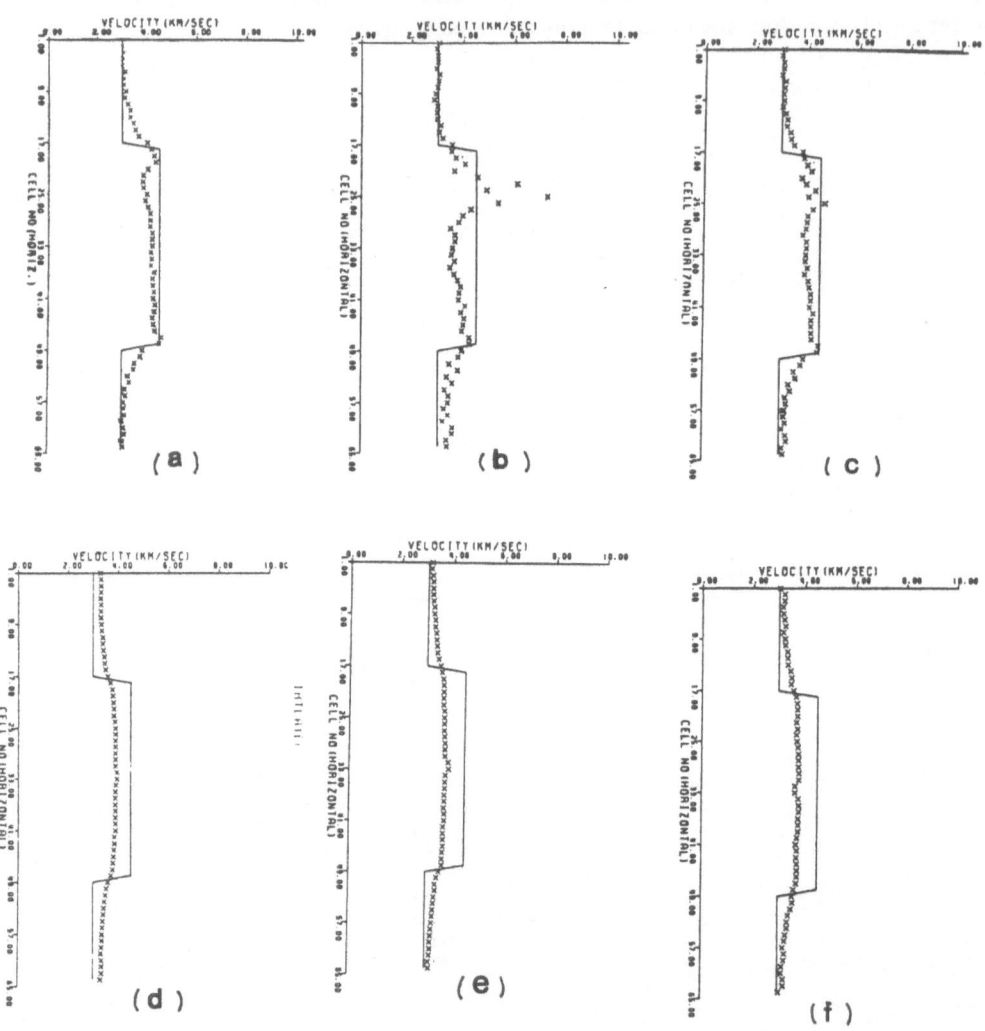

Figure 5: Line plot of reconstructed velocity fields (z=0.36 km) with (a) layered, (b) estimated, and (c) averaged initial models for ART and showing reconstructions with (d) layered, (e) estimated, (f) averaged initial model at the 5 th iteration for SIRT. True(model) velocities are represented by solid lines, reconstructed velocities are represented by x symbol.

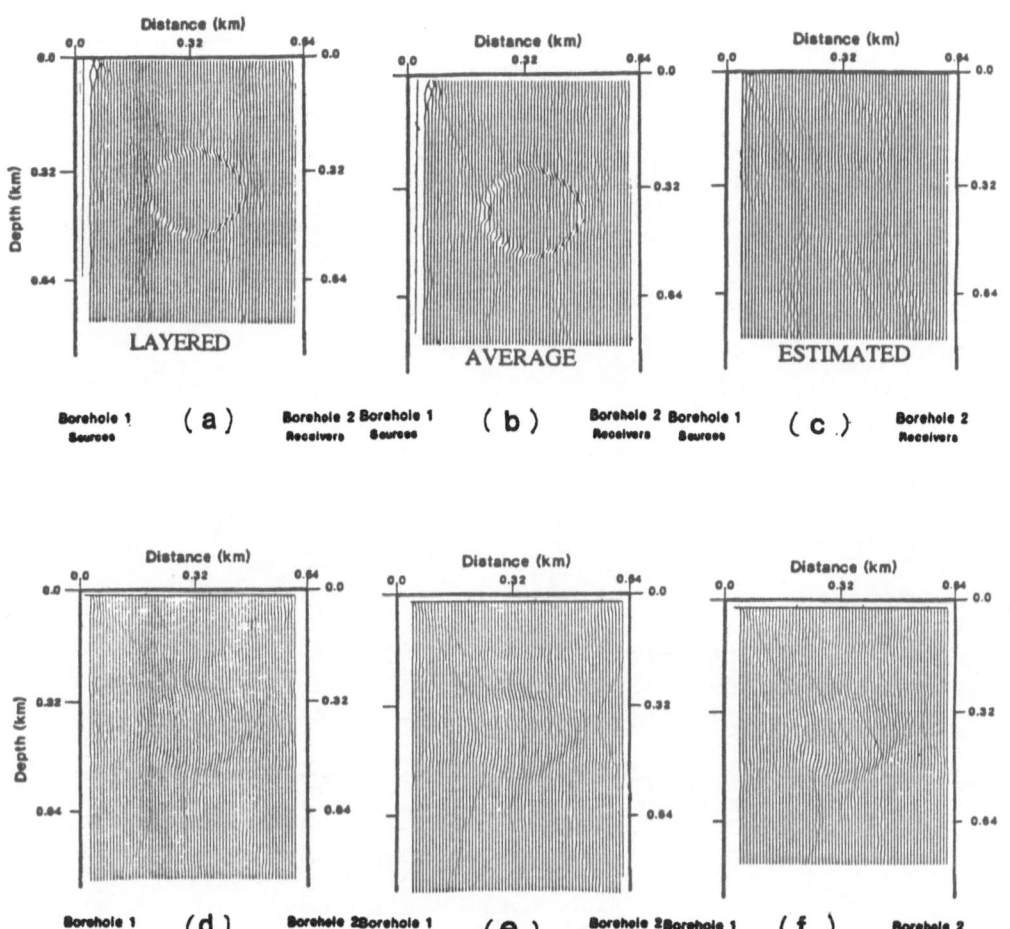

Figure 6: Reconstructions of a circular velocity model with three different initial models; (a) layered, (b) average, (c) estimated at the 10 th iteration with ART and reconstructions for (d) layered, (e) estimated and (f) averaged for the SIRT at the 10 th iteration.

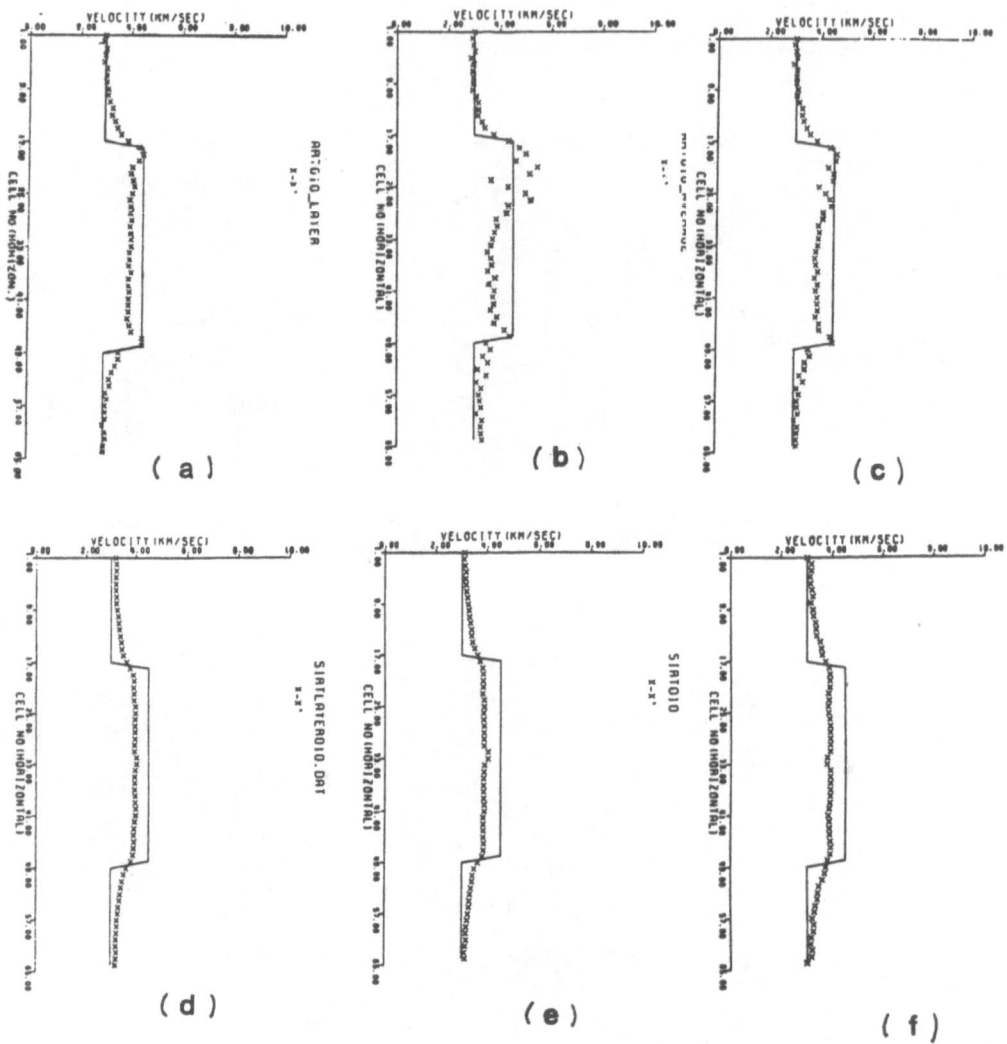

Figure 7: Line plot of reconstructed velocity fields(z=0.36 km) with (a) layered, (b)estimated, and (c) averaged initial models for ART showing reconstructions with (d) layered, (e) estimated, (f) averaged initial model at the 10 th iteration for SIRT. True(model) velocities are represented by solid lines, reconstructed velocities are represented by x symbol.

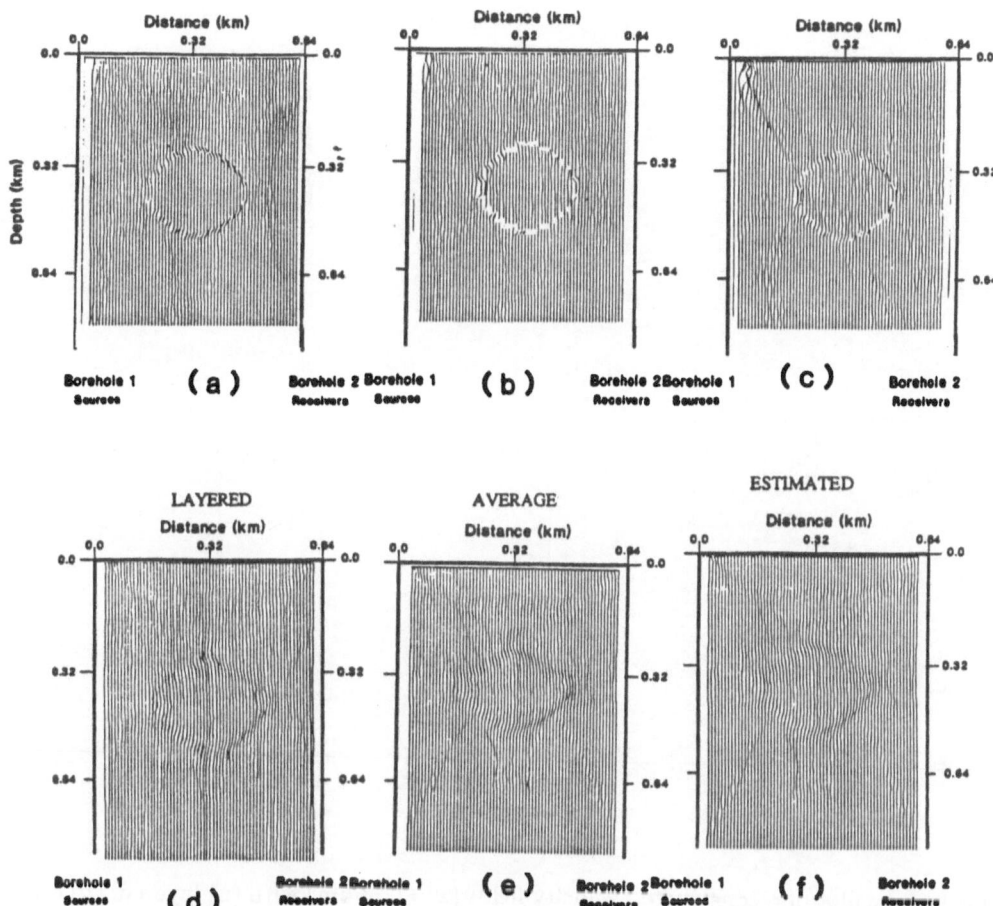

Figure 8: Reconstructions of a circular velocity model with three different initial models; (a) layered, (b) average, (c) estimated at the 20 th iteration with ART and reconstructions for (d) layered, (e) estimated and (f) averaged for the SIRT at the 20 th iteration.

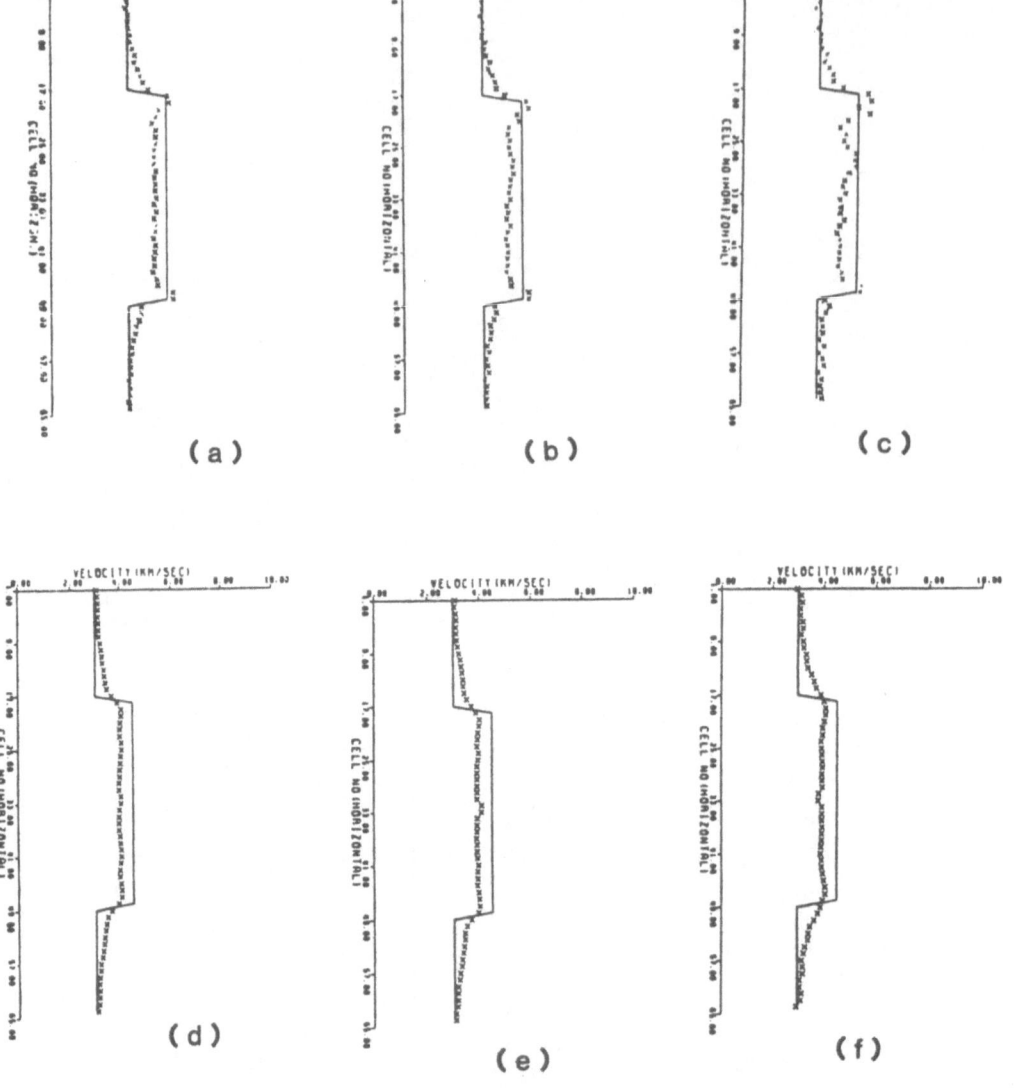

Figure 9: Line plot of reconstructed velocity fields ($z = 0.36km$) with (a) layered, (b) estimated and (c) averaged initial models for ART showing reconstructions with (d) layered, (e) estimated, (f) averaged initial model at the 20 th iteration for SIRT. True(model) velocities are represented by solid lines, reconstructed velocities are represented by x symbol.

4.2 Reconstruction with Noise

To create synthetic data that are similar to real data I have introduced Gaussian random noise with a zero mean and a standard deviation up to 10 percent. The behavior of the algorithms (ART, SIRT) towards noisy data has been examined in this section. A noisy data set (Figure 10) has been generated using a random number generator subroutine. Updating the slowness field was continued up to the 5 th iteration. Layered initial velocity

204

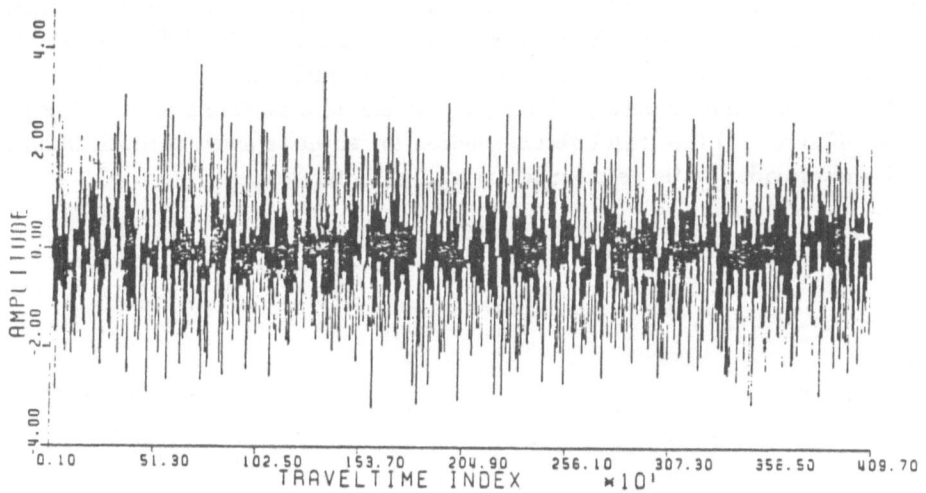

Figure 10: Random noise with zero mean and 0.98 standard deviation.

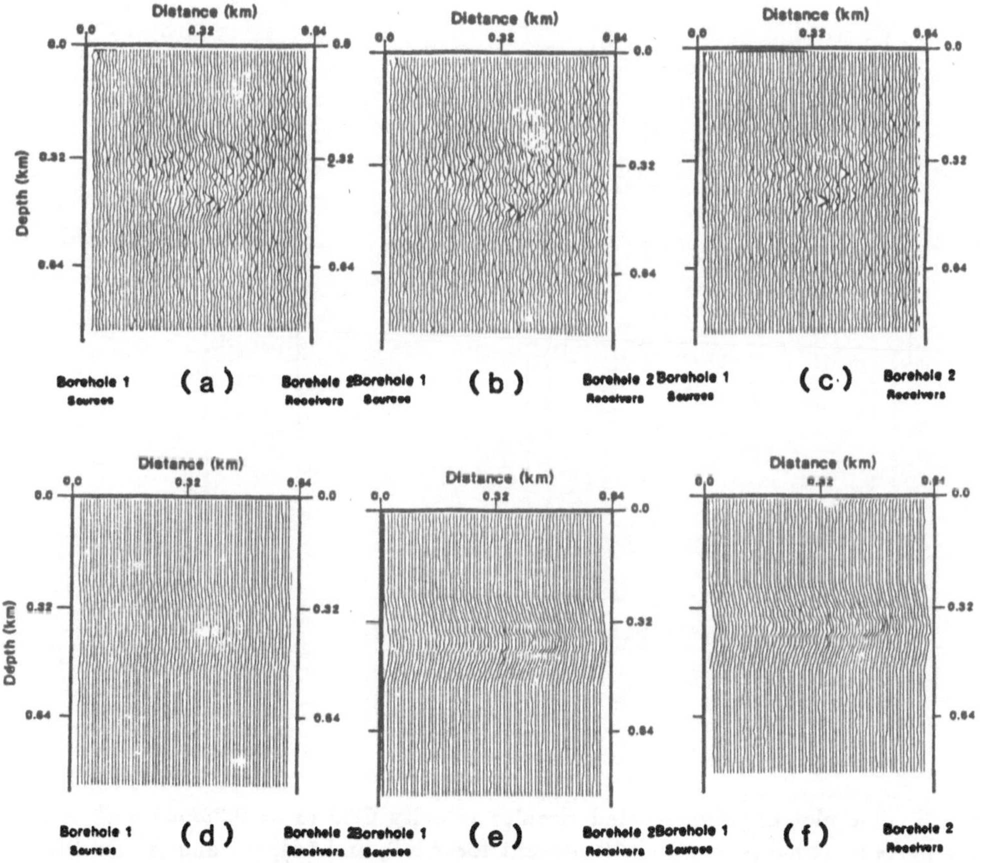

Figure 11: Reconstruction of a circular velocity model with (a) 1 percent (b) 5 percent and (c) 10 percent noise for ART, (d), (e) and (f) for SIRT at iteration 1.

estimate was used to reconstruct the slowness field, since this initial velocity estimate was shown to work the best in the previous section. Figure 11and Figure 13 show the reconstructions for both algorithms from data that included 1 percent noise (a,d), 5 percent noise (b,e) and 10 percent noise (c,f). It can also be seen from the line plots (Figure 12, Figure 14) that SIRT reconstruction for all noise levels is much smoother that ART and provides better reconstruction at higher iterations.

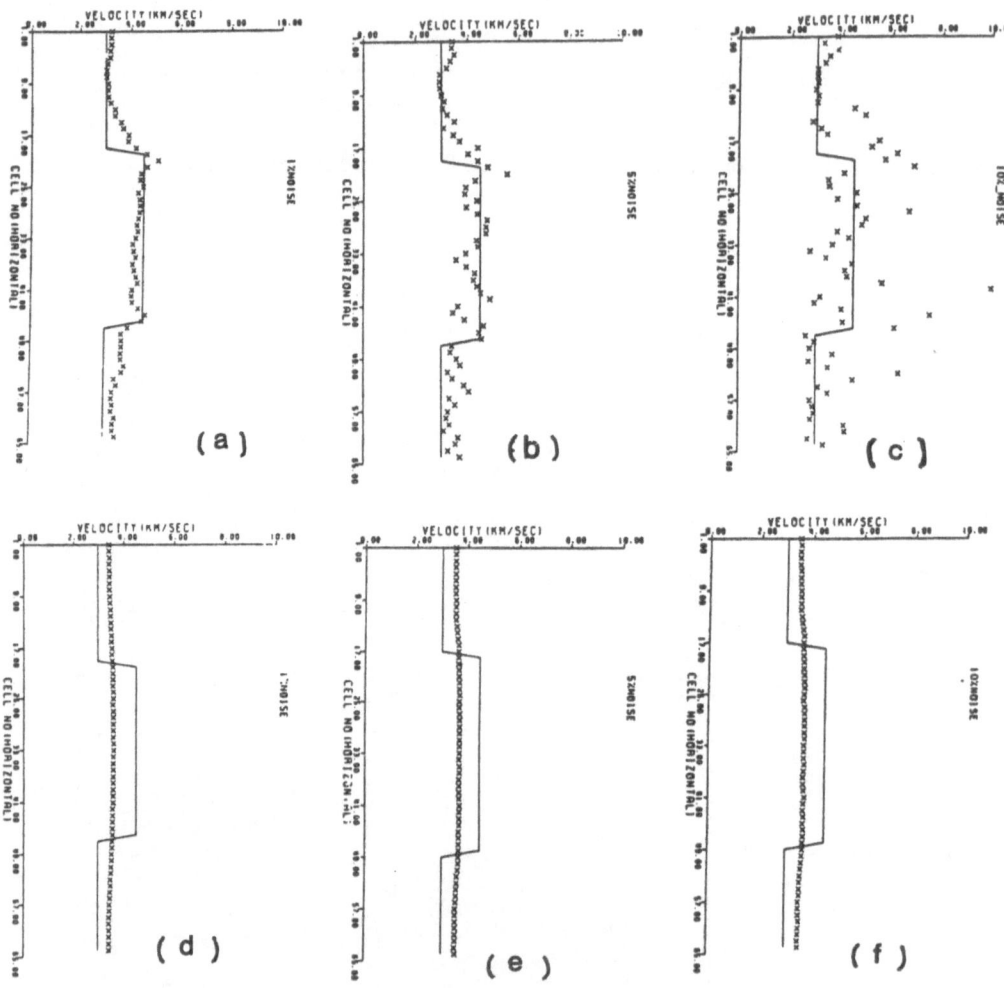

Figure 12: Line plot of reconstructed circular velocity field ($z = 0.36km$) with noise of (a) 1 percent, (b) 5 percent,(c) 10 percent for ART, and (d), (e) and (f) for SIRT at 1 iteration. True(model) velocities are represented by solid lines and reconstructed velocities are represented by x symbol.

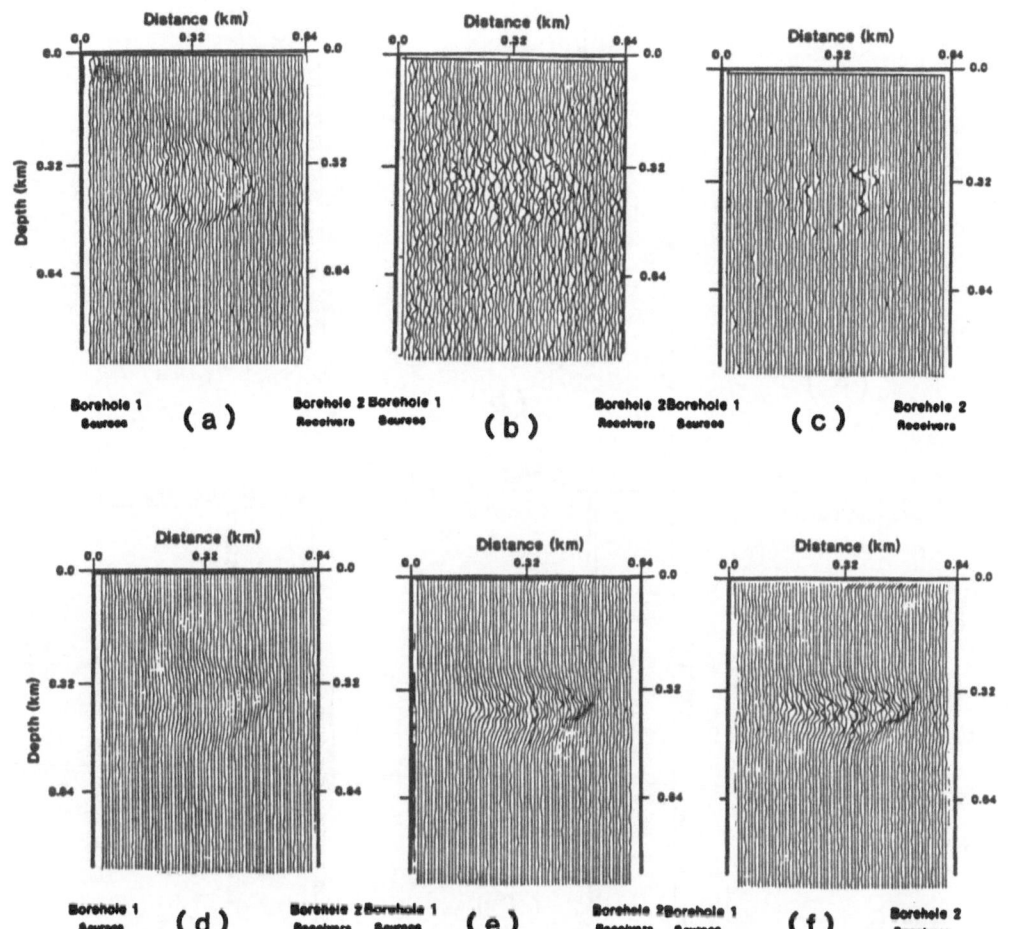

Figure 13: Reconstruction of a circular velocity model with (a) 1 percent (b) 5 percent and (c) 10 percent noise for ART, (d), (e) and (f) for SIRT at iteration 5.

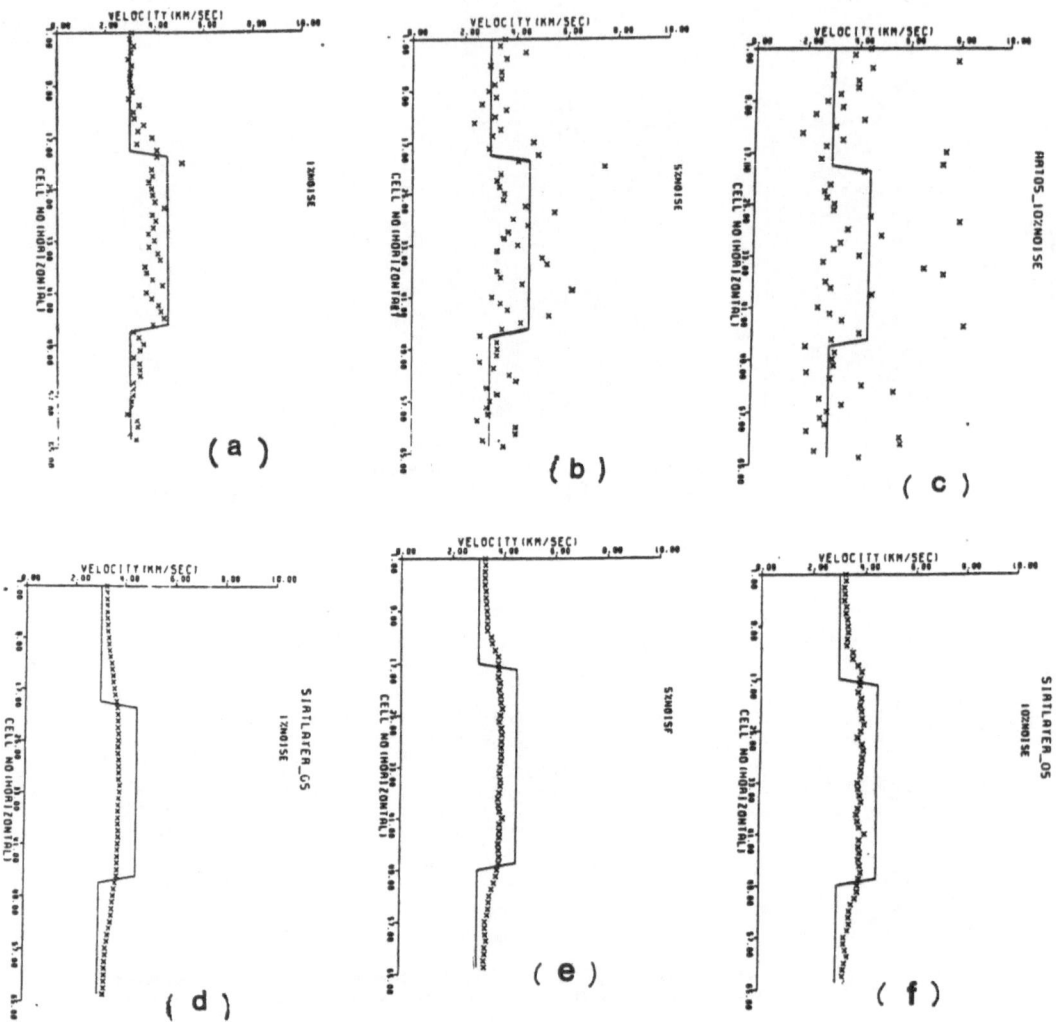

Figure 14: Line plot of reconstructed circular velocity field with noise of (a) 1 percent, (b) 5 percent,(c) 10 percent for ART, and (d), (e) and (f) for SIRT at 5 iteration. True(model) velocities are represented by solid lines and reconstructed velocities are represented by x symbol.

5 Conclusions

Crossborehole seismic tomography is a good technique for determining the velocity structure between parallel boreholes. Two iterative reconstruction techniques (ART and SIRT) have been employed to image the velocity field. The main reason for using an iterative reconstruction technique rather than a direct matrix inversion to reconstruct the velocity is that large matrix problems are difficult to solve.

ART and SIRT share the feature of iteratively adjusting the reconstruction elements (slownesses) with the corresponding observed data. Reconstructing a known velocity model helps to understand the nature of the reconstruction algorithms before applying them to field data. A new approach has been followed in calculating the initial velocity estimates required in the iterative reconstruction techniques. Three different initial velocity models, averaged, layered, and estimated, were calculated by using total ray segment lengths and traveltimes. The layered velocity resulted in the best image reconstruction for a circular velocity model compared to other initial model estimation methods. The initial estimates play a very important role in image quality and convergence rates.

It has been shown that traveltime tomography using iterative reconstruction techniques provides good velocity estimation. ART gives better estimates near the boundaries. SIRT gives better estimates in the presence of noise, though it may require more iterations than ART. Because the SIRT updates the slowness vector based upon the information contained within the entire set of projections per iteration (averaging slowness), the velocity estimates are smoother, yielding more stability in the presence of noise. If the input data are free of noise, convergence to the solution occurs faster with an ART algorithm due to the fact that slowness vectors are being updated ray by ray.

The velocity calculated with the two iterative reconstruction techniques can be used as input for seismic migration to improve the positioning of reflectors representing the velocity contrasts. These velocities can also be used to calculate the total acoustic wave field using finite difference modeling. The finite-difference process generates the direct wave field along with the diffracted wave field. Iteratively updating the velocities with the traveltimes from the total wavefield helps to correct for the ray bending effect. At each iteration step velocities are updated and input to the finite-difference algorithm to select the traveltime. Choosing the ray perpendicular to the wavefront with consideration of the shot and receiver position, the traveltime of this particular ray will take into account the ray bending effects through the area. This process is continued until the difference between calculated traveltime and observed traveltime is close to zero.

References

[1] Aki, K., and Lee, W., 1976, *Determination of three dimensional velocity anomalies under a seismic array using first p arrival times from local earthquakes, a homogeneous initial model.* J. Geophys. Res., 81, 4381-4399.

[2] Backus, G., and Gilbert, F.,1970, *The resolving power of gross earth data.* Geophysics, J. R. Astr. Soc., 16, 169-205.

[3] Benson, E. F. ,1990 *Reverse time migration of crosshole data using tomographic velocities.* Master Thesis The University of New Orleans, New Orleans, Louisiana.

[4] Claerbout, J., F., 1976, *Fundamentals of geophysical data processing*. Blackwell Scientific Publication, London.

[5] Dines, K. A., and Lytle, R., J., 1979, *Computerized tomography*. Proc. IEEE, 67, 1065-1073.

[6] Gilbert, P., 1972, *Iterative methods for the three-dimensional reconstruction of an object from projections*. J. Theor. Biol., 36, 105-117.

[7] Gordon, R., Bender, R., and Herman, G., T., 1970,*Algebraic reconstruction techniques(ART) for three dimensional electron microscopy and x-ray photography*. J.,Theor., Biol., 29, 471-481.

[8] Herman, G., T., 1980 *Image reconstruction from projection*. New York, NY, Academic Press.

[9] Kacmarz, S., 1937 *Angenaherte auflosung von systemen linearer gleichungen*. Bull, Acad. Pol. Sci. Lett.,a, 355-357.

[10] McMechan, G., 1987*Seismic tomography in boreholes.*Geophysics J.,Roy.Astr.Soc.,74,601-612.

[11] Mora, P.,1987, *Nonlinear two dimensional elastic inversion of multioffset seismic data*. Geophysics, 52, 1211-1228.

[12] Tanabe, K., 1971, *Projection method for solving a singular system*. Numer., Math., 17, 203-214.

[13] Tarantola, A., and Valette, B.,1982 b *Generalized nonlinear inverse problems solved using the least square criterion*. Rev., Geophys.,Space phys., 20, no. 2, 219-232.

[14] Wiggins, R., 1972, *The general linear inverse problem: implication of surface waves and free oscillations for earth structure*. Rev. of Geophysics and Space Physics, 10, 336-350.

Reverse-time Migration of Crosshole Data Using Tomographic Velocities

E. F. Benson, R. Guney, A. K. M. Sarwar

University of New Orleans, New Orleans, LA 70148, USA

Abstract

Crosshole data acquisition is a unique way of imaging the subsurface in that the sources are located in one borehole while the receivers are located in another. The zero-offset imaging condition does not apply to this forum. A new imaging condition must be implemented to perform migration. For the crosshole situation, the imaging condition involves the traveltime from the source to the diffractor and the traveltime from the diffractor to the receiver. Since the imaging condition is based so heavily on time, it is only appropriate to migrate the data using time migration.

Modeling crosshole data has been done using explicit second-order finite-difference schemes. Explicit fourth-order schemes are introduced here to show the differences between the two schemes. The data modeled by finite differences are used to test the reverse-time migration algorithm.

Reverse-time migration of crosshole data consists of several steps. First, the imaging condition must be computed (traveltime from a source to a diffractor). This is done using the same finite-difference schemes as those mentioned above instead of using ray tracing. The traveltime for each diffractor (grid point) is stored for migration. The next step involves separating the total wave field (recorded data) into the scattered field and the unperturbed direct field. The scattered wave field is extrapolated backward in time (at each time step) using finite differencing. The final step involves application of the image condition. At each time step, the extrapolated wave field is examined. If the amplitude at a specific grid point has the same traveltime as that of the imaging condition then it is extracted. The final migrated image consists of all the extracted amplitudes. The velocities that are needed for construction of the unperturbed direct field and for migration are calculated using ray tomography.

1 Introduction

Migration of seismic data is a technique used to *move* data recorded in time to its *true* position. There are several types of migration, including phase shift, Stolt, frequency, depth, and reverse-time. Reverse-time migration of normal moveout (NMO) corrected, stacked (approximate zero-offset), vertical seismic profile (VSP), and crosshole data has been presented by several authors, including Baysal et al. (1983), Loewenthal and Mufti (1983), McMechan (1983, 1985), Whitmore (1983), Levin (1984), Chang and McMechan (1986), Hu and McMechan (1988), and Zhu and McMechan (1988). Reverse-time migration extraplolates recorded data backward in time to the initiation of a reflector or diffractor. For zero-offset surface data, extrapolation is performed back to time zero. Extrapolation of the source field and recorder field can be done simultaneously since the waves travel along the same upcoming and downgoing paths. This imaging condition cannot be applied to VSP or crosshole data because the sources and receivers are not located in the same position and the waves do not propagate along the same paths. The imaging condition for VSP and crosshole data, outlined by Chang and McMechan (1986) and by Hu and McMechan (1986), considers the traveltime from a source to a diffractor and the time from the diffractor to a receiver. Thus extrapolation of the source field and recorder field must be done separately. The finite-difference schemes outlined by Alford et al. (1974) are good approximations for the wave equation, which are used for both modeling and migrating. Chang and McMechan (1986) used explicit second-order finite-differences to extrapolate the recorder field and ray tracing for extrapolation of the source field.

In this study, extrapolation of the source (i.e. traveltime computing) and recorder fields are done using explicit second- and fourth-order finite differences. Accuracy of the finite-difference schemes are dependant upon the time sampling rate and the grid spacing. A prior knowledge of the velocity field is necessary for extrapolating both the source and recorder fields. Having a correct velocity field improves the traveltime computing and migration algorithms. The velocities are calculated using ray tomographic methods. Tomography is an image reconstruction technique that has a wide variety of applications. It has been used for imaging in the medical field as well as in astronomy. Several authors, including Lines and LaFehr (1989), McMechan (1986, 1988), and Mora (1989), have incorporated migration with tomography for crosshole data as well as surface reflection data.

This paper focuses on the modeling and migration of synthetic crosshole data using second- and fourth-order finite-difference techniques. The velocities are calculated using straight ray tomography (Guney et al., 1990). The algorithms are outlined using a simple point diffractor model and then are applied to more complex synthetic data.

2 Crosshole Seismic Surveys

2.1 Modeling Crosshole Data

To date there have been very few real data crosshole surveys reported. One study, by Beydoun et al. (1989), dealt with the elastic migration/inversion of crosshole data in the Paris Basin. Another was completed by Lines and LaFehr (1989) concerning tomographic imaging using direct arrival times to compute the velocity field. Applying the algorithms to synthetic data allows for controlled testing before real data is used.

Generating synthetic model data can be thought of as the inverse of migration. Migration extrapolates recorded data backward in time and space to create the subsurface image. Modeling does the opposite. Data are *extrapolated* forward in time and space using some geologic model (including properties of velocity, density, porosity, etc.) to calculate the recorded field, which consists of amplitudes and traveltimes. Generating synthetic model data and migration both involve solving the acoustic and elastic wave equations. The wave equations describe the motion of particles in a given medium.

Several methods have been developed to solve the wave equation. These include the Kirchhoff integral (Hilterman, 1970), finite differencing (Kelly et al., 1976) , and $f - k$ domain solutions (Baysal et al. , 1983). In this study, synthetic model data are created using explicit second- and fourth-order finite-difference techniques (Alford et al. 1974), to compare the differences between the two. Finite-difference methods are very efficient in solving the wave equation.

2.1.1 Finite Differencing

Finite-difference methods have proven to be excellent techniques for solving differential equations, such as the acoustic and scalar wave equations (Alford et al. , 1974; Kelly et al. , 1976; and Alterman and Karal, 1968). These methods work well when modeling or migrating complex geologic structures that generate diffractions. There are two general categories of finite-difference equations. One, an explicit scheme, solves the wave equation by determining the motion at a particular grid point using previously calculated values. The other, an implicit scheme, determines the motion of all grid points from previous values using a matrix inversion technique.

These schemes can be used to solve the 2-D acoustic wave equation (Claerbout, 1976) given by

$$\frac{\partial^2 U}{\partial t^2} = K/\rho \left(\frac{\partial^2 U}{\partial x^2} + \frac{\partial^2 U}{\partial z^2} \right) \tag{1}$$

which can also be written as,

$$V^{-2}(x, z)U_{tt} = U_{xx} + U_{zz}, \tag{2}$$

where U is the acoustic pressure, $K/\rho = V^2(x, z)$, K is the incompressibility, ρ is the density, $V(x,z)$ is the velocity, and the subscripts denote partial derivatives with repsect to x (horizontal coordinate), z (depth coordinate), or t (time).

The second-order approximation written in the form given by Alford et al. (1974) as

$$\begin{aligned} U(x_k, z_j, t_{i+1}) = & \ 2(1 - 2A^2)U(x_k, z_j, t_i) - U(x_k, z_j, t_{i-1}) + A^2 \{ U(x_{k+1}, z_j, t_i) + \\ & \ U(x_{k-1}, z_j, t_i) + U(x_k, z_{j+1}, t_i) + U(x_k, z_{j-1}, t_i) \}. \end{aligned} \tag{3}$$

Higher order approximations can be obtained by using the more accurate fourth-order representation for the Laplacian given by Abramowitz and Stegun (1964) which is rewritten by Alford et al. (1974) as

$$\begin{aligned} U(x_k, z_j, t_{i+1}) = & \ (2 - 5A^2)U(x_k, z_j, t_i) - U(x_k, z_j, t_{i-1}) + (A^2/12) \{ 16[U(x_{k+1}, z_j, t_i) \\ & + U(x_k, z_{j+1}, t_i) + U(x_{k-1}, z_j, t_i) + U(x_k, z_{j-1}, t_i)] - U(x_{k+2}, z_j, t_i) + \\ & \ U(x_k, z_{j+2}, t_i) + U(x_{k-2}, z_j, t_i) + U(x_k, z_{j-2}, t_i) \}, \end{aligned} \tag{4}$$

where $A = V(x_k, z_j)\Delta t/h$, $V(x_k, z_j)$ is the velocity at each x-z grid point, Δt is the time step between successive U wave fields, and h is the distance between the horizontal and vertical grid points.

2.1.2 Stability

In order for finite-difference equations to work properly, they must remain stable as the differencing star moves through the grid. The second-order equation, 3, will remain stable if $\Delta t < hV^{-1}2^{-1/2}$ while the fourth-order equation, 4, will remain stable if $\Delta t < hV^{-1}\sqrt{3/8}$ (Alford et al. 1974). Stablity is dependent on the coarseness of the grid, h, being considered. The grid coarseness is equal to the number of grid points per half-power wavelength of the source. The source field is calculated by the Gaussian function

$$f(t) = te^{-\alpha t^2}, \tag{5}$$

where α is a constant related to the time interval, w, between a negative and a positive peak, as shown in Figure 1. The half-power wavelength corresponds to the frequency of the upper half-power of the source field power spectrum, shown in Figure, 2. The half-power wavelength is calculated by dividing the velocity by the half-power frequency. Then the number of grid points per wavelength is found by dividing the half-power wavelength by the grid spacing. It is imperative that the number of grid points per wavelength be greater than 10 for second-order schemes and greater than 5 for fourth-order schemes. These values were determined to be stable in a study by Alford et al. (1974).

Computing synthetic data causes a problem when an ideal rectangular grid size is used. In order to compute a rectangular shaped grid, artificial boundaries must be used

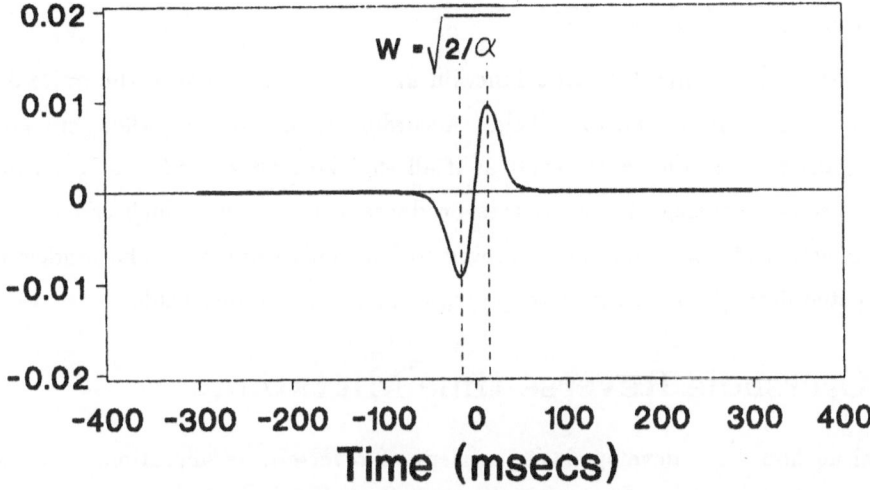

Figure 1: Source distribution time function (Alford et al., 1974)

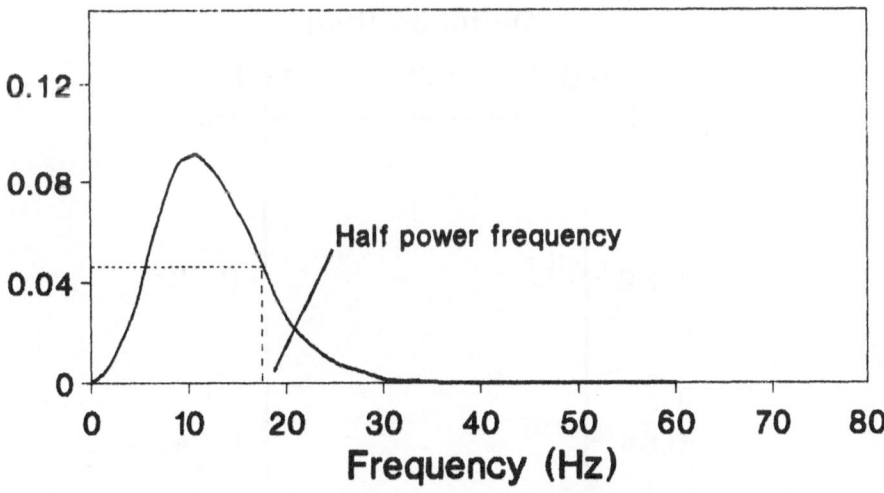

Figure 2: Source field power spectrum. The upper half-power frequency, at ≈ 17.0 hertz, is used to calculate the grid coarseness.

to simulate the edges of the grid so the differencing star can pass over all the values. The problem with this concept is that reflections occur at these boundaries. The boundary conditions absorb incoming energy thus suppressing unwanted reflections. Those used here were developed by Clayton and Enquist (1977).

2.1.3 Crosshole Setup

The geometry for crosshole data acquisition is different from surface seismic surveys. The sources are located in one borehole while the receivers are located in another. It is not necessary for the wells to be vertical and the depths of each well do not have to be the same. In this paper, the sources, each 10.0 m apart, are located in well 1 ($x = 0.0$ km) while the receivers, each 10.0 m apart, are located in well 2 ($x = 0.64$ km) as seen in Figure 3.

The source, a simple Gaussian function $u_t = te^{-\alpha t^2}$, is put into the finite-difference algorithms to compute the total field. According to McMechan, 1988, stacking point source gathers gives the same result as if all the shots were excited at the same time. Simultaneously exciting all the sources produces a line source which can be simulated by a collection of point sources according to Huygen's principle. The models that are constructed here use line sources to generate the total recorded fields.

3   Crosshole Reverse-time Migration

Several authors have developed the concept of reverse-time migration as it applys to zero-offset data, VSP, and crosshole data, including Baysal et al. (1983), Loewenthal

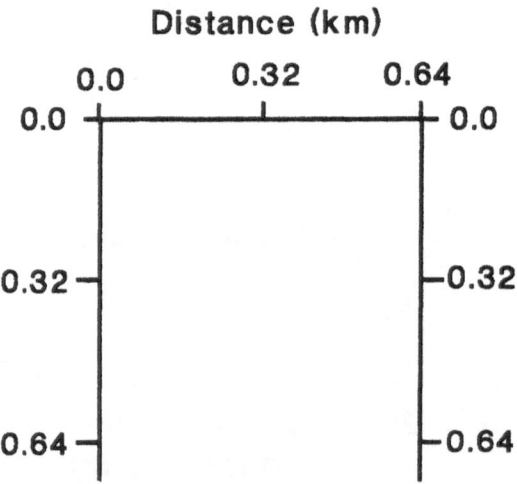

Figure 3: Crosshole geometry with the sources located in borehole 1 at $x = 0.0$ km and the receivers located in borehole 2 at $x = 0.64$ km.

and Mufti (1983), and Levin (1984), McMechan (1983), Hu and McMechan (1986), Zhu and McMechan (1988), and Hu et al. (1988). Reverse-time migration is simply the extrapolation of a recorded wave field backward in time to the initiation of a reflector or a diffractor. This traveltime is called the imaging condition. It is the time at which a diffractor is "exploded" during recording of the total wave field. This total field can then be separated into its component parts: the scattered and direct (transmitted) wave fields. Migration consists of extrapolating the scattered field backward in time to generate the final section.

A model of a point diffractor model in a constant velocity background, shown in Figure 4, is used to illustrate the techniques outlined above. A point diffractor is sufficient for testing the reverse-time migration algorithm, since a more complex structure can be considered the same as a collection of many point diffractors. A line source is located in well 1 and the receivers are located in well 2. The other necessary parameters have been given earlier.

Figures 5 and 6 show how the source field propagates through the finite-difference mesh creating the total recorded wave field (several time steps are shown for both second- and fourth-order finite differencing). Notice at $t = 0.084$ seconds the diffractor is excited. As the wave continues through the mesh, the diffractor initiates a new wave field. This new wave field is the scattered wave field.

Throughout the rest of this work, comparisons are made between the second- and fourth-order finite-difference techniques and will be shown side by side in the figures. In addition, more complex structures are shown in a later section.

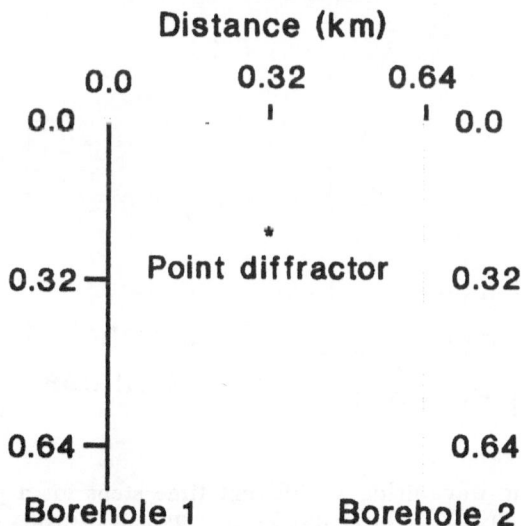

Figure 4: Model of a point diffractor located at $x = 0.32$ km and $z = 0.20$ km in a constant velocity background (3.0 km/sec).

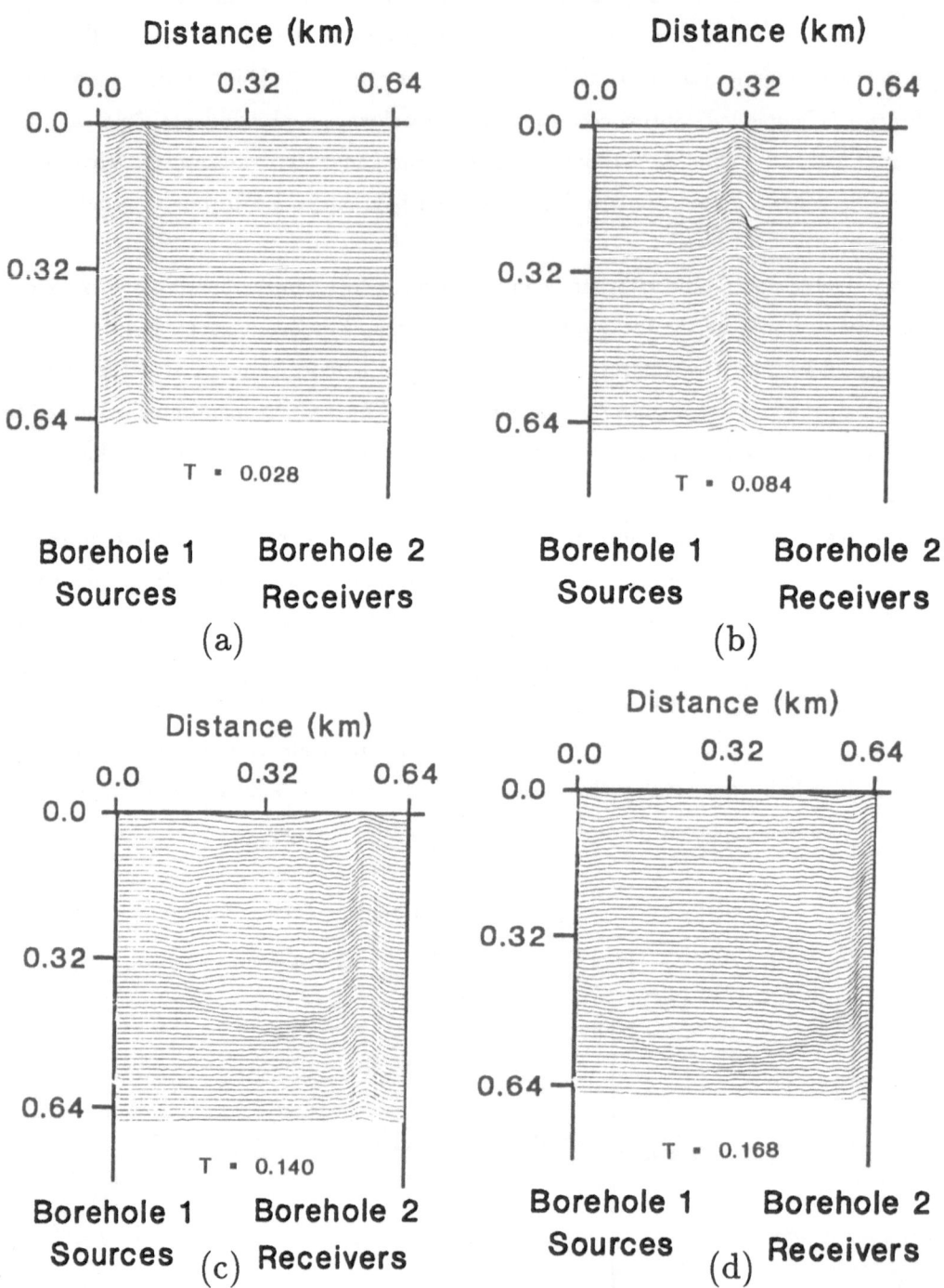

Figure 5: Propagating wave fields at different time steps for a point diffractor model computed using second-order finite differences. The traveltime for each wave field is shown at the bottom.

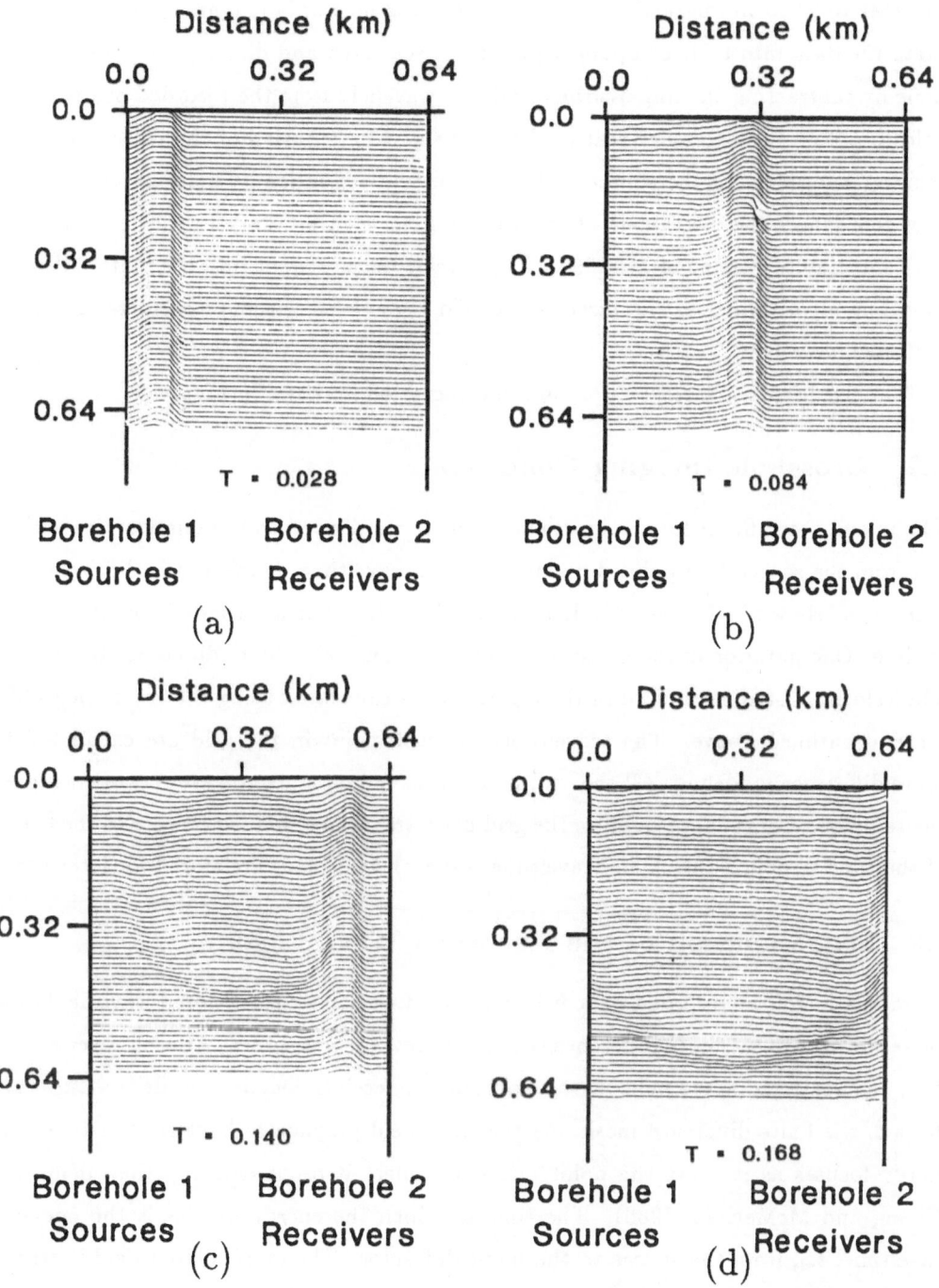

Figure 6: Propagating wave fields at different time steps for a point diffractor model computed using fourth-order finite differences. The traveltime for each wave field is shown at the bottom.

3.1 Separation of Wave Fields

In order to perform migration on the recorded crosshole data (Figure 7) one must separate the data into their component parts: the scattered and direct wave fields. This is done by subtracting the unperturbed or direct wavefield from the recorded or total wave field to obtain the necessary scattered wave field. Computation of the direct wave field requires knowledge of the source and the velocity distribution. The method developed here reproduces the direct field as shown in Figure 8 using the velocities calculated by ray tomography (Guney et al. 1990). These velocities are input into the finite-difference modeling algorithms and the direct wave field is modeled without consideration of the scattered wave field.

The resulting scattered section used for migration is shown in Figure 9.

3.2 Crosshole Imaging Condition

The imaging condition for the crosshole geometry is defined as the one-way traveltime, t_{sd}, from the source to a point diffractor and the traveltime, t_{dr}, from the diffractor to a receiver as shown in Figure 10. It is at the time, t_{sd}, that a secondary point source is excited. Computation of the imaging condition is done using finite-difference techniques. The velocity distribution used in the algorithms is calculated using the ray-tomographic method outlined above. The traveltimes for each point in the grid are calculated by finite-difference modeling. All the grid points that fall on a given wavefront are assigned the same traveltime. Each point in the grid has a traveltime that is accurate to the extent of the time sampling rate. The traveltimes for each grid point are stored for later use.

3.3 Extrapolation of the Recorded Wave Field

Extrapolation of the recorded wave field consists of inserting the scattered wave field from the receivers into the finite-difference schemes in reverse-time order. At each reverse-time step, the scattered wave field, at its appropriate receiver location, is driven *backward* through the finite-difference mesh. As the wave field propagates backward in time, the energy focuses as it nears the point diffractor where it originated and then defocuses (Chang and McMechan, 1986). The time at which the energy focuses is the one-way traveltime, t_{sd}, from the source to the point diffractor. The extrapolated field is stored in a separate array for the final step. Several time snapshots showing the extrapolated field are shown in Figure 11.

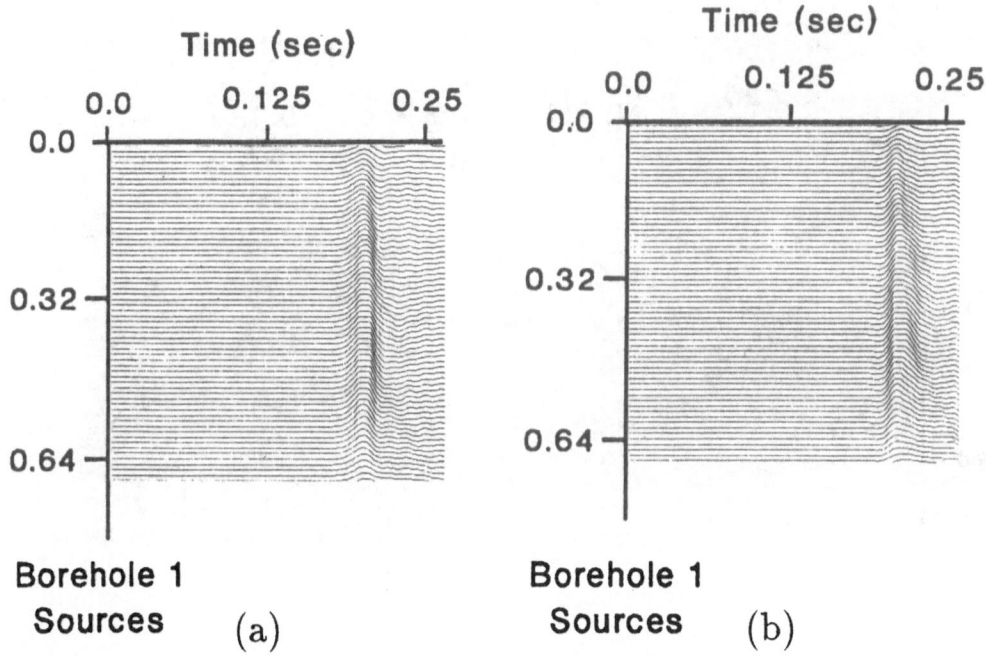

Figure 7: Total recorded wave field: (a) second-order finite differencing, (b) fourth-order finite differencing.

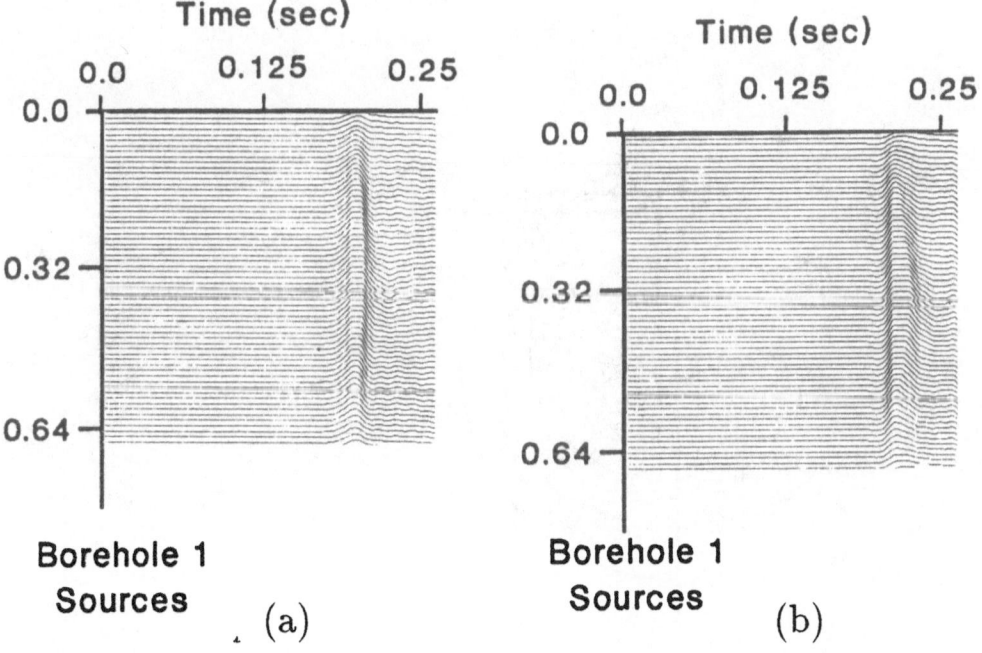

Figure 8: Direct wave field for a point diffractor model using a line source: (a) second-order , (b) fourth-order.

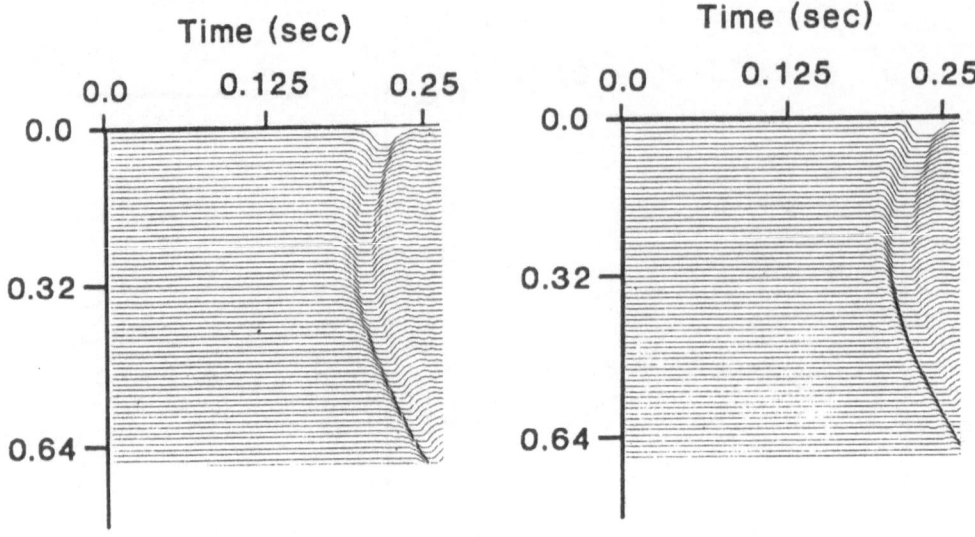

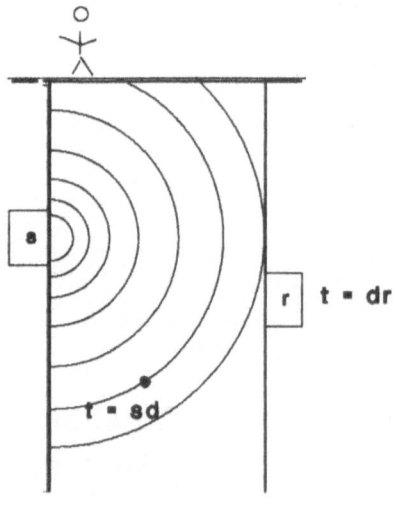

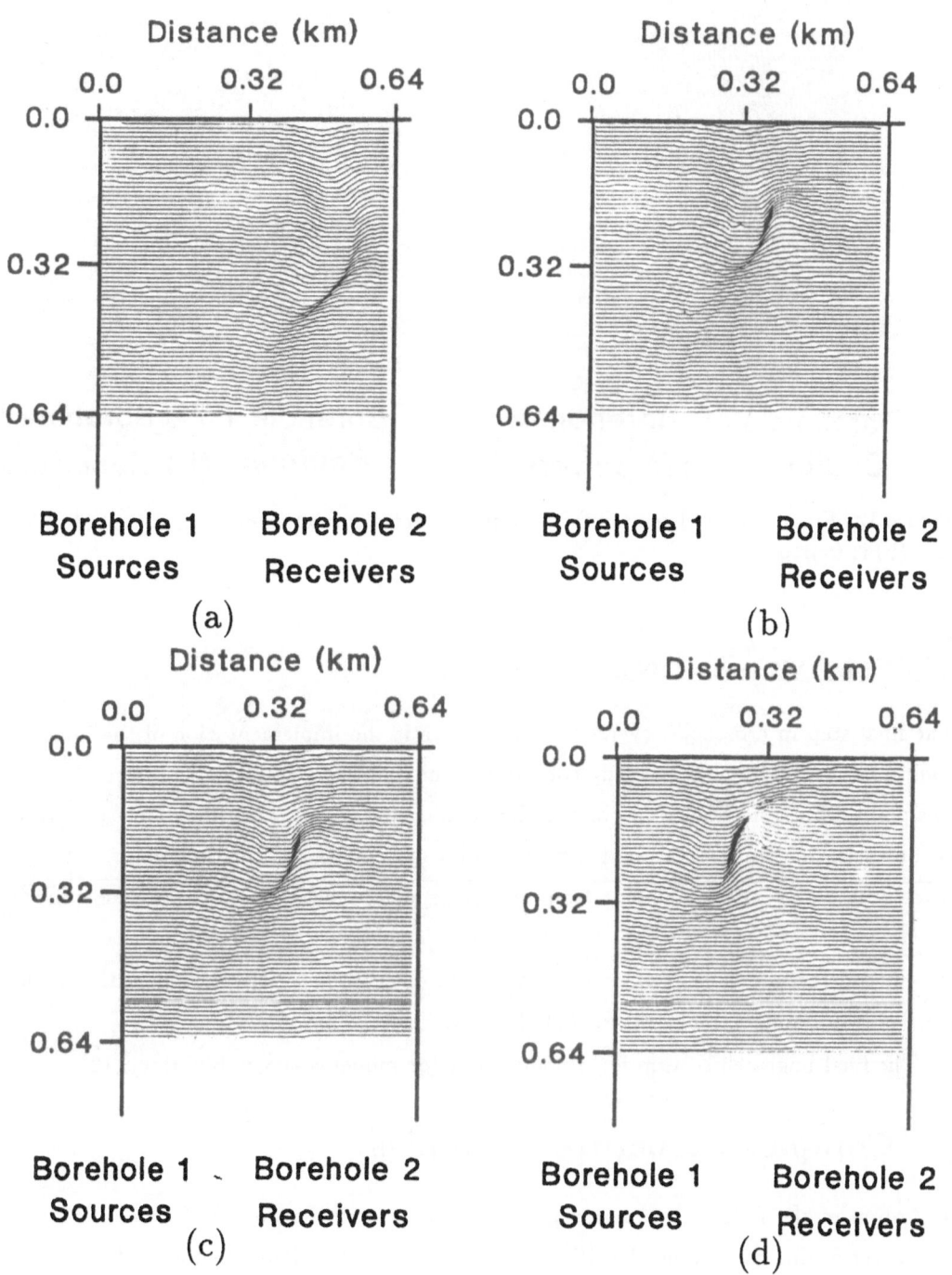

Figure 11: Extrapolation of the scattered wave field at several time snapshots (a,b,c,d) for a second-order approximation.

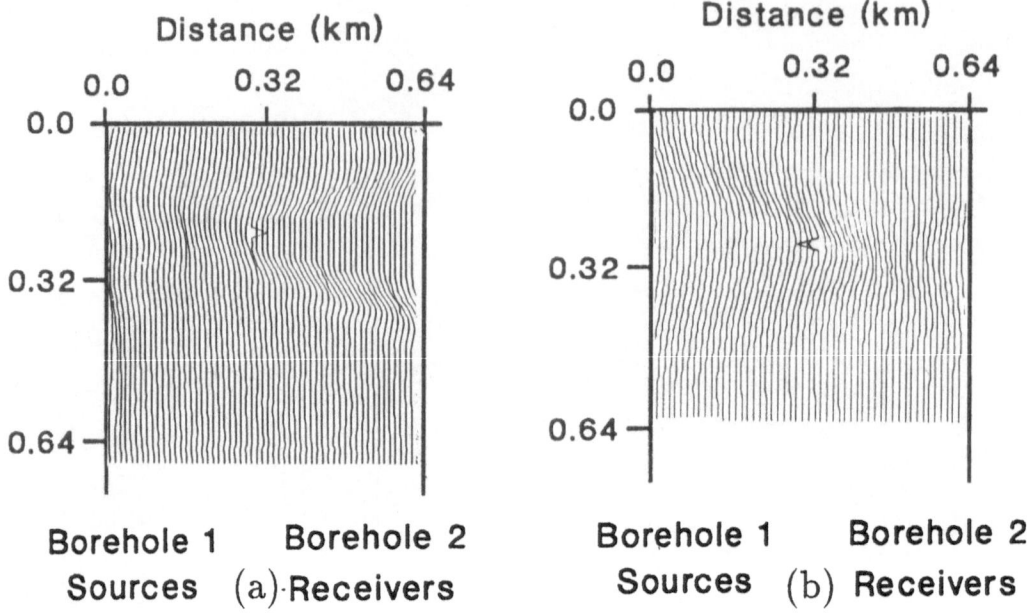

Figure 12: Final migrated section for a point diffractor model where: (a) is second-order and (b) is fourth-order.

3.4   The Migrated Image

The final step in crosshole reverse-time migration is the implementation of the imaging condition. Recall that the imaging condition is the one-way traveltime, t_{sd}, from a source to a point diffractor. Each grid point in the mesh is assigned a traveltime. This traveltime is the excitation time for a point diffractor that may be located at that point. At each reverse-time step any grid point from the extrapolated field sharing the same time is extracted. The extracted values are stored in a separate array, at the same location, as the imaged section. The end result is the migrated section. In this example, the model is a point diffractor located at $x = 0.32$ km and $z = 0.20$ km.

The final migrated section for a point diffractor model is shown in figure, 12.

4   Complex Structure Examples

Migration of the following complex geologic models uses the finite-difference modeling and reverse-time migration algorithms outlined in the previous chapters. The soures located in borehole 1, with the receivers in borehole 2, are used to model the crosshole data. All other necessary variables are outlined above. The velocity distributions used for migrating the scattered field are calculated using ray tomography (Guney et al. 1990).

4.1 Ray Tomography

Tomography is an image reconstruction technique that uses the observed traveltimes to compute the velocity distributions. Much of the theory of tomography was developed for imaging in the medical field, but has been introduced into geophysics by Aki et al. (1977), Spencer and Gubbins (1980), Dines and Lytle (1979, 1981), McMechan et al. (1987), and Wu and Toksöz (1987).

The ray-tomographic methods used here are straight ray methods developed by Guney et al. (1990). The basic principle behind straight ray tomography is reconstruction of the image using line integrals that calculate traveltimes. The required line integral is given by

$$b = \int x \cdot dA, \tag{6}$$

where b is the traveltime, x is the slowness or inverse of velocity, and dA is the ray segment length. This equation can also be written in matrix form:

$$Ax = b. \tag{7}$$

Equation 7 is solved for b using some initial slowness estimate (inverse velocity), x. The traveltimes calculated are compared to traveltimes obtained by ray-tracing methods or the traveltimes from the recorded finite-difference field. The latter comparison is made here. If the traveltimes match, then the slowness estimate is correct. If not, then a new slowness correction, calculated by

$$\Delta x_{ij} = \frac{r_i A_{ij}}{\sum_j A_{ij}^2}, \tag{8}$$

where i and j are the ray number and cell number, and r_i is the residual (traveltime difference), is input into equation 7. This process is continued until the difference between the calculated traveltimes and the observed traveltimes, r_i, is as close to zero as possible. The output is the velocity distribution field, b, used in the migration of crosshole data.

4.2 Synthetic Models and Migrated Results

The first model consists of a horizontally layered medium, shown in Figure 13 a. The recorded wave fields for second- and fourth-order finite differences were calculated by a line source located in borehole 1 with a grid spacing of $h = 0.01$ km. A time sampling rate of $\Delta t = 0.0014$ seconds was used for the second-order schemes and $\Delta t = 0.0012$ for the fourth-order schemes. The time sampling rates are different due to the stability of each approximation. The final migrated section is shown in Figure 14.

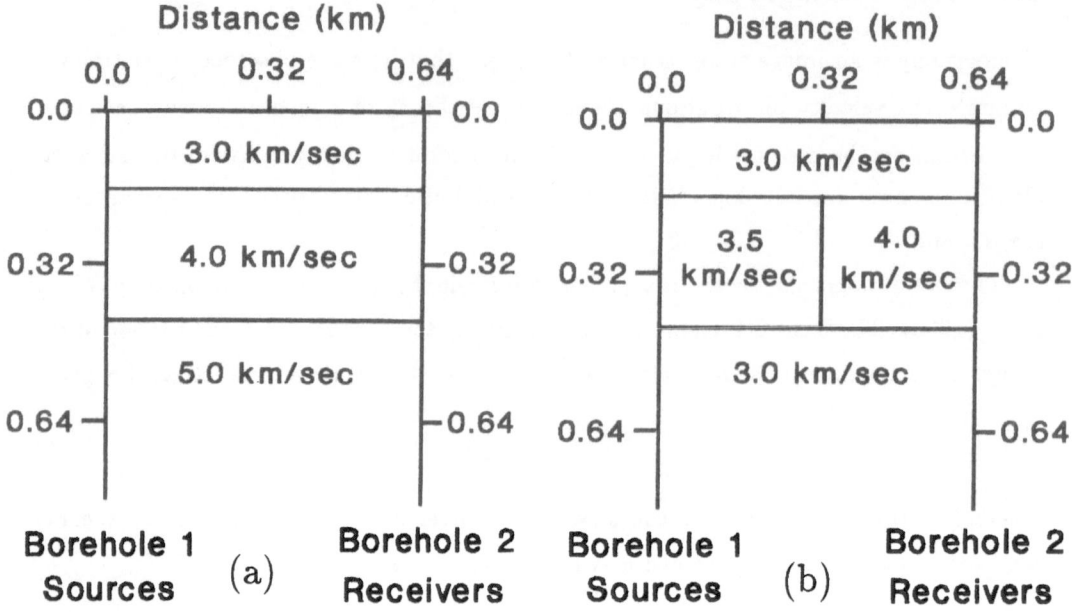

Figure 13: (a) Horizontally-layered velocity model, (b) Horizontally-layered velocity model with a vertical fault in the middle layer.

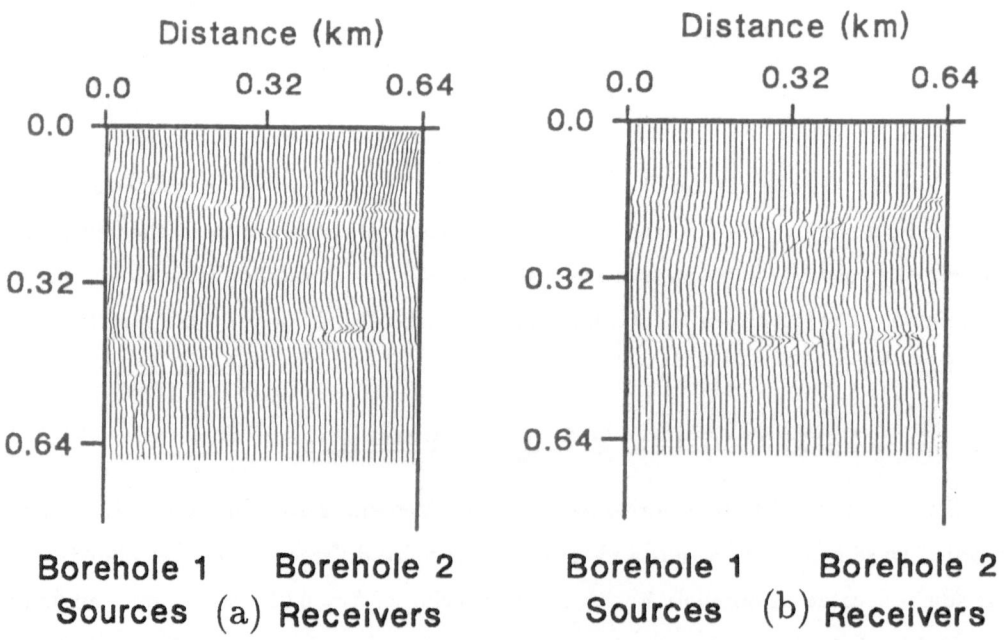

Figure 14: Migrated sections for a horizontally-layered medium, Fig. 13 a, using a line source where: (a) second-order finite-differences, and (b) fourth-order finite-differences.

The next example is also a horizontally layered medium with a fault located in the middle layer, as shown in Figure 13 b. Again a line source is used to generate the total field. The same grid spacing is used but a a time sampling rate of $\Delta t = 0.001 sec.$ is used here. The velocities used to calculate the direct arrival field and perform migrate have and tomographic techniques depend on a wide angle of view for best results. A crosshole survey could be improved by adding sources at the surface, such as in vertical seismic profiling.

Source generation also plays an important role in the modeling and migration of crosshole data. If a line source is used, the wave will tend to propagate parallel to the reflectors (assuming relatively horizontal beds). Those beds that are normal to the planar wavefront will be difficult to image, as some of the energy is reflected back towards the source instead of being diffracted toward the receiver. Computing the scattered wave field by subtracting the direct field from the total field could prove to be an important tool. The main problem occurs in the amplitude differences between the direct and total fields. Finite- difference approximations to the acoustic wave equation, while very efficient for calculating traveltimes, tend to be weak in calculating true amplitudes. As reflections and diffractions occur, the strength of the propagating wave is reduced. The difficulty in calculating amplitude strength could be reduced by scaling the total field in relation to the direct field to remove the direct arrivals. Furthermore, the finite-difference

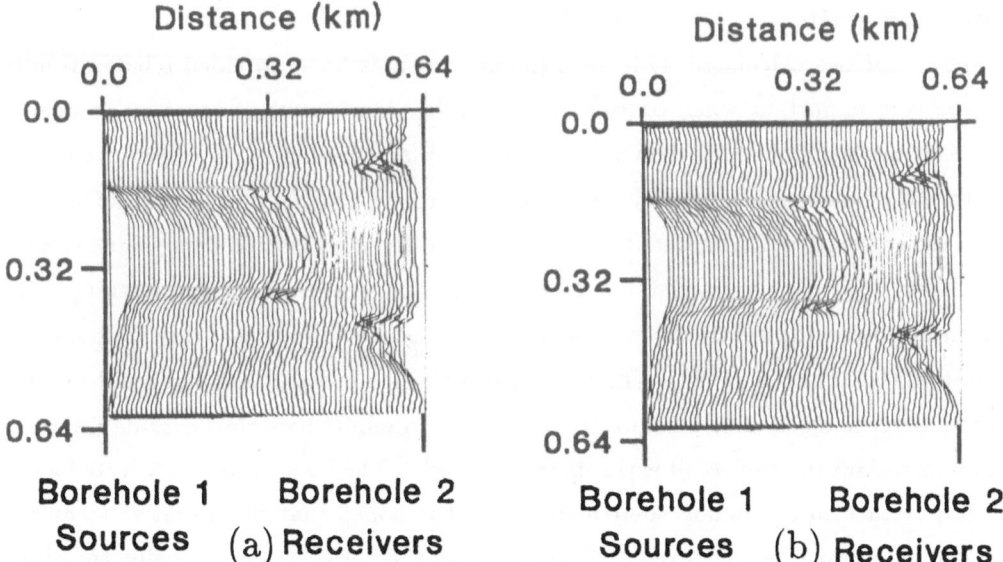

Figure 15: Migrated section for a horizontally layered medium with a fault located in the center, Fig. 13 b, using tomographic velocities where: (a) is second-order and (b) is fourth-order.

approximation to the elastic wave equation is much better for amplitude reconstruction since it takes into consideration both p-wave and s-wave propagation. Thus imaging of been calculated using the tomographic techniques outlined in the previous chapter. The results of this migration are shown in Figure 15. The tomographic velocities are very close to the same velocities used to model the data. Notice, however, that the amplitudes of the reflectors die out as the distance from the source increases. This may be corrected by applying a gain or scaling factor. It may be the result, though, of some source field remnants which mask the results. Another way to correct for this problem would be to increase the view angle of the source/receiver configuration. This can be accomplished by combining VSP methods with crosshole methods. This will provide a greater area of coverage and improve the final migrated image.

5   Conclusions

Reverse-time migration using second- and fourth-order approximations to the acoustic wave equation is a good method for determining the structure between two boreholes. Fourth-order approximations are more accurate because they sample more grid points. The added time for these computations can be reduced by using a coarser grid spacing. Fourth-order approximations will remain stable using fewer grid points than will second-order approximations. Migration results are limited by the view angle between boreholes, the source field generation, and the wavefield separation. Crosshole migration structures that are perpendicular to wave propagation may be better imaged with the elastic wave equation.

The velocities calculated with tomographic methods have provided reliable results. This is very important when considering real data. Much work is spent on recovering the velocities using well-log data and uphole time measurements. A new procedure to improve the tomographic velocity distribution is currently being developed by Guney et al. (1990). Iterative Wavefront Reconstruction (IWR) is an iterative technique created to improve the velocity distribution used in migration and tomographic imaging. The traveltimes from the recorded wave field are selected and input to a ray- tomography algorithm (Guney et al., 1990). The velocity distribution is calculated and input into the finite-difference algorithm. The direct wave field is computed without considering point diffractors, since the end result is the direct wave field. The traveltimes are selected again and input back to the tomographic algorithm. It is hoped that this iterative technique will compute velocities comparable to those used in the initial models. Finally, the direct wave field is subtracted from the total field, resulting in the scattered wave field. Both the direct wave field and the scattered wave field are used later for traveltime picking and migration. I would like to apply the methods of reverse-time migration using finite differences approximations to the elastic wave equation. A two-component receiver in the borehole will be necessary to record both p- and s-waves. With improvements made in tomography, I hope to eventually combine diffraction tomography methods with migration and extend these methods into 3-dimensional space. These techniques will prove to be very useful in the future for exploration for oil and gas, as well as for hydrogeological exploration.

References

[1] Abramowitz, M., and Stegun, I.A., 1965, *Handbook of mathematical functions*, Dover Publishing Co., New York, p. 885.

[2] Aki, K., Christofferson, A. and Husebye, E. S., 1977, *Determination of the three-dimensional seismic structure of the lithosphere*, J. Geophys. Res., 82, 277-296.

[3] Alford, R.M., Kelly, K.R., and Boore, D.M., 1974, *Accuracy of finite-difference modeling of the acoustic wave equation*, Geophysics, 39, 834-842.

[4] Alterman, Z. and Karal, F.C., 1968, *Propagation of elastic waves in layered media by finite difference methods*, Bulletin of the Seismological Society of America, 58, 367-398.

[5] Baysal, E., Kosloff, D.D., and Sherwood, J.W.C., 1983, *Reverse time migration*, Geophysics, 48, 1514-1524.

[6] Beydoun, W.B., Delvaux, J., Mendes, M., Noual, G., and Tarantola, A., 1989, *Practical aspects of an elastic migration/inversion of crosshole data for reservoir characterization: A Paris basin example*, Geophysics, 54, 1587-1595.

[7] Chang, W.F. and McMechan, G.A.,1986,*Reverse-time migration of offset vertical seismic profiling data using the exitation-time imaging condition*, Geophysics, 51, 67-84.

[8] Claerbout, J.F., 1976, *Fundamentals of geophysical data processing*, Blackwell Scientific Publicationsm London.

[9] Claerbout, J.F., 1985, *Imaging the Earth's Interior*, Blackwell Scientific Publications, London.

[10] Clayton, R.W. and Engquist, B., 1977, *Absorbing boundary conditions for acoustic and elastic wave equations* , Bulletin of the Seismological Society of America, 67, 1529-1540.

[11] Clayton, R.W. and Engquist, B., 1980, *Absorbing boundary conditions for wave-equation migration*, Geophysics, 45, 895-904.

[12] Devaney, A.J., 1982, *A filtered back propagation algorithm for diffraction tomography*, Ultrasonic Imaging, 4, 336-350.

[13] Devaney, A.J., 1983, *A computer simulation study of diffraction tomography*, Inst. Electr. Electron. Eng., Trans. Biomed. Eng., BME-30, 377-386.

[14] Devaney, A.J.,1984, *Geophysical diffraction tomography*, Trans., Inst. Electr. Electron. Eng., GE-22, 3-13.

[15] Dines, K.A. and Lytle, R.J., 1979, *Computerized geophysical tomography*, Proc. IEEE, 67, 1065-1073.

[16] Dines, K.A. and Lytle, R.J., 1981, *Analysis of electrical conductivity imaging*, Geophysics, 46, 1025-1036.

[17] Engquist, B. and Majda, A., 1977, *Absorbing boundary conditions for the numerical simulation of waves*, Mathematics of Computation, 31, 629-651.

[18] Guney, R., Benson, E.F., and Sarwar, A.K.M., 1990 *Imgaging of the earth with a new iterative wave reconstruction (IWR) method*, in Monograph, Methods of Applied Geophysics, 8, Friedrick Vieweg and Sohn, Wiesbaden, Germany.

[19] Hilterman, F.J., 1970, *Three-dimensional seismic modeling*, Geophysics, 35, 1020.

[20] Hu, L., and McMechan, G.A., 1986, *Migration of VSP data by ray equation extrapolation in 2-D variable velocity media*, Geophysical Prospecting, 34, 704-734.

[21] Hu, L., McMechan, G.A., and Harris, J.M., 1988, *Acoustic prestack migration of cross-hole data*, Geophysics, 53, 1015-1023.

[22] Kak, A.C., 1985, *Tomographic imaging with diffracting and non-diffracting sources* in Haykin, S., Array processing systems: Prentice Hall.

[23] Kaveh, M., Mueller, P.K., and Greenleaf, J.F., 1983, *Acoustical imaging*, 13: Plenum Press.

[24] Kelly, K.R., Ward, R.W., Treitel, S., Alford, R.M., 1976 *Synthetic seismograms: A finite-difference approach*, Geophysics, 41, 2-27.

[25] Levin, S.A., 1984, *Principles of reverse time migration*, Geophysics, 49, 581-583.

[26] Lines, L.R., and LaFehr, E.D., 1989, *Tomographic modeling of a cross-borehole data set*, Geophysics, 54, 1249-1257.

[27] Lowenthal, D., and Mufti, I.R., 1983, *Reverse time migration in spatial frequency domain*, Geophysics, 48, 627-635.

[28] McMechan, G.A., 1983, *Migration by extrapolation of time- dependent boundary values*, Geophysical Prospecting, 31, 413-420.

[29] McMechan, G.A., 1985, *Synthetic finite-offset vertical seismic profiles for laterally varying media.*, Geophysics, 50, 627-636.

[30] McMechan, G.A., Harris, J.M., and Anderson, L., 1987, *Cross-hole tomography for strongly variable media with applications to scale model data*, Bull., Seis. Soc. Am., 77, 1945-1960.

[31] Mora, P. , 1989, *Inversion = migration + tomography*, Geophysics, 54, 1575-1586.

[32] Mueller, R.K., Kaveh, M., and Wade, G., 1979, *Reconstructive tomography and applications to ultrasonics*, Proc. Inst. Elsctr. Electron. Eng., 67, 567-587.

[33] Mueller, R.K., 1980, *Diffraction tomography I: the wave equation*, Ultrasonic imaging, 2, 213-222.

[34] Mueller, R.K., Kaveh, M., and Iverson, D., 1980, *A new approach to acoustic tomography using diffraction techniques* in Metherell, A. F., Acoustic holography, 8,: Plenum Press, 615-628.

[35] Nahamoo, D., Pan, S.X., and Kak, A.C., 1984, *Synthetic aperture diffraction tomography and its interpolation-free computer implenentation*, Inst. Electr. Electron. Eng. Trans. sonics and ultrasonics, SU-31, 218-229.

[36] Pratt, R.G. and Worthington, M.H., 1988, *The application of diffraction tomography to cross-hole seismic data*, Geophysics, 53, 1284-1294.

[37] Spencer, C. and Gubbins, D., 1980, *Travel-time inversion for simultaneous earthquake location and velocity structure determination in laterally varying media.*, Geophys. J. R. astr. Soc., 63, 95-116.

[38] Whitmore, N.D., 1983, *Iterative depth migration by backward time propagation*, Presented at the 53rd Ann. Mtg. and Expos., Soc. Explor. Geophysics, Las Vegas.

[39] Wu, R-S. and Toksöz, M.N., 1987, *Diffraction tomography and multisource holography applied to seismic imaging*, Geophysics, 52, 11-25.

[40] Yilmaz, Ö., 1987, *Seismic data processing*, Society of Exploration Geophysicists, Tulsa, Oklahoma.

[41] Zhu, X. and McMechan, G.A., 1988, *Acoustic modeling and migration of stacked cross-hole data*, Geophysics, 53, 492-500.

From Radon to Kirchhoff Migration

M. Novotný

Geofyzika Brno, Ječná 29a, 612 46 Brno, Czechoslovakia

Abstract. The interrelation between the Radon wavefield extrapolation and
integral solutions of boundary value problems posed for scalar wave equation is studied. Using the stationary phase method it is found that the
Radon extrapolator degenerates into the Rayleigh-Sommerfeld solution of
the Dirichlet problem in the far-field zone. The derivations are done for
the 3-D case, but the 2-D analogues are also mentioned. The practical consequences and advantages of slant-stack representation of space-undersam-
pled noisy data are discussed.

1. Introduction

Radon-like transformation was first applied in seismics at the beginning
of the 1970s when the tau method for solving the inverse kinematic seismological problem appeared (Bessonova 1974). The tau method further developed
by many authors (Milkereit 1987, Diebold and Stoffa 1981, Stoffa et al.
1981) dealt mainly with the kinematic aspect of the solved problem. In short,
seismic data in the form of time-distance curves were Legendre transformed
to the tau representation which allows a certain generalized solution of
the inverse kinematic problem. From the kinematic ray concept it was only
one step to slant-stack which transforms the complete wavefield to the tau-p
domain. This was achieved by Schultz and Claerbout (1978) who noticed the
similarity between the slant-stack and the plane wave decomposition of the
wavefield proposed by G. Müller (1971) for the reflectivity method of computing synthetic seismograms.

Chapman (1978, 1981) studied the mathematical aspect of the direct and
inverse problems in seismics based on slant-stack. He referred to the original Radon publication (Radon   1917). He was the first to point out the
fact that the slant-stack or Radon transformation does not change the basic
attribute of the wavefield, i.e. it satisfies the wave equation. Later, the

231

plane wave representation was used to solve migration seismic problems for horizontally stratified media (Hubral 1981, Levin 1980, Clayton and McMechan 1981 , Tygel and Hubral 1990 , Brysk 1986).

In a recent paper (Novotný 1990) I compared the slant-stack migration formula with the Kirchhoff integral solution of wave equation in the 2-D case. In the present paper the comparison is done for the 3-D case. Similarly, I will show that the Rayleigh-Sommerfeld integral solution of the Dirichlet boundary value problem represents the high-frequency asymptotic of Radon wavefield extrapolator.

2. Formulation of problems

In a homogeneous medium the propagation of the compressional wave $u(x,y,z,t)$ is described by the scalar wave equation

$$\Box u = u_{xx} - u_{yy} - u_{zz} - u_{tt}/c^2 = 0 \tag{1}$$

where c stands for the constant propagation velocity. The wave equation (1) allows to extrapolate the surface data $u(x,y,0,t)$ to the depth $z > 0$. Naturally the surface data are recorded in a finite region with sampling steps Δx, Δy and Δt. The used coordinate system is depicted in Fig.1.

The seismic migration or the inverse problem involves just reconstruction of the wavefield $u(x,y,z,0)$ in the lower half-space for t=0. The term wavefield extrapolation is used for reconstruction of the wavefield in the lower half-space $z > 0$ and $t > 0$. The opposite reconstruction $u(x,y,z,0) \rightarrow u(x,y,0,t)$ is called the direct problem or seismic modelling. As a rule, the wavefield extrapolation is performed with a depth step Δz in a recur-

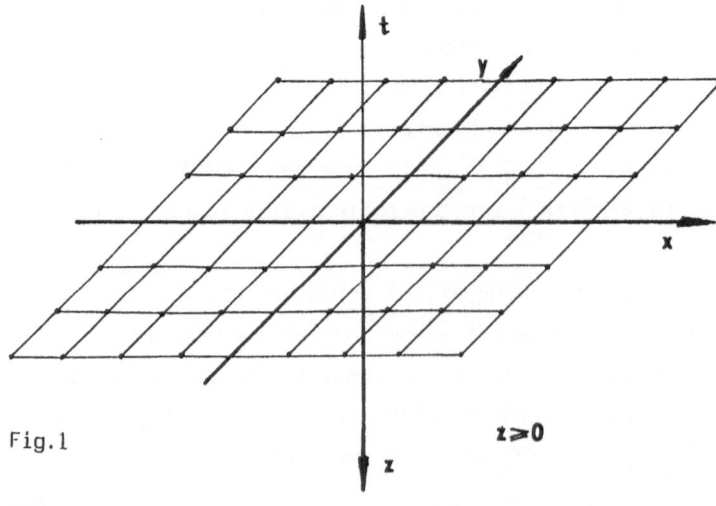

Fig.1

sive way. In further, we shall deal with such depth recursive wavefield extrapolators, for simplicity setting $\Delta z=z$. Thus, one depth recursion step corresponds to the wavefield extrapolation through a homogeneous layer with velocity c and thickness $\Delta z=z$. We shall investigate only the wavefield extrapolators of integral type that can also be employed for resampling of input data, the so called trace interpolation problem.

The most straightforward integral solution of wave equation (1) can be derived by the Fourier transform technique. By the Fourier transform of the surface data $u(x,y,0,t)$

$$U(k_x,k_y,0,\omega) = \frac{1}{(2\pi)^3} \iiint u(x,y,0,t)e^{i(xk_x+yk_y-i\omega t)}dxdyd\omega \qquad (2)$$

we shall find a solution in the spectral domain in the form

$$U(k_x,k_y,z,\omega) = e^{izk_z} U(k_x,k_y,0,\omega) \qquad (3)$$

with

$$k_z = \mp\sqrt{\frac{\omega^2}{c^2} - (k_x^2 + k_y^2)} . \qquad (4)$$

If not specified otherwise, integration in all formulas will be assumed from $-\infty$ to ∞.

In solving the inverse (direct) problem we engage the upgoing(downgoing) waves specified in (4) by the upper (lower) sign. This convention will be held also in the next. Transforming (3) into the space-time domain (Berkhout 1980), we shall obtain the Rayleigh-Sommerfeld integral solution of a certain boundary value problem - see eq.(9) in the next section.

3. Integral solutions of boundary value problems

The classical approach employs the Green function technique (Morse, Feshbach 1953). Comparison and classification of integral solutions can be found in (Kuhn, Alhilali 1977). For our purposes we recapitulate the main results following from the monograph by Morgan (1983).

Consider the Green function $G(\vec{X},t\,|\,\vec{X}_0,t_0)$ satisfying the inhomogeneous wave equation for the variables with zero indices.

$$\Box_0\, G(\vec{X},t\,|\,\vec{X}_0,t_0) = -4\pi\,\delta(\vec{X}-\vec{X}_0)\delta(t-t_0) . \qquad (5)$$

Then, the Green function describes the propagation of the δ-impulse initiated at $t=t_0$ at the point $\vec{X}=\vec{X}_0$. For the homogeneous free space it holds

$$G_c(\vec{X},t\,|\,\vec{X}_0,t_0) = \delta(\vec{R},t_0-t\mp R/c) \qquad (6)$$

where $\vec{R} = \vec{X} - \vec{X}_o$ and R/c is the traveltime between $\vec{X}_o$ and $\vec{X}$. Under certain assumptions - for details see e.g.(Morgan 1983)- the wavefield $u(\vec{X},t)$ can be expressed using the Green theorem in the form of surface integral

$$u(\vec{X},t) = \frac{1}{4\pi} \int_0^t dt_o \iint_S d\vec{A}_o (G\nabla_o u - u\nabla_o G). \tag{7}$$

The integration is referred to the boundary plane $z=0$ where the surface data are recorded - see Fig.2. Expression (7) represents solution of the homogeneous wave equation (1) for zero initial condition.

Thus, according to (7) the wavefield $u(\vec{X},t)$ can be expressed by means of the Green function and the wavefield, and their normal derivatives at the boundary S. The exact form of the Green function depends on the type of the prescribed boundary condition. The three boundary value problems are summarized in Table 1.

TABLE 1 BOUNDARY CONDITIONS

	Cauchy problem	Dirichlet problem	Neumann problem
Wavefield	free space	free surface	rigid surface
$u(\vec{X}_o,t)$	yes	yes	no
$\nabla_o u(\vec{X}_o,t)$	yes	no	yes
Green function	G_C	G_D	G_N
$G(\vec{X},t\|\vec{X}_o,t_o)$	G_C	0	$2G_C$
$\nabla_o G(\vec{X},t\|\vec{X}_o,t_o)$	$\nabla_o G_C$	$2\nabla_o G_C$	0

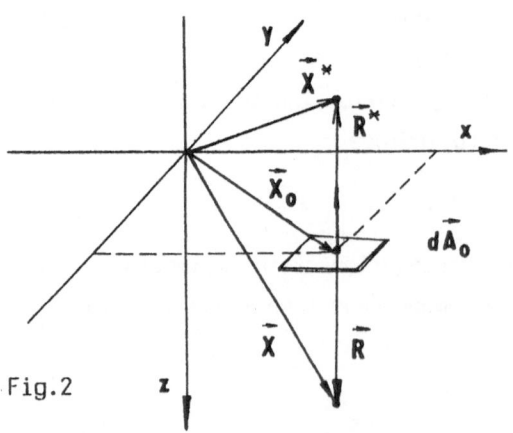

$\vec{X}_o = (x_o,y_o,0)$

$\vec{X} = (x,y,z)$

$\vec{X}^* = (x,y,-z)$

$\vec{R} = \vec{X} - \vec{X}_o, \quad \vec{R}^* = \vec{X}^* - \vec{X}_o$

$R = \sqrt{(x-x_o)^2 + (y-y_o)^2 + z^2}$

Fig.2

The Cauchy problem requires specification of both the value $u(\vec{X}_0, t)$ and of the normal derivative $\nabla_0 u(\vec{X}_0, t)$ of the wavefield at the boundary surfaces, while the Dirichlet and Neumann problems require either the boundary values of the wavefield or its normal derivative.

The remaining two conditions are known as the Rayleigh-Sommerfeld conditions (Rayleigh 1897, Sommerfeld 1912). They introduced the boundary surface S as perfectly reflecting. The Dirichlet problem relates to the free surface condition with the reflecting coefficient $r = -1$ on S, while the Neumann problem establishes the rigid surface condition with $r = +1$.

The Green function for the Cauchy problem is given by (6). The Green functions corresponding to the reflecting boundary surface in the Dirichlet and Neumann problems can be constructed by the method of images (Fig.2).

$$G_D = \delta(\vec{R}, t_0 - t \mp R/c) - \delta(\vec{R}^*, t_0 - t \mp R^*/c) , \tag{8}$$

$$G_N = \delta(\vec{R}, t_0 - t \mp R/c) + \delta(\vec{R}^*, t_0 - t \mp R^*/c) . \tag{9}$$

Evidently,

$$G_C = \frac{1}{2} (G_D + G_N). \tag{10}$$

We used the notations depicted in Fig.2. Useful interrelations between the above Green functions and their normal derivatives are reviewed in Table 1.

Now, let us express the integral solution for the Dirichlet problem. Evaluating the normal derivative $\nabla_0 G_D = 2 \nabla_0 G_C$ and inserting it in (7), we obtain

$$u_D(x,y,z,t) = \frac{1}{2\pi} \iint \left(\frac{z}{R^3} - \frac{z}{R^2} \frac{\partial}{\partial R} \right) u(x_0, y_0, 0, t \pm \frac{R}{c}) \, dx_0 dy_0 \tag{11}$$

The far-field approximation $R\omega/c \gg 1$ yields

$$U_D(x,y,z,t) \approx \frac{\mp 1}{2\pi} \iint \frac{z}{cR^2} u_t(x_0, y_0, 0, t \pm \frac{R}{c}) \, dx_0 dy_0 \tag{12}$$

The upper sign again holds for the upgoing waves. The hyperbolic travel-time t+R/c and the geometric quantities R, x_0, $\cos \phi = z/R$ for $y_0 = 0$ are depicted in Fig.3a. Relation (11) can also be obtained by integral transforms of one-way extrapolator $\exp(izk_z)$ in (3) – see Berkhout (1980). Thus, the spectral domain solution (3) relates to the Dirichlet problem with the prescribed free surface condition.

Without derivation we write the 2-D analogue of (12) – see (Novotný 1990, Berkhout 1980)

$$u_D(x,z,t) \approx d(t) \circledast \int u(x_0, t \pm \frac{R}{c}) R^{-3/2} dx_0 . \tag{13}$$

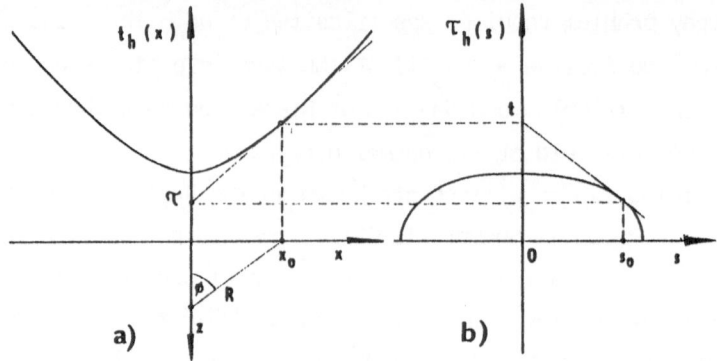

Fig.3 a) Hyperbolic traveltime $t_h(x)$ used in the Kirchhoff integral (12)

b) Legendre transform $\tau_h(p)$ used in the Radon wavefield extrapolation

The convolution operator $d(t)$ is defined by its spectrum $D(\omega)$

$$d(t) = \frac{1}{2\pi} \int D(\omega)e^{-i\omega t}d\omega, \quad D(\omega) = ze^{\pm \frac{i\pi}{4}} \left(\frac{\omega}{2\pi c}\right)^{1/2} . \tag{14}$$

4. Radon wavefield extrapolation

Since the early 1970s, the Radon transform has received increasing attention in the kinematic analysis of reflection and refraction data (Bessonova et al. 1974, Diebold et al. 1981, Stoffa et al. 1981, Milkereit 1987). Schultz and Claerbout (1978) first applied the slant stacking to derive the plane-wave decomposition of reflection wavefields. The thorough mathematical theory for extrapolation of slant-stacked wavefields is due to Chapman (1978) who also recalled the original work by Radon (1917). Later, the Radon transform was applied for seismic migration in horizontally layered media (Levin 1980). Clayton and McMechan (1981) have originally applied Radon migration to velocity analysis of refraction data.

In this section, we recall the basic relations for Radon wavefield extrapolation in 3-D case. According to (Brysk,McGowan 1986) the slant stack of areal seismic data $u(x_0,y_0 0,t)$ can be defined as a simple extension of the 2-D case, i.e.

$$\tilde{u}(p_x,p_y,\tau) = f_2(t) \bullet \iint u(x_0,y_0 0,\tau+x_0 p_x+y_0 p_y)dx_0 dy_0 \tag{15}$$

where $f_2(t)$ denotes the convolution filter

$$f_2(t) = \frac{1}{2\pi} \int F_2(\omega)e^{-i\omega t} d\omega, \quad F_2(\omega) = \frac{\omega^2}{4\pi^2} . \tag{16}$$

The inverse transform is then

$$u_{RT}(x,y,0,t) = \iint \tilde{u}(p_x,p_y,t-xp_x-yp_y)dp_x dp_y . \tag{17}$$

To satisfy the wave equation (1), we must delay each plane-wave component $\tilde{u}$ in (17) by the time

$$\tau_h(p_x,p_y) = \frac{-z \cos \phi}{c} = -zp_z \tag{18}$$

with p_z given by the dispersion relation

$$p_z = \mp \sqrt{1/c^2 - (p_x^2+p_y^2)} . \tag{19}$$

The upper sign again stands for upgoing waves. The delay (18) corresponds to the traveltime across the homogeneous layer $\Delta z = z$ in the direction (p_x, p_y) - see Fig.4.

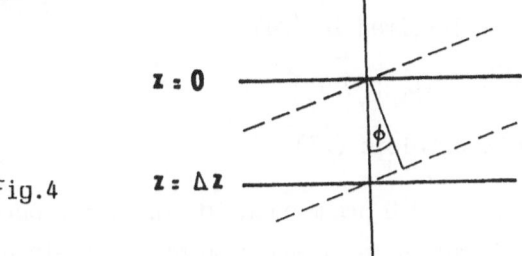

Fig.4

Thus, the wavefield extrapolation through a layer $\Delta z = z$ can be written as the sum of delayed plane-wave components (17)

$$u_{RT}(x,y,z,t) = \iint \tilde{u}(p_x,p_y,t-xp_x-yp_y-zp_z)dp_x dp_y . \tag{20}$$

For completness note that the 2-D analogues can be derived from the previous relations by omitting the p_y component and replacing $F_2(\omega)$ with $F_1(\omega) = |\omega|/(2\pi)$ - see Novotný (1990).

Let us discuss formula (20). The pure plane-wave superposition is involved in (20) for z=0. In this case, the slant stack $\tilde{u}(p_x,p_y,\tau)$ needed in (20) includes the slanting planes $(x_0-x)p_x + (y_0-y)p_y + t$ in (15) with the common point $(x_0,y_0,0)$. With $z > 0$, the term $-zp_z$ in (20) represents the delay time (18). For downward continuation employing the minus sign in (19) we get

$$\tau_h(p_x,p_z) = -zp_z = z\sqrt{1/c^2 - (p_x^2+p_y^2)} \tag{21}$$

or more concisely with $s^2 = p_x^2+p_y^2$

$$\tau_h(s) = z\sqrt{1/c^2-s^2} . \tag{22}$$

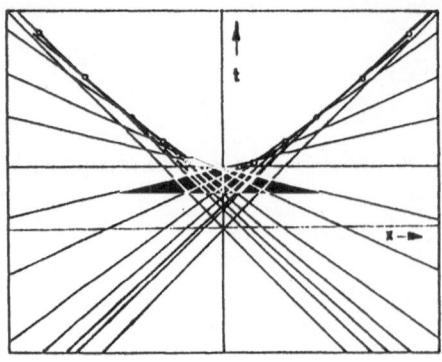 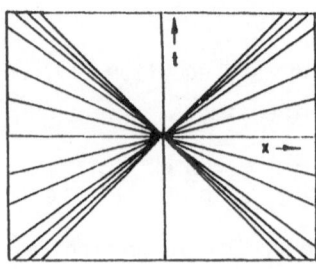

Fig.5

This expression represents intercept times for the rotating hyperboloid (Fig.3b),

$$t_h(x_o,y_o) = \frac{1}{c} \sqrt{(x_o-x)^2 + (y_o-y)^2 + z^2} = \frac{1}{c} R .$$ (23)

Thus, the slanting planes involved in (20)

$$(x_o-x)p_x + (y_o-y)p_y + \tilde{\tau}_h(p_x,p_y) + t = 0$$ (24)

envelope the rotating hyperboloid (23).

The situation is depicted in 2-D projection in Fig.5. The pure plane-wave supersposition can be viewed as the limit case of wavefield extrapolation $z \rightarrow 0$ when the hyperboloid approaches its asymptotic cone.

5. High-frequency asymptotic

The aim of the present paper is to establish the interrelation between the classical integral solutions of the wave equation and of the Radon wavefield extrapolation. The key to such comparison provides the stationary-phase method. In agreement with Papoulis (1968) the following integral may be estimated for ω large enough

$$\iint g(p,q)e^{i\omega\mu(p,q)}dpdq \approx \frac{2\pi i g(p_o,q_o)}{\omega\sqrt{K(p_o,q_o)}} e^{i\omega\mu(p_o,q_o)}$$ (25)

where (p_o,q_o) is the only stationary point S_o in region R with disappearing first derivatives of the phase function $\mu(p,q)$, i.e.

$$\mu_p = 0, \quad \mu_q = 0 .$$ (26)

$K(p_o,q_o)$ stands for the Gauss curvature of the phase function at the stationary point S_o - see (Born and Wolf 1975),

$$K(p_o,q_o) = \mu_{pp}\mu_{qq} - \mu_{pq}^2 .$$ (27)

Approximation (21) holds under the assumption that $\mu_{qq} \neq 0$ and $K(p_0,q_0) \neq 0$ at the stationary point S_0.

Let us apply the above high-frequency asymptotic to the Radon extrapolation formula (20). First let us express the surface wavefield in (15) by its Fourier transform (2). After inserting into the inverse slant-stack and utilizing (16), we have

$$U_{RT}(x,y,z,t) = \frac{1}{(2\pi)^5} \iint dx_0 dy_0 \iiint d\omega dk_x dk_y \, \omega^2 U(k_x,k_y, \,) \cdot$$

$$\cdot e^{i(x_0 k_x + y_0 k_y - i\omega t)} \iint dp_x dp_y \, e^{-i\omega\mu(p_x,p_y)} \tag{28}$$

We denoted $\mu(p_x,p_y)$ the resulting phase function

$$\mu(p,q) = \tau_h(s) + (x_0-x)p + (y_0-y)q \tag{29}$$

with the notation $p = p_x$, $q = p_y$, $s^2 = p^2 + q^2$. For certainty, we again consider the delay time $\tau_h(s)$ according to (22) defined for inverse modelling. From equation (22) we get the stationary point

$$p_o = \frac{x_0-x}{cR}, \qquad q_o = \frac{y_0-y}{cR} \tag{30}$$

where again $R = \sqrt{(x_0 - x)^2 + (y_0- y)^2 + z^2}$ as in (23). Further, at the stationary point $S_0(p_0,q_0)$ we determine

$$s_0 = \frac{1}{c}\left(1 - \frac{z^2}{R^2}\right), \quad \tau_h(s_0) = \frac{z^2}{cR}, \quad \mu(p_0,q_0) = \frac{R}{c} \, . \tag{31}$$

The geometric situation in 2-D projection is illustrated in Fig.3. Ellipse $\tau_h(s)$ represents the Legendre transform of the hyperbola $l_h(x)$.

Further derivation provides

$$K(p_0, q_0) = \mu_{pp}\mu_{qq} - \mu_{pq}^2 = \ldots = \frac{\tau_h'(s_0)\, \tau_h''(s_0)}{s_0} = \frac{c^2 R^4}{z^2} \, . \tag{32}$$

Thus, the asymptotic estimation (25) of the p_x, p_y integration in (28) yields

$$U_{RT}(x,y,z,t) \approx \frac{z}{c(2\pi)^4} \iint \frac{dx_0 dy_0}{R^2} \iiint i\omega U(k_x,k_y,\omega) \cdot$$

$$\cdot e^{i(x_0 k_x + y_0 k_y - i\omega t - R/c)} \, dk_x dk_y d\omega \, . \tag{33}$$

After the inverse Fourier transform we finally get the high-frequency asymptotic of the Radon wavefield extrapolator

$$u_{RT}(x,y,z,t) \approx -\frac{1}{2\pi} \iint \frac{z}{cR^2}\ u_t(x_o,y_o,0,t + \frac{R}{c})dx_o dy_o \qquad (34)$$

This is the exact form of the far-field approximation of the Rayleigh-Sommerfeld integral solution (10) for the Dirichlet problem. The derivation was done for the 3-D case, but it can be transferred to the 2-D case (Novotný 1990).

6. Discussion and conclusions

The Radon wavefield extrapolator exhibits some convenient properties as compared to the conventional integral solutions. First of all, any depth step Δz can be used for the Radon extrapolator while the Kirchhoff type solutions fail for $\Delta z=0$ owing to diverging amplitude factors: R^{-2} or $R^{-3/2}$ in the 3-D case and the 2-D case, respectively. For $\Delta z=0$ the Radon extrapolator yields the familiar plane-wave superposition formula (Treitel et al. 1982, Claerbout 1985).

The Radon extrapolator exhibits another favourable property in processing noisy and spatially undersampled data. The migration algorithms based on Kirchhoff-type summation include just one data sample per input trace. This is obviously insufficient in the presence of noise. The situation is still worse if the summing runs over traces with stepouts close to the spatial Nyquist limit. As a rule, it happens in the boundary region of the migration aperture.

The Radon extrapolator works with more competent representation of noisy data. Instead of one sample per trace the resulting integral (20) comprises the double slant-stack performed along the slanting plane (24). Slant stacking also reduces spatial undersampling of coherent events with "quasi-linear" course over several traces. Thus, steep plane horizons can be readily represented under the assumption that inner spatial aliasing in the integral (20) does not take place. It means that the events should have mutual stepouts smaller than half the dominant period. This was proved in the 2-D case for the trace interpolation problem solved by means of the Radon extrapolator (Novotný 1990).

The revealed relationship of asymptotic representations of Kirchhoff-type and Radon extrapolation gives a theoretical justification to replacing infinite slant-stacks by slant-stacking over a finite trace gather localized at the stationary point. To determine gather sizes, the concept of Fresnel zones can be utilized. Note that finite slant-stacks were introduced before for velocity interpretation of seismic data (McMechan 1983, Estevez 1977, Milkereit 1987). The results of the Radon trace interpolation (Novotný 1990) show that even a narrow slant-stack gather can successfully supply infinite slant-stack in many practical problems.

References

Berkhout,A.J., 1980: Seismic migration - Imaging of acoustic energy by wavefield extrapolation. Amsterdam/New York, Elsevier (north Holland Publ.Co.)

Bessonova,E.N.,V.M.Fishman,V.Z.Ryaboi and G.A.Sitnikova, 1974: The Tau method for inversion of traveltimes - I. Deep sounding data.Geophys.J.R. Astr.Soc.,36,377-398.

Born,M. and E.Wolf, 1975: Principles of optics, 5th ed.,New York,Pergamon Press.

Brysk,H. and D.W.McGowan, 1986: A slant-stack procedure for point-source data,Geophysics 51,1370-1386.

Chapman,C.H., 1978: A new method for computing synthetic seismograms, Geophys.J.R.Astr.Soc. 54,481-518.

Chapman,C.H., 1981: Generalized Radon transform and slant stacks,Geophys. J.R.Astr.Soc. 66,445-453.

Claerbout,J.F., 1985: Imaging the Earth's Interior. Blackwell scientific publications.

Clayton,R.W. and A.McMechan, 1981: Inversion of reflection data by wavefield continuation. Geophysics 46,860-868.

Diebold,J.B. and P.L.Stoffa, 1981: The traveltime equation, tau-p mapping, and inversion of common midpoint data. Geophysics 46,238-254.

Estevez,R., 1977: Slant stacks and interval of optimum stacking. SEP Report No.11,Stanford University,Stanford,California.

Hubral,P., 1981: Slant-stack migration. In Festschrift Theodor Krey, Prakla-Seismos,Hannover,72-78.

Kuhn,M.J. and K.A.Alhilali, 1977: Weighting factors in the construction and reconstruction of acoustical wavefields. Geophysics 42,6,1183-1198.

Levin,S., 1980: A frequency-dip formulation of wave-theoretic migration in stratified media. Acoustical Imaging 9,Plenum Press,681-697.

McMechan,G.A., 1983: P-x imaging by localized slant-stacks of T-x data, Geophys.J.R.Astr.Soc. 72,213-221.

Milkereit,B., 1987: Decomposition and inversion of seismic data - an instantaneous slowness approach. Geophysical prospecting 35,875-894.

Morgan,T.R., 1983: Foundations of wave theory for seismic exploration.Boston, International Human Resources Development Corporation.

Morse,P.M. and H.Feshbach, 1953: Methods of Theoretical Physics, New York. McGraw-Hill Book company.

Müller,G., 1971: Direct inversion of seismic observations.Zeits.Geophys., 37,225-235.

Novotný,M., 1990: Trace interpolation by slant-stack migration.Geophysical Prospecting, to be published.

Papoulis,A., 1968: Systems and transforms with applications in optics.New York,McGraw-Hill Book Company.

Radon,J., 1917: Über die Bestimmung von Funktionen durch ihre Integralwerte längs gewisser Mannigfaltigkeiten.Ber.Vehr.Akad.Wiss.,69,262-277.

Rayleigh, 1897: On the passage of waves through apertures in plane screens and allied problems.Phil.Mag. 43,259.

Sommerfeld,A., 1912: Die Greensche Funktion der Schwingungsgleichungen. Jahresber.Deut.Math.Ver.,21,309.

Schultz,P.S. and J.F.Claerbout, 1978: Velocity estimation and downward continuation by wavefront synthesis,Geophysics,43,691-714.

Stoffa,P.L., P.Buhl,J.B.Diebold and F.Wenzel, 1981: Direct mapping of seismic data to the domain of intercept time and ray parameter - A plane wave decomposition.Geophysics 46,255-267.

Treitel,S., P.R.Gutowski and D.E.Wagner, 1972: Plane-wave decomposition of seismograms.Geophysics 47,1375-1401.

Tygel,M. and P.Hubral, 1989: Constant velocity migration in the various guises of plane-wave theory, to be published in Surveys of Geophysics, Kluwer Academic Publishers.

An Integral Equation Method for Seismic Modelling and Inversion

H. Freter

Institut für Geophysik, Bundesstraße 55, D-2000 Hamburg 13, Germany

Abstract

In this paper a Fredholm integral equation of the second kind for the Green's function associated to a heterogenous medium is derived. This approach is based on the idea to interpret the space dependent propagation speed of the wave equation as a perturbation of a constant reference velocity. The integral equation can be solved by standard quadrature methods. Once the Green's function is calculated, the seismic response to an arbitrary source function can be calculated by a simple convolution. Also, since the calculations are performed in the frequency domain, an incorporation of attenuation mechanisms can be applied easily. Some numerical examples for one-dimensional acoustic and viscoacoustic media are presented. The derived integral equation can also be interpreted as first kind integral equation for the perturbation potential. Thus it can be applied to the inverse problem as well. In order to treat the ill-posedness of this problem it is necessary to apply certain regularization methods such as truncated singular value decomposition or Tikhonov regularization. Some numerical results for Born inversion applied to synthetic data obtained by the integral equation modelling are presented.

1. Introduction

At present the more common methods for calculating exact seismograms for laterally inhomogeneous media are finite difference techniques (e.g.

Dablain, 1986), pseudo-spectral methods like the Fourier method (e.g. Fornberg, 1987) or finite element techniques (e.g. Marfurt, 1984). In all these cases numerical solutions for the equations of motion are calculated for a given particular source function.

In practical seismic forward modelling where the exact wavelet used in field experiments is unknown it is sometimes necessary to repeat model computations with different appropriate estimations of the source function in order to achieve a congruency between real and synthetic data.

We suggest an approach which is based on the calculation of the Green's function in the frequency domain. Thus the calculation of seismic responses for different source functions can be reduced to a simple convolution of the Green's function and the considered wavelet. Hence it is not necessary to repeat the entire numerical procedure for a different source function once the Green's function for the medium is known.

The main objective of this paper is to describe the derivation of a Fredholm integral equation of the second kind for the Green's function associated to the wave equation with a space dependent velocity. This derivation is based on the idea to interpret the space dependent propagation speed as a perturbation of a constant reference velocity. The solution of the integral equation can be calculated by solving a linear system which has been obtained by using a quadrature rule. Numerical examples for one-dimensional heterogenous media are presented.

A generalization to viscoacoustic media is possible since all calculations are performed in the frequency domain. In particular, an application of the theory of Liu et al. (1976) leads to a model rheology with a constant Q factor. Numerical examples are presented.

Given the Green's function of the heterogenous medium at the surface, the derived integral equation can be regarded as Fredholm integral equation of the first kind for the perturbation potential. Such problems are known to

244

be ill-posed and hence the system of linear equations received by applying a quadrature rule becomes ill-conditioned. In order to obtain appropriate results the problem has to be regularized. We apply the two well-known methods of truncated singular value decomposition and of Tikhonov regularization. Numerical examples for a linearized inversion applied to synthetic data obtained by the integral equation modelling are presented.

2. Green's function

We consider the initial value problem for the wave equation of a one-dimensional heterogeneous medium with constant density:

$$LU(z,t) = -F(z,t), t < 0, -\infty < z < \infty,$$
$$U(z,0) = f(z), U_t(z,0) = g(z), \tag{1}$$

where $U(z,t)$ denotes the acoustic wavefield and $F(z,t)$ is an arbitrary space and time dependent source function. L represents the one-dimensional wave equation operator

$$L = \frac{\partial^2}{\partial z^2} - \frac{1}{c^2(z)} \frac{\partial^2}{\partial t^2}. \tag{2}$$

Using this formulation the vertical incidence of seismic waves to a horizontally layered acoustic medium with constant density can be described.

We note that (1) is a whole space problem, i.e. no surface effects are considered.

The intention is now to produce a general solution of the initial value problem (1) in which the acoustic wavefield $U(z,t)$ is explicit in the initial values $U(z,0) = f(z), U_t(z,0) = g(z)$ and the inhomogeneous term $F(z,t)$.

Let now (ζ, τ) be the arbitrary but fixed point of the space-time-domain where the solution $U(\zeta, \tau)$ of (1) should be determined. The value of U at this point depends only on the values of F inside the domain $D(\zeta, \tau)$ bounded by the two characteristics through the point (ζ, τ) and the initial curve as shown in Fig. 1. $D(\zeta,\tau)$ is called the domain of dependence of the point (ζ,τ). The characteristics can be interpreted as the space-time diagram of a point moving with the velocity $c(z)$. As described e.g. in Courant and Hilbert (1968) it can be shown that the characteristics are represented by the two equations

$$t \pm \int_0^z \frac{dz'}{c(z')} = const. \qquad (3)$$

In the special case of constant velocity $c(z) \equiv c_0$ it is well known that the characteristics are two lines represented by the equations $z \pm c_0 t = \zeta \pm c_0 \tau$. As demonstrated in Bleistein (1984), the following expression for the solution $U(\zeta,\tau)$ can be obtained:

$$U(\zeta,\tau) = \int_{z_1-}^{z_2+} \frac{1}{c^2(z)} \left[G\left(\zeta,\tau;z,0\right) g(z) - G_t\left(\zeta,\tau;z,0\right) f(z) \right] dz$$
$$+ \int_D G\left(\zeta,\tau;z,t\right) F(z,t) dz dt. \qquad (4)$$

Here $G(\zeta,\tau;z,t)$ denotes the Green's function which is defined by the initial value problem

$$LG(\zeta,\tau;z,t) = -\delta(z-\zeta)\delta(t-\tau), t < \tau^+, -\infty < z < \infty,$$
$$G(\zeta,\tau;z,\tau) = 0, G_t(\zeta,\tau;z,\tau) = 0. \qquad (5)$$

It describes the time-reverse response to a δ-perturbation.

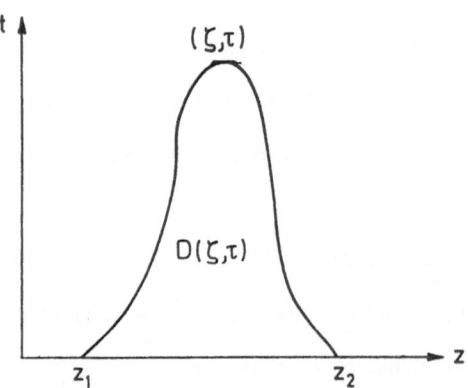

Fig.1 Domain of dependence $D(\zeta,\tau)$.

Due to the homogenity of the wave equation with respect to time it holds

$$G(\zeta, \tau; z, t) = G(\zeta, 0; z, t - \tau). \tag{6}$$

Thus the integral over the domain D in (4) has the form of a convolutional integral with respect to time. By a Fourier transformation of (4) to the frequency domain we obtain for vanishing initial conditions $f = g = 0$ the expression

$$\hat{U}(\zeta, \omega) = \int_{-\infty}^{\infty} \hat{G}(\zeta, z; \omega) \hat{F}(z, \omega) dz. \tag{7}$$

since a convolution in the time domain corresponds to a multiplication in the frequency domain. The hat denotes the Fourier transform with respect to time. In particular, for a point explosion source of the form

$$F(z, t) = \delta(z - z_s) f(t) \tag{8}$$

we obtain

$$\hat{U}(\zeta; \omega) = \hat{G}(\zeta, z_s; \omega) \hat{F}(\omega) \tag{9}$$

where $\hat{F}(\omega)$ denotes the Fourier transform of $f(t)$. Thus, by determining the Green's function $\hat{G}$ associated to the heterogenous medium, the seismic response to an arbitrary source can be calculated. The whole method can be summarized as follows:

(i) Determine the Green's function in the frequency domain (by solving the integral equation derived in the subsequent section)

(ii) Multiply the Green's function by the Fourier transform of the source signal

(iii) Backtransformation to the time domain gives the seismogram

If we take the wave operator with constant velocity $L = L_0 = \frac{\partial^2}{\partial z^2} - \frac{1}{c_0^2} \frac{\partial^2}{\partial t^2}$ and transform (5) to the frequency domain, we obtain

$$\frac{\partial^2 \hat{G}_0}{\partial z^2} + \frac{\omega^2}{c_0^2} \hat{G}_0 = -\delta(z - \zeta), \tag{10}$$

247

where $\hat{G}_0(\zeta, z; \omega)$ denotes the Fourier transform of the Green's function $G_0(\zeta, \tau; z, t)$ corresponding to the operator L_0. $\hat{G}_0$ can be expressed explicitly (e.g. Stakgold, 1979):

$$\hat{G}_0(\zeta, z; \omega) = \frac{c_0}{2i\omega} e^{-i\omega \frac{|z - \zeta|}{c_0}}. \tag{11}$$

We observe that $\hat{G}_0$ only depends on the difference $|z - \zeta|$ as it is true in general for linear partial differential equations with constant coefficients (Gustafson, 1987). Hence, in addition to (6), it also holds

$$\hat{G}_0(\zeta, z; \omega) = \hat{G}_0(\zeta - z, 0; \omega). \tag{12}$$

In the following section we shall relate the solution of the problem with space dependent velocity to the analytically known solution of the one with constant velocity.

3. Fredholm integral equation of the second kind for G

Based on the theory presented in the previous section we now derive an integral equation for the Green's function associated to the one-dimensional wave equation for a heterogenous medium.

Let $G(z_s, 0; z, t)$ be the solution of the initial value problem

$$LG = -\delta(z - z_s)\delta(t),$$
$$G(z_s, 0; z, 0) = G_t(z_s, 0; z, 0) = 0 \tag{13}$$

where z_s denotes the position of the source.

This corresponds with the equation

$$L_0 G = -\delta(z - z_s)\delta(t) + \frac{\alpha(z)}{c_0^2} \frac{\partial^2 G}{\partial t^2}, \tag{14}$$

which follows by writing

$$\frac{1}{c^2(z)} = \frac{1}{c_0^2}(1 + \alpha(z)) \tag{15}$$

(see e.g. Cohen and Bleistein, 1979), i.e. we consider the spatially varying propagation speed as a perturbation of a constant reference velocity.

After a transformation of (14) to the frequency domain we receive

$$\frac{\partial^2 \hat{G}}{\partial z^2} + \frac{\omega^2}{c_0^2}\hat{G} = -\delta(z - z_s) - \frac{\omega^2}{c_0^2}\alpha(z)\hat{G}, \qquad (16)$$

where $\hat{G}(z, z_s; \omega)$ is the Fourier transform of $G(z_s, 0; z, t)$. By means of Equation (7) we obtain

$$\hat{G}(z, z_s; \omega) = \int\limits_{-\infty}^{\infty} \hat{G}_0(z, z'; \omega)\left(\delta(z' - z_s) + \frac{\omega^2}{c_0^2}\alpha(z')\hat{G}(z', z_s; \omega)\right) dz'$$

$$= \hat{G}_0(z, z_s; \omega) + \frac{\omega^2}{c_0^2}\int\limits_{-\infty}^{\infty} \alpha(z')\hat{G}_0(z - z', 0; \omega)\,\hat{G}(z', z_s; \omega)\, dz'. \qquad (17)$$

For simplicity we consider only the case $z_s = 0$ and write $\hat{G}_0(z; \omega)$ and $\hat{G}(z; \omega)$ instead of $\hat{G}_0(z, 0; \omega)$ and $\hat{G}(z, 0; \omega)$, respectively. Thus we obtain a Fredholm integral equation of the second kind for $\hat{G}(z; \omega)$:

$$\hat{G}(z; \omega) = \hat{G}_0(z; \omega) + \frac{\omega^2}{c_0^2}\int\limits_{-\infty}^{\infty} \alpha(z')\hat{G}_0(z - z'; \omega)\,\hat{G}(z'; \omega)\, dz'. \qquad (18)$$

Note that this integral equation has to be solved for each frequency. Doing this, the Green's function $\hat{G}(z; \omega)$ associated to a heterogenous medium can be calculated. It may be pointed out that this approach can be generalized to higher dimensions.

4. Numerical solution and examples

A simple method for the solution of integral equations is the so-called Nystrom method (e.g. Delves and Walsh, 1974). Assuming $\alpha(z) \equiv 0$ for $z < 0$ and $z > z_{max}$ with z_{max} sufficiently large, the limits of integration become finite. The interval $[0, z_{max}]$ is divided into $N - 1$ subintervals

$$[z_k, z_{k+1}], k = 1, \ldots, N - 1, z_k = (k - 1)h$$

of equal length $h = \frac{z_{max}}{N-1}$.

Rewriting the integral equation (18) for fixed ω as

$$u(z) = g(z) + \int\limits_{-\infty}^{\infty} K(z, z')u(z')dz', \qquad (19)$$

249

where
$$u(z) = \hat{G}(z;\omega),$$

$$g(z) = \hat{G}_0(z;\omega), \tag{20}$$

$$K(z,z') = \frac{\omega^2}{c_0^2}\alpha(z')\hat{G}_0(z-z';\omega),$$

and applying a quadrature rule of the type

$$\int f(z)dz \approx \sum_{m=1}^{N} w_m f_m \tag{21}$$

we obtain the following linear system of equations with the N unknowns u_i:

$$u_i = g_i + \sum_{j=1}^{N} K_{ij}w_j u_j, \qquad i = 0, 1, \ldots, N-1, \tag{22}$$

where
$$u_i = u(z_i),$$

$$g_i = g(z_i), \tag{23}$$

$$K_{ij} = K(z_i, z_j).$$

This system of equations can be solved by Gaussian elimination using standard library routines.

For the following three examples we calculate the seismic response by a convolution of the numerically generated Green's function $G(z;\omega)$ with a Ricker wavelet

$$\hat{F}(\omega) = \frac{\sqrt{2\pi}}{\kappa^{\frac{3}{2}}}\omega^2 e^{-\frac{\omega^2}{2\kappa}+i\omega t_s} \tag{24}$$

as the souce function; compare Sheriff and Geldart (1983,p.125), where $\kappa := 2\pi^2\nu_M^2$.

ν_M denotes the dominant frequency and has been chosen to be $20 Hz$, t_s is an arbitrary time shift which has been chosen to be 0.4 s.

The shape of the seismic response in the 1D case is expected to have the form of the integral of the source function (Chester, 1971). The seismograms computed by the integral equation method are represented by continuous lines. For comparison seismograms computed by the reflectivity method are shown in dashed lines. Source and receiver are both located at $z = 0$, i.e. on the "surface".

250

In the first example (Fig. 2a) we chose the lithological parameters covering a sufficiently large velocity contrast. We consider a layer of thickness 400 m with a high velocity contrast of 2000 m/s : 4000 m/s giving a reflection coefficient of 1/3. In Fig. 2b a comparison of synthetic seismograms calculated by the integral equation approach (continuous line) and the reflectivity method (dashed line) is shown. We see a reflection from the top of the layer with correct reflection coefficient, a reflection from the bottom of the layer with slightly smaller amplitude due to transmission losses, and a first order multiple. The second example (Fig. 3a) displays a layered medium consisting of five reflectors which is expected to produce a relatively complex wave field (Fig. 3b). We see five reflections each of which belongs to one of the five layers. Finally, in the third example a thin layer is considered (Fig. 4a). Here the well-known property of thin layers to act as a differentiator to the signal is reproduced (Fig. 4b).

All results of the presented three examples show an excellent agreement between the integral equation method and the reflectivity method.

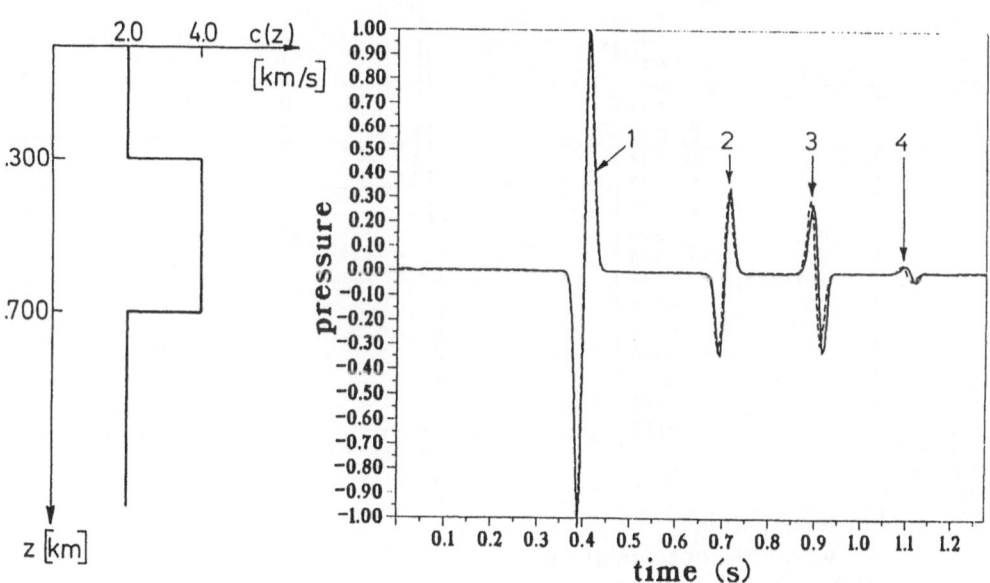

Fig.2 Two plane reflector model
(a) Velocity-depth relation
(b) Synthetic seismograms

(1=direct wave; 2,3= reflections from the upper and lower interface; 4 = multiple).

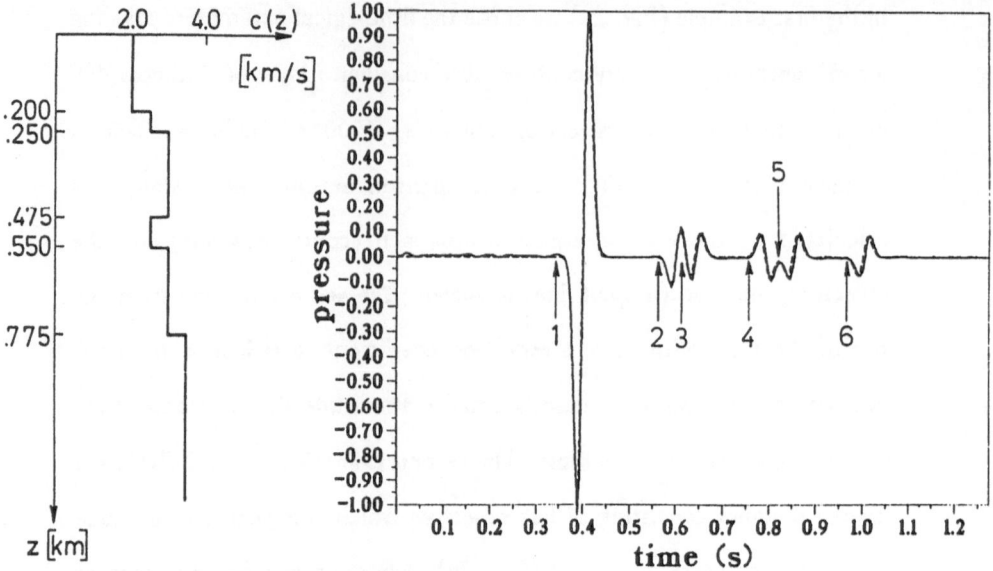

Fig. 3 Layered medium with five reflectors
    (a) Velocity-depth relation
    (b) Synthetic seismograms
        Five reflections (2) – (6) after the direct wave (1)

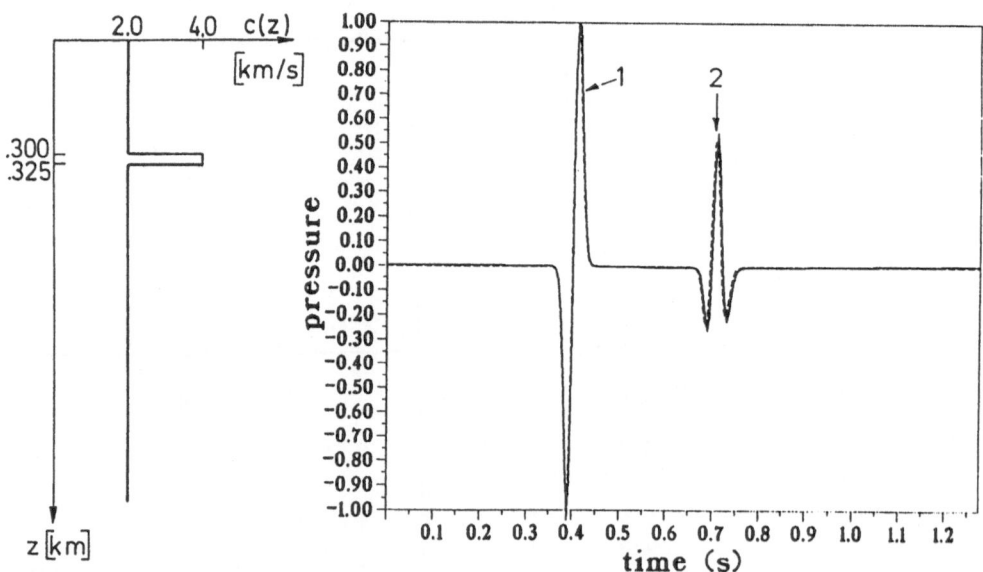

Fig. 4 Thin layer model
    (a) Velocity-depth relation
    (b) Synthetic seismograms

        The heterogenity acts as a differentiator (2) on the in
        cident wave (1)

5. Generalization to viscoacoustic media

A simulation of wave propagation in a viscoacoustic medium should be able to reproduce the effects of attenuation and dispersion. In seismics it is of particular importance to model a material rheology which gives an approximately constant Q factor in the seismic frequency band. As it has been shown by Liu et al. (1976) this can be achieved by a generalized standard linear solid rheology. With a suitable choice of material parameters, constant Q values can be obtained (Carcione et al., 1988).

It would be beyond the scope of this paper to review this theory in detail. This can be found in the paper of Carcione et al. (1988). Here we just state the result.

The frequency dependent complex phase velocity can be expressed as

$$c(\omega) = c_0(1 - L + \sum_{l=1}^{L} \frac{1 + i\omega\tau_{\epsilon l}}{1 + i\omega\tau_{\sigma l}})^{1/2} \tag{25}$$

where $\tau_{\sigma l}(z)$ and $\tau_{\epsilon l}(z)$ denote material relaxation times of the lth mechanism and L is the number of relaxation mechanisms. The $\tau_{\sigma l}, \tau_{\epsilon l}$, $l=1,...,L$ have to be chosen appropriately in order to simulate a constant Q factor (Carcione et al., 1988).

We now start from Equ. (16) where, applying the correspondence principle (Carcione et al., 1988), we substitute c_0 by $c(\omega)$ using (25). In order to be able to apply (7) we put all contant terms to the left and all perturbation and frequency dependent terms to the right. Thus we get the equation

$$\frac{\partial^2 \hat{G}}{\partial z^2} + \frac{\omega^2}{c_0^2}\hat{G} = -\delta(z - z_s) - \frac{\omega^2}{c_0^2} \frac{\alpha(z) + L - \sum_{l=1}^{L} \frac{1 + i\omega\tau_{\epsilon l}}{1 + i\omega\tau_{\sigma l}}}{1 - L + \sum_{l=1}^{L} \frac{1 + i\omega\tau_{\epsilon l}}{1 + i\omega\tau_{\sigma l}}}\hat{G}. \tag{26}$$

Applying (8) we obtain the integral equation

$$\hat{G}(z,\omega) = \hat{G}_0(z,\omega) + \frac{\omega^2}{c_0^2} \int_{-\infty}^{\infty} \beta(z';\omega)G_0(z - z';\omega)G(z';\omega)dz' \tag{27}$$

where

$$\beta(z;\omega) = \frac{\alpha(z) + L - \sum_{l=1}^{L} \frac{1 + i\omega\tau_{\epsilon l}(z)}{1 + i\omega\tau_{\sigma l}(z)}}{1 - L + \sum_{l=1}^{L} \frac{1 + i\omega\tau_{\epsilon l}(z)}{1 + i\omega\tau_{\sigma l}(z)}} \tag{28}$$

denotes the frequency dependent perturbation potential. For simplicity, as in the derivation of (18), we only considered the case $z_s = 0$.

This integral equation can be solved by the same method as (18). We consider two numerical examples with Q=100 and Q=15, respectively (Fig. 5). The two effects of viscous media on wave propagation, attenuation and dispersion, can be seen clearly by a comparison of the seismic response to the viscoacoustic medium (continuous line) with the one to a pure acoustic medium (dashed line). The response to the viscous medium is damped and faster due to the frequency dependent phase velocity.

6. Integral equation inversion

The integral equation (18) has been derived for the treatment of the direct problem. But it can also be interpreted as Fredholm integral equation of the first kind for the perturbation potential $\alpha(z)$.

Observing the seismograms at $z = 0$, (19) simplifies to

$$\frac{\omega^2}{c_0^2} \int_{-\infty}^{\infty} \alpha(z') \hat{G}_0 (-z'; \omega)\, \hat{G} (z'; \omega)\, dz' = \hat{G}(0; \omega) - \hat{G}_0 (0; \omega). \qquad (29)$$

This integral equation is nonlinear since $\hat{G}(z; \omega)$ depends on $\alpha(z)$ by means of (19). Inserting the Born approximation ($\hat{G} \approx \hat{G}_0$) to the integrand leads to the linear first kind integral equation

$$\frac{\omega^2}{c_0^2} \int_{-\infty}^{\infty} \alpha(z') \hat{G}_0^2 (z'; \omega)\, dz' = \hat{G}(0; \omega) - \hat{G}_0 (0; \omega). \qquad (30)$$

Using the same assumptions for α as in Section 4, applying a quadrature rule like (21) and collocating at M frequencies $\omega_1,...,\omega_M$ leads to an $M \times N$ system of linear equations (M frequencies, N integration points)

$$G\underline{\alpha} = \underline{g} \qquad (31)$$

where

$$G_{mn} = \frac{\omega_m^2}{c_0^2} w_n \hat{G}_0^2 (z_n; \omega_m), m = 1, \ldots, M, n = 1, \ldots, N,$$

$$\alpha_n = \alpha(z_n), n = 1, \ldots, N, \qquad (32)$$

$$g_m = \hat{G}(0; \omega_m) - \hat{G}_0(0; \omega_m), m = 1, \ldots, M.$$

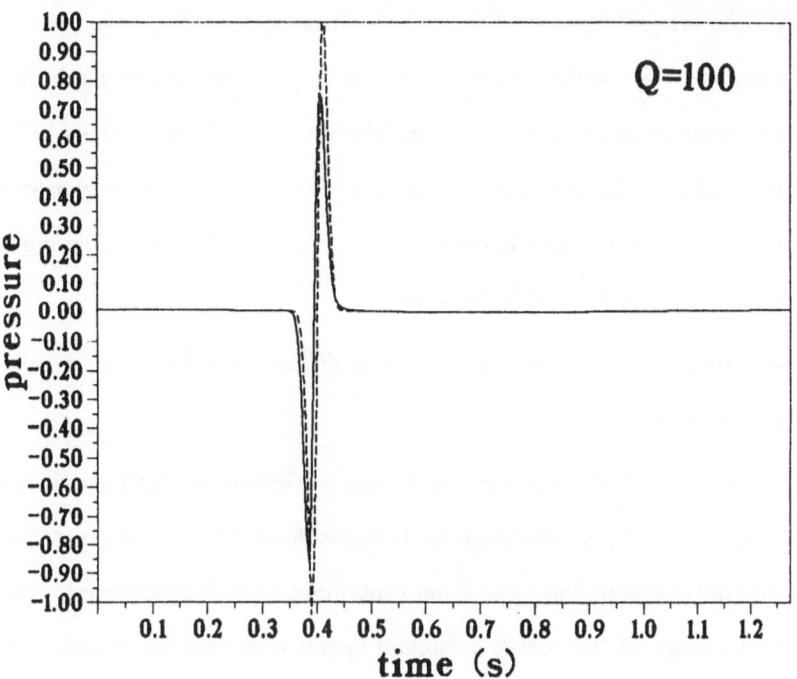

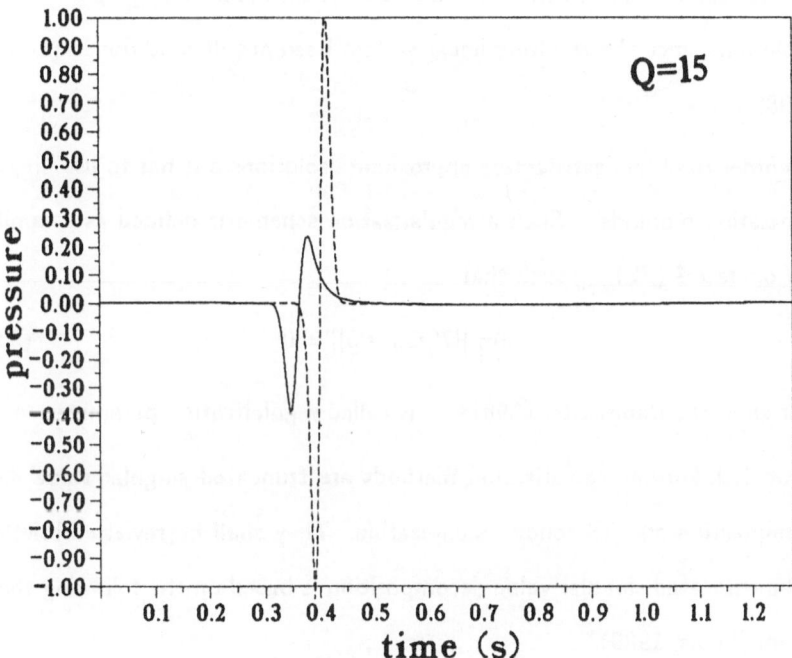

Fig. 5 Viscoacoustic medium
$(a)Q = 100$
$(b)Q = 15$

Both the integral equation (30) and the linear system (31) may not have a solution. We therefore have to look for a "best-approximate solution" which leads to the concept of a generalized solution (Baumeister (1987), Louis (1989)). The generalized solution $\alpha^\dagger$ is the least squares solution of minimal norm of the equation $G\underline{\alpha} = \underline{g}$, i.e. it has minimal norm among all minimizers of the residual $\|G\underline{\alpha} = \underline{g}\|$.

The corresponding operator $G^\dagger$ which is defined by $G^\dagger g = \alpha^\dagger$ is called generalized inverse of G.

It is well-known that the problem of solving a Fredholm integral equation of the first kind with square-integrable kernel is ill-posed, i.e. the generalized inverse of the integral operator is not continuous. This follows from the fact that the range of the compact integral operator defined by the left hand side of (30) is not closed if it has infinite dimension (Groetsch, 1977).

Hence simple discretization methods do not give satisfactory approximate solutions since the resulting linear system becomes ill-conditioned (Louis, 1989).

In order to obtain satisfactory approximate solutions one has to use "regularization methods". Such a regularization scheme is defined as a family of operators $(T_\gamma)_{\gamma > 0}$ such that

$$\lim_{\gamma \to 0} \|T_\gamma Gg - g\| = 0, \tag{33}$$

for all g, see Baumeister (1987). γ is called regularization parameter.

Two well-known regularization methods are truncated singular value decomposition and Tikhonov regularization. They shall be reviewed briefly. The truncated singular value decomposition is based on the following theorem (Louis, 1989).

Let G be a $M \times N$ matrix. Then there exists a unitary $M \times M$ matrix U and a unitary $N \times N$ matrix V such that

$$G = U \begin{bmatrix} S & 0 \\ 0 & 0 \end{bmatrix} V^* \tag{34}$$

with $S = diag(\sigma_1, ..., \sigma_r)$ where σ_i are the singular values of G, defined by

$$Gv_i = \sigma_i u_i$$

$$i = 1, \ldots, r, \tag{35}$$

$$G^* u_i = \sigma_i v_i$$

where u_i and v_i denote the column vectors of the matrices U and V, respectively, and r denotes the rank of G. From this definition it can be seen that the singular values are the square roots of the positive eigenvalues of G^*G or GG^*, respectively, while the vectors u_i and v_i are the eigenvectors of these matrices.

The generalized inverse can be expressed as

$$G^\dagger g = \sum_{\sigma_n > 0} \frac{1}{\sigma_n} < g, u_n > v_n, \tag{36}$$

see Louis (1989). From this expression it can be seen that small singular values may amplify corresponding errors in the data. This leads to the idea of truncated singular value decomposition which is defined by the operator T_γ by means of

$$T_\gamma g = \sum_{\sigma_n > \gamma} \frac{1}{\sigma_n} < g, u_n > v_n. \tag{37}$$

It can be shown that the truncated singular value decomposition defines a regularization of $G^\dagger$ (e.g. Louis, 1989) with regularization parameter $\gamma > 0$.

The other widely used method is Tikhonov regularization. Here α_γ is determined by minimizing the functional

$$F_\gamma(\alpha) = ||G\alpha - g||^2 + \gamma^2 ||\alpha||^2. \tag{38}$$

This defines a regularization of $G^\dagger$ (e.g. Louis, 1989) with regularization parameter $\gamma > 0$.

7. Numerical Examples

In order to test the inversion method synthetic data are calculated using the integral equation method. Thus the values $\hat{G}(0; \omega)$ which are needed for the right hand side of the integral equation (30) are obtained. The kernel of the integral equation is given via the explicit expression (11).

In order to find the regularized solution $\alpha^\dagger$ via the truncated singular value decomposition the eigenvalues and eigenvectors of the matrix G^*G are determined using standard library routines. Then the expression (37) for the truncated expansion is used. Eventually the velocity-depth relation can be reconstructed using the definition of the perturbation potential (15).

The regularization parameter γ is found by trial and error. In this context this is an appropriate method since the results can be immediately viewed and optimized until they appear reasonable.

For obtaining the Tikhonov regularized solution, i.e. for minimizing the functional (38) the system of normal equations

$$(G^*G + \gamma^2 I)\alpha_\gamma = G^*g \tag{39}$$

has to be solved (Louis, 1989). Again the regularization parameter is found by trial and error.

Both the regularization by truncated singular value decomposition and by Tikhonov's method are tested on synthetic data for three model cases. They cover a structure of four layers with a very low velocity contrast, a layer with 10% velocity contrast and a stair stepping velocity-depth relationship used recently by Rajan and Frisk (1989).

The results are shown in Figs. 6 - 8 where the continuous line shows the velocity function to be reconstructed and the dashed line shows the reconstructed function by truncated singular value decompostion (left) and by Tikhonov regularization (right), respectively.

In all cases reasonable results are obtained taking into account that the underlying integral equation has been received by linearization.

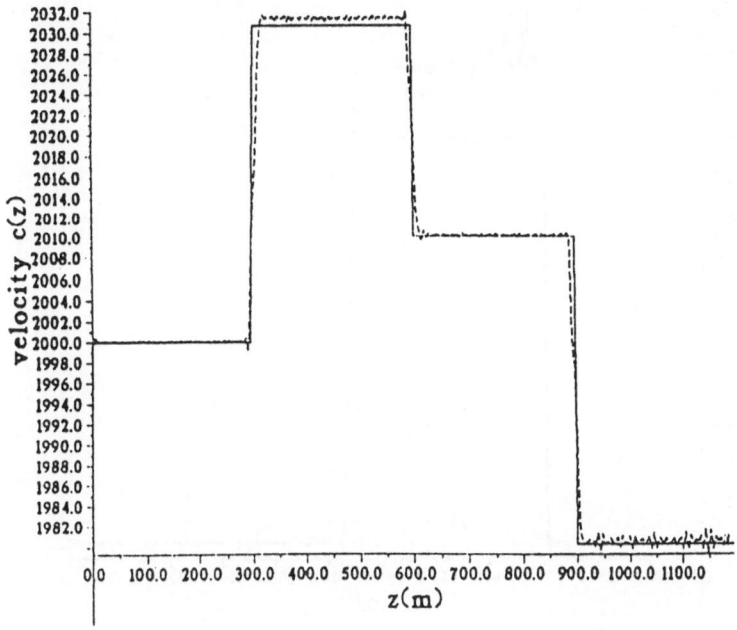

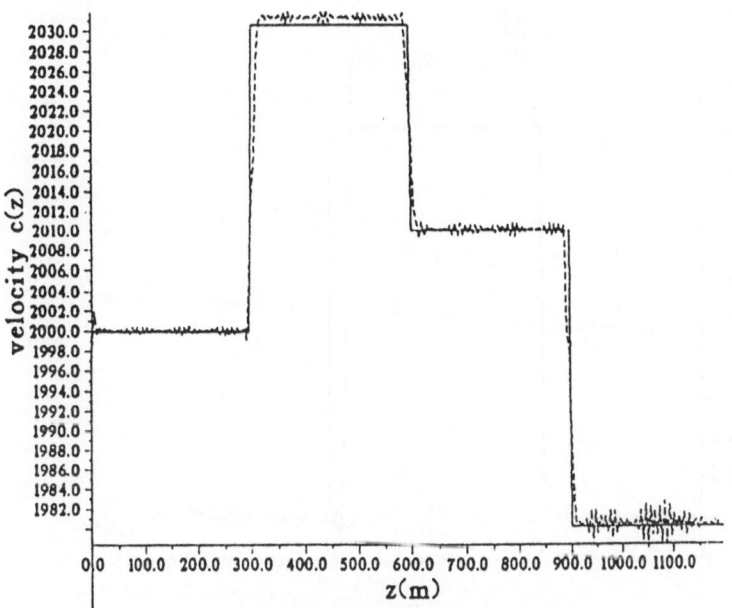

Fig. 6 Inversion of a small velocity contrast
(*a*) Truncated singular value decomposition
(*b*) Tikhonov regularization

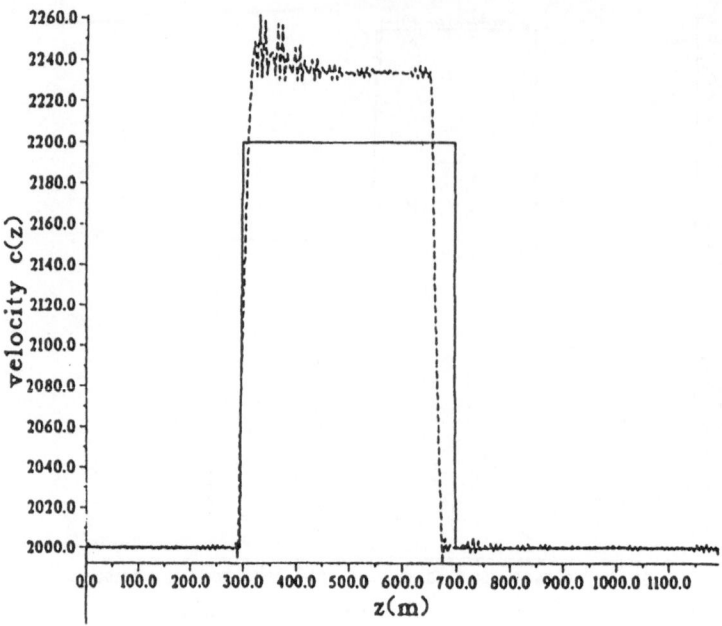

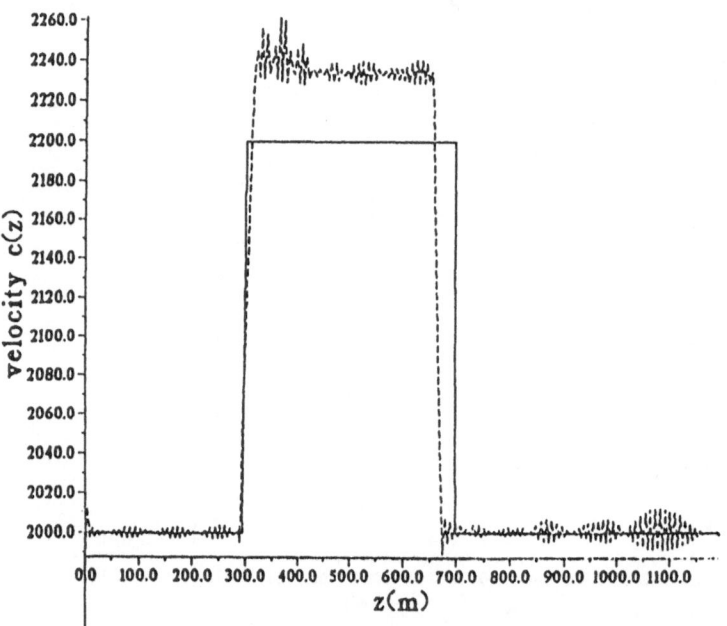

Fig. 7 Inversion of a 10% velocity contrast
(a) Truncated singular value decomposition
(b) Tikhonov regularization

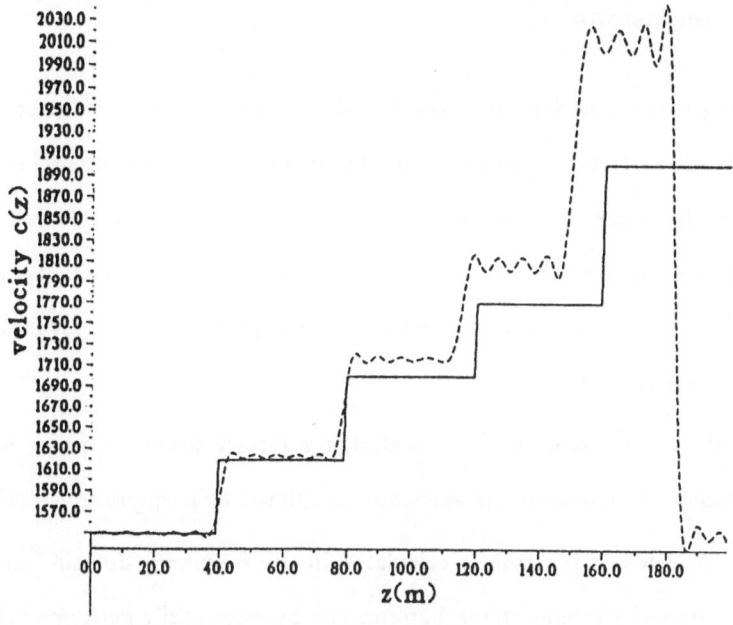

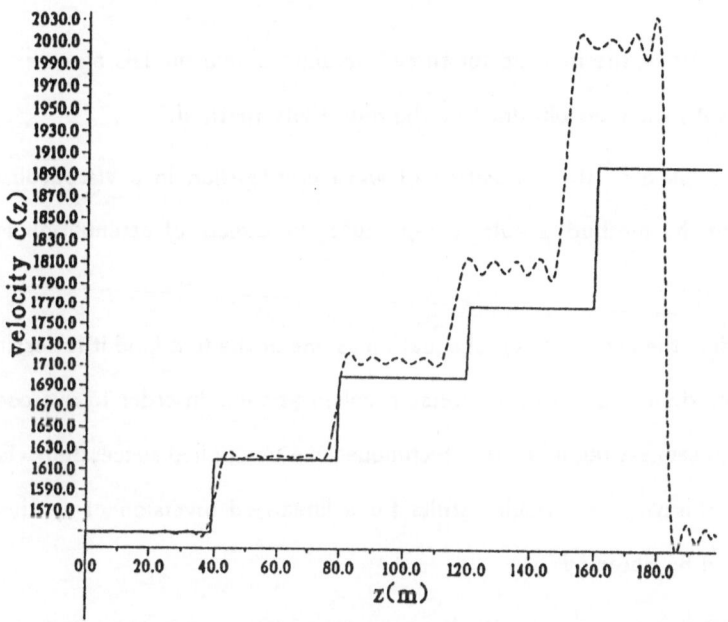

Fig. 8 Inversion of a stair stepping velocity function
(*a*) Truncated singular value decomposition
(*b*) Tikhonov regularization

8. Conclusions

In this paper a new seismic forward modelling method has been presented which calculates the seismic response by solving a Fredholm integral equation of the second kind for the Green's function. The derivation of this equation is based on the idea to interpret the space dependent propagation speed as a perturbation of a constant reference velocity. The advantages of this approach are:

Once the Green's function is calculated an arbitrary seismic response for a given source function can be efficiently calculated by a simple convolution.

Since the Green's function is calculated in the frequency domain, an incorporation of attenuation mechanisms can be more easily performed than in approaches which produce solutions of the wave equation in the time domain.

The results of the method for three one-dimensional models agree excellently with the ones obtained by the reflectivity method.

When applied to the simulation of wave propagation in a viscoacoustic medium the method is able to reproduce the effects of attenuation and dispersion.

Regarding the derived integral equation as one of the first kind it is possible to apply the method to the inverse problem as well. In order to overcome the ill-posedness regularization techniques can be applied successfully. Following this way reasonable results for a linearized inversion of synthetic data can be obtained.

In future it is intended to develop an iterative inversion scheme by a combination of the integral equation modelling with the integral equation inversion. In this method the approximations for the velocity-depth function are used to calculate new approximations for the Green's function which are better than the simple Born approximation used here.

Also some investigations on an automatic choice of the regularization parameter should be performed as an alternative to the trial and error method used so far.

References

Baumeister, J., 1987, *Stable solution of inverse problems*, Vieweg Verlag, Braunschweig.

Bleistein, N., 1984 *Mathematical methods for wave phenomena*, Academic Press, Orlando.

Carcione, J., Kosloff, D., and Kosloff, R., 1988, Wave propagation simulation in a linear viscoacoustic medium, Geophysical Journal, 93, 393-407.

Chester, R., 1971, *Techniques in partial differential equations*, McGraw Hill, New York.

Cohen, J.K. and Bleistein, N., 1979, Velocity inversion procedure for acoustic waves, Geophysics 44, 1077-1087.

Courant, R. and Hilbert, D., 1968, *Methoden der mathematischen Physik, Vol.2*, Springer Verlag, Berlin.

Dablain, M.A., 1986, The application of high-order differencing to the scalar wave equation, Geophysics, 51, 54-66.

Delves, L. and Walsh, J., 1974, *Numerical methods for integral equations*, Clarendon Press, Oxford.

Fornberg, B., 1987, The pseudospectral method: Comparisons with finite differences for the elastic wave equation, Geophysics 52, 483-501.

Groetsch, C.W., 1977, *Generalized inverses of linear operators*, Marcel Dekker Inc., New York.

Gustafson, K.E., 1987, *Introduction to partial differential equations and Hilbert space methods*, 2nd edition, John Wiley and Sons, New York.

Liu, H.P., Anderson, D.L., and Kamamori, H., 1976, Velocity dispersion due to anelasticity; implications for seismology and mantle composition, Geophys. J. R. astr. Soc., 47, 41-58.

Louis, A.K., 1989, *Inverse und schlecht gestellte Probleme*, Teubner Verlag, Stuttgart.

Marfurt, K.J., 1984, Accuracy of finite-difference and finite-element modelling of the scalar and elastic wave equation, Geophysics, 49, 533-549.

Rajan, S.D., and Frisk, G.V., 1989, A comparison between the Born and Rytov approximation for the inverse backscattering problem, Geophysics, 54, 864-887.

Sheriff, R.E. and Geldart, L.P., 1983, *Exploration seismology, Vol.1*, Cambridge University Press, Cambridge.

Stakgold, I., 1979, *Green's functions and boundary value problems*, John Wiley, New York.

Inverse Scattering for Goupillaud Horizontally Layered Earth Model

V. Bardan

Computer Center of IPGG, Str. Coralilor 20, Sect. 1, Bucharest 78449, Rumania

Abstract

The Goupillaud horizontally layered earth model and its inverse scattering problem is considered in this paper: the determination of the reflection coefficents from an excitation-response pair.This problem is the prototype for the inverse scattering problems arising in various fields, obtaining the equations of the inverse scattering problem and mentioning that these equations are direct consequences of the causality property of the signal propagation model.The discrete analogues of the classical equations of Gelfand-Levitan, Marchenko and Krein are obtained with special choices of scattering data pairs for free-surface and non-free-surface reflections.

With the Goupillaud model, it can be seen that an efficient layer-recursive inverse scattering procedure arises more directly than do the linear equations; this result only invokes the causality of signal propagation and a simple rule for recursively computing the discrete waveforms at increasing depths in the scattering medium.This approach provides fast recursive inversion procedures, that at the beginning were called dynamic predictive deconvolutions and afterwards direct layer-peeling or Schur-type algorithms.

1. INTRODUCTION

The inverse problem in geophysics is to determine the subsurface makeup on the basis of wave motion observed at the surface of the earth. In many other real-world problems, it must also infer the size, shape, and texture of remote objects via the "scatter" traveling waves. For this reason, beside geophysicists, numerous physicists, electrical engineers and mathematicians have studied a wide range of inverse scattering problems. It turns out that inverse spectral problems for Schrodinger operators and vibrating strings, algorithms for transmission-line synthesis, determination of the vocal tract area function in speech research, the design of digital filters in cascade form, the derivation of fast lattice-form linear least-squares prediction algorithms and, of course, the identification of the acoustic impedance in layered-earth models all share a common mathematical foundation. They all require a procedure that identifies the parameters of a highly structured signal propagation model from scattering data measured at the boundary, or from excitation-response pairs, whose prototype for geophysicists is the Goupillaud horizontally layered earth model.

The important work of Gelfand and Levitan, published in 1951, reduced the solution of the Schrodinger inverse scattering problem to the solution of a parametrized set of linear integral equations. Subsequent papers of Marchenko, Krein and others, further strengthened the belief that inverse scattering is equivalent to solving linear integral or, in the discrete case, matrix equations (see, e.g., Agranovich et al, 1963 and Chadan et al, 1977). However, in geophysical research a different approach emerged. This approach, more directly concerned with a local analysis of acoustic wave propagation, provided recursive inversion procedures that were called dynamic deconvolution, or downward continuation, methods (see Robinson, 1975 and Bardan, 1977).

In this paper we use the Goupillaud horizontally layered earth model to present a unified, straightforward approach for a wealth of discrete inverse scattering problems arising in various fields; i.e., we show how it can obtain the classical equations of discrete inverse scattering problems (Gelfand-Levitan, Marchenko and Krein) and direct fast recursive inversion algorithms. For this study we had as an example the paper of Bruckstein et al (1984) on the inverse scattering problem for discrete transmission-line models.

2. THE GOUPILLAUD HORIZONTALLY LAYERED EARTH MODEL

The determination of the properties of the earth from waves that have been reflected from the earth is the classic problem of reflection seismology. As a first step in mathematical analysis, the problem is

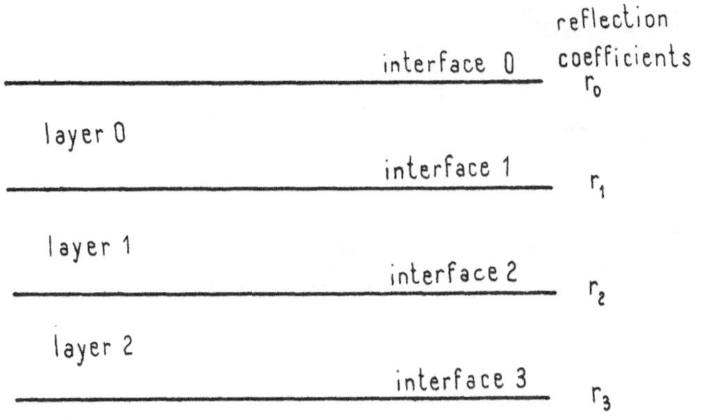

reflection
interface 0   coefficients
r_0

layer 0

interface 1 r_1

layer 1

interface 2 r_2

Fig. 1.

layer 2

The Goupillaud model.

interface 3 r_3

usually simplified by assuming that the earth's crust is made up of a sequence of sedimentary layers. The Goupillaud model (Figure 1) approximates the heterogeneous earth with a sequence of horizontal layers, each of which is homogeneous, isotropic, and nonabsorptive. This model was introduced by Goupillaud (1961) but important contributions for the description and solution of its inverse scattering problem were made by Robinson (1975 and 1982). The Goupillaud model is subject to vertically traveling plane compressional waves, and thus it is a normal incidence model. It is assumed that the two-way travel time in each layer is the same and is equal to one time unit. In other words, the one-way travel time in each layer is taken to be one-half of the discrete unit of time. The first layer is called layer 0, the next layer is called layer 1, and so on. Interface 0 is the interface at the top of layer 0, interface 1 is the interface at the top of layer 1, and so on. Let r_k be the reflection coefficient for downgoing waves striking interface k. The reflection coefficient for upgoing waves striking interface k is thus equal to $-r_k$. We will assume that the amplitudes of our waves are measured in units such that squared amplitude is proportional to energy. Then the transmission coefficient through interface k is equal to $(1-r_k^2)^{1/2}$ for either upgoing or downgoing waves. All waves are digitized with unit time spacing. Although the waves exist throughout the layers, we will only be concerned with them as measured at the tops of the layers (see Figure 2). For example, if a downgoing impulse is introduced at time 0 at the top of layer 0, then it arrives at the top of layer 1 at time 0.5, at the top of layer 2 at time 1 and so on. The downgoing wave at the top of layer k is denoted by $d_k(n)$ if k is even, and by $d_k(n+0.5)$ if k is odd, where n is an integer. The respective z-transform is $D_k(z) = \sum_n d_k(n) z^{-n}$ if k is even, or $D_k(z) =$

267

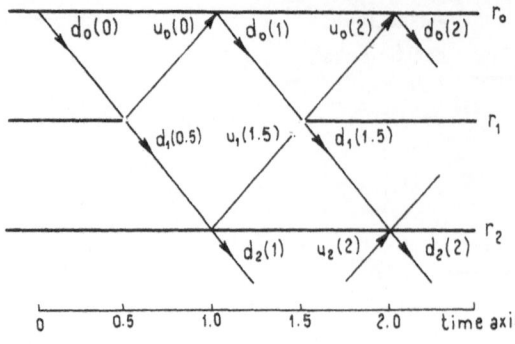

Fig. 2.

Space-time lattice diagram.

$\sum_n d_k (n+0.5) z^{-(n+0.5)}$ if k is odd. Similarly, the upgoing wave $u_k (n)$ for k even, and $u_k (n+0.5)$ for k odd, is also measured at the top of the layer k. Its z-transform is $U_k (z)$.

We now consider an idealized seismic experiment. The source is a downgoing unit impulse introduced at the top of layer 0 at time instant 0. This pulse proceeds downward where it undergoes multiple reflections and refractions within the layered system. Some of the energy is returned to the top of layer 0, where it is recorded in the form of a seismic trace, which we denote by the sequence $x_0, x_1, x_2, \ldots$, where the subscript indicates the discrete time index. Its z-transform is $X(z)$.

There are two types of boundary conditions commonly imposed on the top interface (interface 0 with reflection coefficient r_0). One is the free-surface condition, which says that interface 0 is a perfect reflector; that is, the free-surface condition is that $|r_0|=1$. The free-surface condition approximately holds in the case of a marine seismogram, when the surface of the water (interface 0) is virtually a perfect reflector and the hydrophones record the downgoing wave at the top of the layer 0 (see Figure 3). In this case we have $x_0=1$, $x_1=r_1$,

$$X(z)=D_0(z)=1+r_1 z^{-1} +x_2 z^{-2} +x_3 z^{-3} +\ldots$$

and

$$U_0 (z)=r_1 z^{-1} +x_2 z^{-2} +x_3 z^{-3} +\ldots$$

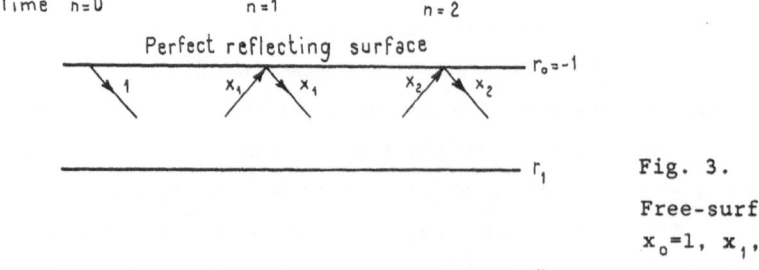

Fig. 3.

Free-surface seismogram:
$x_0=1$, $x_1, x_2, \ldots$

268

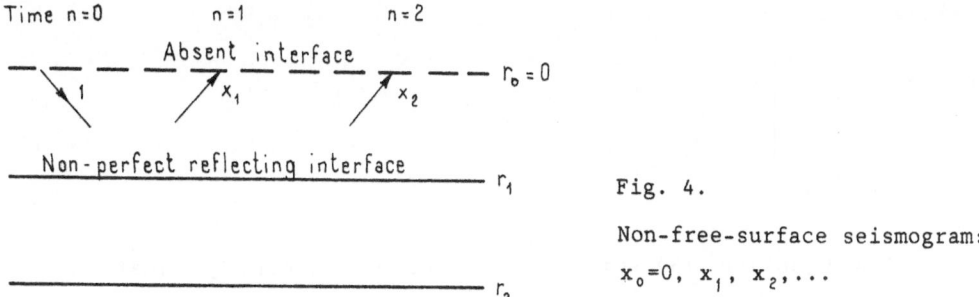

Fig. 4.

Non-free-surface seismogram:

$x_0 = 0$, x_1, x_2, ...

The other condition is the non-free-surface case. For notational convenience, we will choose the non-free-surface as interface 1; that is, the air-earth interface is taken as the interface 1, where $|r_1| < 1$. Since the interface 1 represents the surface of the ground, it follows that the interface 0 is not present, that is $r_0 = 0$. Thus the non-free-surface condition is that $r_0 = 0$ and r_1 is arbitrary. In this case we suppose that the geophones record the upgoing wave at the top of the layer 0 (see Figure 4). Therefore, we have $x_0 = 0$, $x_1 = r_1$, $D_0(z) = 1$ and

$$X(z) = U_0(z) = r_1 z^{-1} + r_2 z^{-2} + r_3 z^{-3} + \ldots$$

The relationship between the waves in the Goupillaud model can be written in matrix form as

$$\begin{bmatrix} D_{k+1}(z) \\ U_{k+1}(z) \end{bmatrix} = \theta_{k+1} \begin{bmatrix} z^{-1/2} & 0 \\ 0 & z^{1/2} \end{bmatrix} \begin{bmatrix} D_k(z) \\ U_k(z) \end{bmatrix}, \tag{1}$$

where

$$\theta_k = (1 - r_k^2)^{-1/2} \begin{bmatrix} 1 & -r_k \\ -r_k & 1 \end{bmatrix} \tag{2}$$

(see e.g., Robinson, 1982).

The matrices $\{\theta_k\}$ are often called chain scattering or transfer matrices and their natural cascade rule is clearly the usual matrix multiplication.

The relationship (1) represents a scheme which does not correspond to a causal signal flow. This relationship, by simple algebraic rearrangement, can be written

$$\begin{bmatrix} D_{k+1}(z) \\ U_k(z) \end{bmatrix} = \begin{bmatrix} 1 & 0 \\ 0 & z^{-1/2} \end{bmatrix} \Sigma_k \begin{bmatrix} z^{-1/2} & 0 \\ 0 & 1 \end{bmatrix} \begin{bmatrix} D_k(z) \\ U_{k+1}(z) \end{bmatrix}, \tag{3}$$

where matrix Σ_k is given by

$$\Sigma_k = \begin{bmatrix} (1-r_k^2)^{1/2} & -r_k \\ r_k & (1-r_k^2)^{1/2} \end{bmatrix}. \tag{4}$$

The equation (3) is a causal wave scattering equation. It represents a scheme which corresponds to a causal signal flow.

The matrix Σ_k is called the scattering matrix of the layer k of the model since it describes the interaction between the waves which go into the layer k (d_k and u_{k+1}) and the waves which go out of the layer k (d_{k+1} and u_k). It is easy to check that Σ_k is unitary, i.e.,

$$\Sigma_k^T \Sigma = \Sigma_k \Sigma_k^T = I$$

(I is unit matrix and T denotes transpose), which corresponds to the physical property of energy-conservation or losslessness, i.e. (with $\|f(\cdot)\|^2$ denoting l^2 norm),

$$\| d_k \|^2 + \| u_{k+1} \|^2 = \| d_{k+1} \|^2 + \| u_k \|^2,$$

or the energy which goes into the layer k is equal to the energy which emerges.

3. LINEAR EQUATIONS FOR INVERSE SCATTERING

Suppose we have the matrix transfer function corresponding to the cascade of the first n+1 layers of Goupillaud model. Therefore, we have using the z-transform notation,

$$\begin{bmatrix} D_n(z) \\ U_n(z) \end{bmatrix} = \begin{bmatrix} M_{11}(n,z) & M_{12}(n,z) \\ M_{21}(n,z) & M_{22}(n,z) \end{bmatrix} \begin{bmatrix} D_o(z) \\ U_o(z) \end{bmatrix}. \tag{5}$$

Considering (1) we can obtain

$$\begin{bmatrix} M_{11}(n,z) & M_{12}(n,z) \\ M_{21}(n,z) & M_{22}(n,z) \end{bmatrix} = \prod_1^n [\theta_k \Delta(z)], \tag{6}$$

where

$$\Delta(z) = \begin{bmatrix} z^{-1/2} & 0 \\ 0 & z^{1/2} \end{bmatrix}. \tag{7}$$

There is a complete symmetry in the equation (5), in that we can replace $\Delta(z)$ by $\Delta(z^{-1})$ and D by U without affecting the relationships between the z-transforms (see (1),(2) and (6)). We therefore have the following useful identities

$$M_{22}(n,z)=M_{11}(n,z^{-1}) \quad \text{and} \quad M_{21}(n,z)=M_{12}(n,z^{-1}). \tag{8}$$

Using the polynomials $P_n(z)$ and $Q_n(z)$, defined by Robinson (see, i.e., Robinson, 1982), we can write

$$M_{22}(n,z) = z^{n/2} P_n(z) \prod_1^n (1-r_k^2)^{-1/2} \tag{9}$$

and

$$M_{21}(n,z) = z^{n/2} Q_n(z) \prod_1^n (1-r_k^2)^{-1/2}, \tag{10}$$

where

$$P_n(z) = 1+\ldots+r_1 r_n z^{-n+1} \tag{11}$$

and

$$Q_n(z) = -r_1 z^{-1}+\ldots-r_n z^{-n}. \tag{12}$$

We can rewrite the basic relation (5) as

$$z^{-n/2} P_n(z^{-1}) D_o(z) + z^{-n/2} Q_n(z^{-1}) U_o(z) = D_n(z) \prod_1^n (1-r_k^2)^{1/2} \tag{13}$$

and

$$z^{-n/2} Q_n(z) D_o(z) + z^{n/2} P_n(z) U_o(z) = U_n(z) \prod_1^n (1-r_k^2)^{1/2}. \tag{14}$$

Now we define the function

$$G_n(z)=z^{-n/2} P_n(z^{-1})+z^{n/2} Q_n(z)=g_{no} z^{\frac{n}{2}} +g_{n1} z^{\frac{n}{2}-1} +\ldots+g_{nn} z^{-\frac{n}{2}} \tag{15}$$

and we can note that at time n/2, by causality, we have

$$d_n(t)=0 \quad \text{for} \quad t<n/2 \quad \text{and} \quad u_n(t)=0 \quad \text{for} \quad t<(n+2)/2 \tag{16}$$

while for $t=n/2$

$$d_n(n/2)=d_o(0) \prod_1^n (1-r_k^2)^{1/2}. \tag{17}$$

Adding the equations (13) and (14) we have

$$D_o(z)G_n(z) + U_o(z)G_n(z^{-1}) = [D_n(z)+U_n(z)] \prod_1^n (1-r_k^2)^{1/2}. \tag{18}$$

We equate coefficients on each side of the equation (18) for the powers of z from $-n/2$ to $n/2$. Using the causality relations (16) and (17) we obtain n+1 equations (one for each power of z: $-n/2$, $(-n+2)/2$, .., $n/2$) given by

$$
\begin{bmatrix}
d_o(0) & 0 & & & \\
d_o(1) & d_o(0) & & & \\
& & & & \\
& & & d_o(0) & \\
d_o(n) & & & d_o(1) & d_o(0)
\end{bmatrix}
\begin{bmatrix}
g_{no} \\
g_{n1} \\
\cdot \\
\cdot \\
g_{nn}
\end{bmatrix}
+
\begin{bmatrix}
0 & & & 0 & u_o(0) \\
& & u_o(0) & u_o(1) & \\
0 & u_o(0) & & & \\
u_o(0) & u_o(1) & & & u_o(n)
\end{bmatrix}
\begin{bmatrix}
g_{no} \\
g_{n1} \\
\cdot \\
\cdot \\
g_{nn}
\end{bmatrix}
=
\begin{bmatrix}
0 \\
\cdot \\
\cdot \\
0 \\
d_n(n)\prod_1^n(1-r_k^2)^{1/2}
\end{bmatrix}
\tag{19}
$$

We note that $g_{no}=0$ and $u_o(0)=0$ (see (11), (12),(15) and (16)). The above equation is a general linear equation, which was derived by Bruckstein et al (1984) for the inverse scattering problem of discrete transmission-line models.

For free-surface reflection seismograms of the Goupillaud model we have

$$X(z) = D_o(z) = 1 + x_1 z^{-1} + x_2 z^{-2} + \ldots$$

and

$$U_o(z) = x_1 z^{-1} + x_2 z^{-2} + \ldots ,$$

where the sequence $x_o=1$, x_1, x_2,... represents a seismic trace. In this case the equation (19) reduces to the form

$$
\begin{bmatrix}
1 & 0 & & & 0 \\
x_1 & 1 & 0 & & \\
& & & & \\
& & x_1 & 1 & 0 \\
x_n & & x_2 & x_1 & 1
\end{bmatrix}
\begin{bmatrix}
g_{no} \\
g_{n1} \\
\cdot \\
\cdot \\
g_{nn}
\end{bmatrix}
+
\begin{bmatrix}
0 & 0 & & & 0 \\
0 & & & 0 & x_1 \\
& & & x_1 & x_2 \\
& & 0 & x_1 & \\
0 & x_1 & x_2 & & x_n
\end{bmatrix}
\begin{bmatrix}
g_{no} \\
g_{n1} \\
\cdot \\
\cdot \\
g_{nn}
\end{bmatrix}
=
\begin{bmatrix}
0 \\
\cdot \\
\cdot \\
0 \\
\prod_1^n(1-r_k^2)
\end{bmatrix}
, \tag{20}
$$

a linear system of equations with a Toeplitz+Hankel coefficient matrix, which is a discrete form of the Gelfand-Levitan equation.

For non-free-surface reflection seismograms we have $D(z)=1$ and

$$X(z) = U_o(z) = x_1 z^{-1} + x_2 z^{-2} + \cdots ,$$

where the sequence $x_o=0, x_1, x_2, \ldots$ represents a seismic trace. In this case the equation (19) reduces to the form

$$
\begin{bmatrix} g_{no} \\ g_{n1} \\ \cdot \\ \cdot \\ g_{nn} \end{bmatrix}
+
\begin{bmatrix} 0 & 0 & \cdot & \cdot & 0 \\ 0 & \cdot & \cdot & 0 & x_1 \\ \cdot & \cdot & \cdot & x_1 & x_2 \\ \cdot & \cdot & 0 & x_1 & \cdot \\ 0 & x_1 & x_2 & \cdot & x_n \end{bmatrix}
\begin{bmatrix} g_{no} \\ g_{n1} \\ \cdot \\ \cdot \\ g_{nn} \end{bmatrix}
=
\begin{bmatrix} 0 \\ \cdot \\ \cdot \\ 0 \\ \prod\limits_1^n (1-r_k^2) \end{bmatrix} ,
\tag{21}
$$

which can be recognized as a discrete Marchenko equation.

The Gelfand-Levitan and Marchenko equations have coefficient matrices that are respectively Toeplitz+Hankel and Hankel matrices.

There is a somewhat lesser known formulation due to Krein, which has a purely Toeplitz coefficient matrix. To derive this equation we use the free-surface reflection seismogram and rewrite equation (5) as

$$z^{-n/2} P_n(z^{-1}) + z^{-n/2} [P_n(z^{-1}) + Q_n(z^{-1})] X(z) = D_n(z) \prod\limits_1^n (1-r_k^2)^{1/2} \tag{22}$$

and

$$z^{n/2} Q_n(z) + z^{n/2} [P_n(z) + Q_n(z)] X(z) = U_n(z) \prod\limits_1^n (1-r_k^2)^{1/2}. \tag{23}$$

Now we define the function

$$H_n(z) = z^{-n/2} [P_n(z^{-1}) + Q_n(z^{-1})] = h_{no} z^{\frac{n}{2}-1} + h_{n1} z^{\frac{n}{2}-1} + \ldots + h_{nn} z^{-\frac{n}{2}} \tag{24}$$

and we note that $h_{no}=-r_n$ and $h_{nn}=1$ (see (11) and (12)).

Changing z by z^{-1} in the equation (23) and adding this equation to the equation (22) we have

$$H_n(z) [1+X(z)+X(z^{-1})] = [D_n(z)+U_n(z^{-1})] \prod\limits_1^n (1-r_k^2)^{1/2}. \tag{25}$$

Equating coefficients on each side of the equation (25) for the powers of z from $-n/2$ to $n/2$ we obtain the Krein equation

$$\begin{bmatrix} 1 & x_1 & x_2 & \cdot & x_n \\ x_1 & 1 & x_1 & \cdot & \cdot \\ \cdot & \cdot & \cdot & \cdot & \cdot \\ \cdot & \cdot & \cdot & 1 & x_1 \\ x_n & \cdot & \cdot & x_1 & 1 \end{bmatrix} \begin{bmatrix} h_{no} \\ h_{n1} \\ \cdot \\ \cdot \\ h_{nn} \end{bmatrix} = \begin{bmatrix} 0 \\ \cdot \\ \cdot \\ 0 \\ \overset{n}{\underset{1}{\Pi}}(1-r_k^2) \end{bmatrix} \qquad (26)$$

We recall that in the above equation we again used the causality relations (16) and (17). The linear system of equations (26) has a symmetric Toeplitz matrix of coefficients. It is well-known in linear prediction theory as the "normal equations", and hence the Levinson recursion can be used to solve it, i.e., to determine the coefficients $h_{no}, h_{n1}, \ldots, h_{nn}$. After this we can find the reflection coefficients $r_1, r_2, \ldots, r_n$ from the free-surface reflection seismogram (see, e.g., Robinson, 1967).

4. FAST INVERSE SCATTERING ALGORITHMS

The inverse scattering problem for the Goupillaud model is to determine the reflection coefficients $r_1, r_2, \ldots, r_n$ from the seismic trace $x_o, x_1, \ldots, x_n$. In the previous section we showed that the inverse scattering problem for the Goupillaud model is equivalent to the solution of sets of linear equations, which are discrete analogs of certin well-known integral equations of continuous inverse scattering associated with the names of Gelfand-Levitan, Krein and Marchenko.

The linear equations required to determine the model up to interface N are of size NxN, so that their solution would in general require $O(N^3)$ elementary operations. However the special properties of the Goupillaud model impose further structure on the linear equations, e.g., making their coefficient matrices be symmetric Toeplitz (Krein), Toeplitz+Hankel (Gelfand-Levitan) and Hankel (Marchenko). These special structures can be exploited to reduce the computational burden by an order of magnitude to $O(N^2)$, leading to so-called fast inverse scattering algorithms. It turns out, however, that this is a rather indirect route to fast inverse scattering algorithms, because we obtain, for beginning, the coefficients $g_{no}, g_{n1}, \ldots, g_{nn}$ or the coefficients h_{no}, $h_{n1}, \ldots, h_{nn}$, and after this, the reflection coefficients.

The special properties of the Goupillaud model, analyzed in detail in the second section, suggest a type of direct fast scattering inversion algorithms, which were discovered by geophysicists and called "dynamic deconvolution" algorithms (see Robinson, 1975 and Bardan, 1977). Foias (1977) noted the close connection between the Robinson-Bardan's fast inverse scattering algorithm and a classical algorithm of Schur (see, e.g., Chamfy, 1958) for testing whether a given power series is bounded in the region of analyticity (see Bardan, 1978).

Now we develop this type of algorithms in a general form, which immediately suggests extensions to more general inverse scattering problems. For this we use a function often encountered in the scattering literature, the layer k reflection function, defined as

$$R_k(z) = \frac{U_k(z)}{D_k(z)} .$$

(27)

In particular, at k=0, we have

$$R_0(z) = \frac{U_0(z)}{D_0(z)} = \frac{u_0(1) z^{-1} + u_0(2) z^{-2} + \ldots}{1 + d_0(1) z^{-1} + d_0(2) z^{-2} + \ldots} .$$

(28)

Referring to relation (1) we can, after some easy algebra, obtain the following recursion

$$R_{k+1}(z) = \frac{z R_k(z) - r_{k+1}}{1 - r_{k+1} z R_k(z)} .$$

(29)

This recursion, which is a linear fractional map, provides the evolution of the reflection function as we proceed deeper and deeper into the model, provided we are given the reflection coefficient r_k. But, using the causality relations (16), we have

$$R_k(z) = \frac{u_k\left(\frac{k}{2}+1\right) z^{-\frac{k}{2}-1} + u_k\left(\frac{k}{2}+2\right) z^{-\frac{k}{2}-2} + \ldots \ldots}{d_k\left(\frac{k}{2}\right) z^{-\frac{k}{2}} + d_k\left(\frac{k}{2}+1\right) z^{-\frac{k}{2}-1} + \ldots \ldots} =$$

$$= \frac{u_k\left(\frac{k}{2}+1\right) z^{-1} + u_k\left(\frac{k}{2}+2\right) z^{-2} + \ldots \ldots}{d_k\left(\frac{k}{2}\right) + d_k\left(\frac{k}{2}+1\right) z^{-1} + \ldots \ldots} ,$$

(30)

275

where $\quad d_k(k/2) = \prod_1^n (1-r_k^2)^{1/2}$ and $\quad u_k(k/2+1) = r_{k+1} d_k(k/2)$ (see Figure 2). Using (30) we have

$$r_{k+1} = \lim_{z \to \infty}[zR_k(z)] \cdot \qquad (31)$$

Thus we have an alternative inversion algorithm, that could recursively propagate (29) together with the identification formula (31).

At the initial step, we have

$$r_1 = \lim_{z \to \infty}[zR_0(z)]$$

and

$$R_1(z) = \frac{zR_0(z)-1}{1-r_1 zR_0(z)} = \frac{r_2 z^{-1} + u_0'(2)z^{-2} + u_0(3)z^{-3} + \ldots}{1+ d_0'(1)z^{-1} + d_0'(2)z^{-2} + \ldots} , \qquad (32)$$

where $\qquad r_2 = u_0'(1) = u_1(3/2)(1-r_1^2)^{1/2} , \qquad u_0'(2) = u_1(5/2)/(1-r_1^2)^{1/2} ,$

$u_0'(3) = u_1(7/2)/(1-r_1^2)^{1/2} , \quad d_0'(1) = d_1(3/2)/(1-r_1^2)^{1/2} , \quad d_0'(2) = d_1(5/2)/(1-r_1^2)^{1/2} ,$ and so on.

Therefore, once we have determined the reflection coefficient r_1, we can use the formula (29) to "peel off" the effect of first layer and put ourselves in the same position as before with a model parametrized by $[r_2, r_3, \ldots)$. Indeed, using r_1 and the original scattering data we are able to obtain equivalent "scattering data" for the model extending over $[r_2, r_3, \ldots)$. We can continue in this way, succesively determining a reflection coefficient and peeling off the associated layer to determine the whole model. For these reason, the procedure is called "layer-peeling" inversion algorithm.

The recursion formula (28) can be writen

$$R_{k-1}(z) = \frac{z^{-1}\left[r_k + R_k(z)\right]}{1+r_k R_k(z)} \cdot \qquad (33)$$

We now suppose that the reflection coefficients $r_k = 0$ for $k>s$ and $r_s \neq 0$. In this case $R_{s-1}(z) = r_s z^{-1}$ (see (27) and Figure 5). Starting from $R_{s-1}(z)$ and using the recursion (33) we obtain

$$R_0(z) = \frac{r_1 z^{-1} + \ldots + r_s z^{-s-1}}{1 + \ldots + r_1 r_s z^{-s}} , \qquad (34)$$

provided we are given the reflection coefficients $r_1, r_2, \ldots, r_s$.

276

Supposing that only the reflection coefficient $r_s = \mathcal{E}$, where $|\mathcal{E}| = 1$, we have $R_{s-1}(z) = \mathcal{E}z^{-1}$ (see again (27) and Figure 5). Using the same

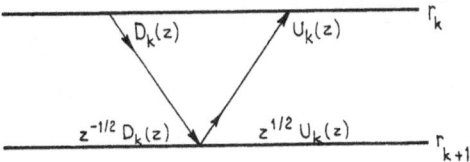

Fig. 5. The layer k reflection function.

recursion (33) we obtain

$$R_0(z) = \frac{A(z)}{B(z)} = \frac{r_1 z^{-1} + \ldots + \mathcal{E}z^{-s-1}}{1 + \ldots + \mathcal{E}r_1 z^{-s}},\tag{35}$$

where

$$B(z) = \mathcal{E}z^{-s+1} A(z^{-1}).\tag{36}$$

To show the close connection between the above inverse scattering procedure and the algorithm of Schur for testing whether a given power series is bounded in the region of analyticity, we present Schur's theorem on power series of meromorphic functions (see, e.g., Chamfy, 1958) in the appendix.

Let

$$R_k(z) = \rho_k(1)z^{-1} + \rho_k(2)z^{-2} + \rho_k(3)z^{-3} + \ldots\tag{37}$$

be the power series expansion (in z^{-1}) of the reflection function $R_k(z)$.

Let us now define a power series as

$$F_k(z) = zR_k(z) = f_{ko} + f_{k1} z^{-1} + f_{k2} z^{-2} + \ldots\tag{38}$$

so that the recursion (28) becomes

$$F_{k+1}(z) = z \frac{F_k(z) - r_{k+1}}{1 - r_{k+1}F_k(z)}.\tag{39}$$

The reason for doing this is that now (39) turns out to be identical to the recursion of Schur (A3) for checking when a power series in z^{-1},

277

$F_k(z)$, is such that

$$|F_k(z)| \leqslant 1 \quad \text{for} \quad |z| > 1. \tag{40}$$

Schur's theorem is precisely that the sequence of number $f_{ko} = r_k$ implicitly defined by the recursion (39)

$$\lim_{z \to \infty} [F_{k-1}(z)] = f_{ko}, \tag{41}$$

should satisfy

$$|f_{ko}| \leqslant 1. \tag{42}$$

From this result we also obtain that the reflection function associated with any layer will be bounded by one outside the unit circle of the complex plane. Here we see that this algorithm may be regarded as describing the propagation of waves into the Goupillaud model and consequently as providing a direct inverse scattering procedure, which is named "Schur-type algorithm".

We now show how the first n reflection coefficients from a seismic trace can be effectively determined.

The reflection function $R_o(z)$ can be expressed as

$$R_o(z) = 1 - 1/X(z)$$

for free-surface reflection seismograms , and

$$R_o(z) = X(z)$$

for non-free-surface reflection seismograms, where $X(z)$ is the z-transform of a seismic trace.

We consider the function

$$R_{on}(z) = \frac{u_o(1) z^{-1} + u_o(2) z^{-2} + \ldots + u_o(n) z^{-n}}{1 + d_o(1) z^{-1} + \ldots + d_o(n-1) z^{-n+1}} . \tag{43}$$

The power series expansion of $R_{on}(z)$ is

$$R_{on}(z) = \rho_{on}(1) z^{-1} + \rho_{on}(2) z^{-2} + \rho_{on}(3) z^{-3} + \ldots \tag{44}$$

and it can be obtained by using the recursive formula

$$\rho_{on}(k) = u_o(k) - \sum_{i=1}^{n-1} d_o(i) \rho_{on}(k-i) \quad \text{for} \quad k > 1 \tag{45}$$

(see, e.g., Bardan, 1977).

The first n terms in the power series expansion of $R_o(z)$ are identical to corresponding terms in expansion of $R_{on}(z)$ (see formula (45)), i.e., $\rho_o(1) = \rho_{on}(1), \rho_o(2) = \rho_{on}(2), \ldots, \rho_o(n) = \rho_{on}(n)$.

To determine the first n reflection coefficients, at the initial step of the algorithm, we can use the function

$$R_{on}^n(z) = R_o^n(z) = \rho_{on}(1)z^{-1} + \rho_{on}(2)z^{-2} + \ldots + \rho_{on}(n)z^{-n} \tag{46}$$

instead of the function $R_o(z)$, because for this determination only the coefficients $\rho_o(1), \rho_o(2), \ldots, \rho_o(n)$ of the power series expansion of $R_o(z)$ are necessary (see (A6)). Of course,

$$r_1 = \lim_{z \to \infty}[zR_{on}(z)] = \rho_{on}(1) \tag{47}$$

and

$$R_{1n}(z) = \frac{z\,R_{on}^n(n) - r_1}{1 - r_1 z\,R_{on}^n(z)} \quad . \tag{48}$$

Using formula (45) we can obtain

$$R_{1n}(z) = \rho_{1n}(1)z^{-1} + \rho_{1n}(2)z^{-2} + \ldots + \rho_{1n}(n-1)z^{-n+1} \tag{49}$$

with

$$r_2 = \lim_{z \to \infty}[zR_{1n}^n(z)] = \rho_{1n}(2) \quad . \tag{50}$$

We can continue in this way to determine the reflection coefficients $r_3, r_4, \ldots, r_n$.

5. CONCLUSIONS

We conclude that the Goupillaud model is a very interesting model for the discrete inverse scattering problem because using this model we can present a unified, straightforward approach for a wealth of inverse scattering problems arising in various fields.

ACKNOWLEDGMENT

The author is indebted to Prof. Enders A. Robinson who suggested the approach to this subject and helped with some observations and comments.

REFERENCES

Agranovich, Z.S., and V. A. Marchenko, 1963: The Inverse Problem of Scattering Theory. Gordon and Breach, New York.

Bardan, V., 1977: Comments on Dynamic Predictive Deconvolution. Geophysical Prospecting, 25, 569-572.

Bardan, V., 1978: Despre o Problema Inversa in Geofizica. Report of IPGG Bucharest.

Bruckstein, A., M., and T. Kailath, 1984: Inverse Scattering for Discrete Transmission-Line Models. Report of Stanford University, CA 94305.

Chadan, K., and P. C. Sabatier, 1977: Inverse Problems in Quantum Scattering Theory. Springer-Verlag, New-York.

Chamfy, C., 1958: Fonctions Meromorphes dans le Cercle-Unite et leurs Series de Taylor. Ann. Inst. Fourier, 8, 211-251.

Foias, Ciprian, 1977: Private communication, University of Bucharest.

Robinson, E.A., 1967: Multichannel Time Series Analysis with Digital Computer Programs. San Francisco, CA: Holden-Day.

Robinson, E.A., 1975: Dynamic Predictive Deconvolution. Geophysical Prospecting, 23, 779-797.

Robinson, E.A., 1982: Spectral Approach to Geophysical Inversion by Lorentz, Fourier and Radon Transforms. Proc. IEEE, 70/9, 1039-1054.

APPENDIX

SCHUR'S THEOREM ON POWER SERIES OF MEROMORPHIC FUNCTIONS

Let $F(z)$ be a Schur-type function, i.e., a complex function of complex variable, meromorphic and with

$$|F(z)| \leqslant 1 \quad \text{for} \quad |z| \geqslant 1. \tag{A1}$$

Let

$$F(z) = f_{00} + f_{01} z^{-1} + f_{02} z^{-2} + \ldots \tag{A2}$$

be its power series.

Starting from $F_0(z)=F(z)$, by formula

$$F_{k+1}(z)=z\frac{F_k(z)-f_{k0}}{1-f_{k0}f_k(z)} ,$$ (A3)

we can obtain a sequence of Schur-type functions. This sequence is infinite or ends in a constant function whose absolute value is 1. We can associate with the function $F_0(z)$ a sequence of coefficients

$$f_{00},f_{10},f_{20},\ldots$$ (A4)

This sequence is infinite and $\forall k$, $k \geqslant 0$, $|f_{k0}|<1$, or it is finite and ends in f_{s0} with $|f_{s0}|=1$ and $|f_{k0}|<1$ for $k=0,1,\ldots,s-1$. In the second case we say that the order of the function $F(z)$ is s.

Schur's theorem shows that the existence of the sequence (A4), finite or infinite, is a necessary and sufficient condition for the function $F(z)$ to be a Schur-type function.

In other words:

a) - A necessary and sufficient condition to associate with the $F(z)$, by formula (A3), a finite sequence of coefficients $f_{00},f_{10},\ldots,f_{s0}$ with $|f_{k0}|<1$ for $k<s$ and $|f_{s0}|=1$ is as

$$F(z)=\frac{A_s(z)}{B_s(z)} ,$$ (A5)

where

$$A_s(z)=a_{s0}+a_{s1}z^{-1}+\ldots+a_{ss}z^{-s}$$

and

$$B_s(z)=\xi z^{-s}A_s(z^{-1}).$$

ξ is a complex constant whose absolute value is 1 and the zeros of $A_s(z)$ are situated inside the unit circle of the complex plane. The function $F(z)$ is defined if we know the complex constants $f_{00},f_{10},\ldots,f_{s0}$, where $|f_{k0}|<1$ for $k<s$ and $f_{s0}=\xi$.

b) - The coefficients f_{k0} for $k<s$ can be expressed as a function of coefficients $f_{00},f_{01},\ldots,f_{0k}$. Thus we have

$$f_{k0}=\phi_k(f_{00},f_{01},\ldots,f_{0k})$$ (A6)

and, conversly,

$$f_{0k}=\psi_k(f_{00},f_{10},\ldots,f_{k0}) .$$ (A7)

Let $f_{00}, f_{10}, f_{20}, \ldots$ be an infinite sequence of complex constants with $|f_{k0}| < 1$. Then, the series

$$F(z) = \sum_{0}^{\infty} \psi_k(f_{00}, f_{10}, \ldots, f_{k0}) z^{-n} \tag{A8}$$

is convergent outside the unit circle of the complex plane and $F(z)$ is a Schur-type function.

Therefore,

$$|\phi_k(f_{00}, f_{01}, \ldots, f_{0k})| < 1 \quad \text{for} \quad k \geqslant 0$$

or

$$|\phi_k(f_{00}, f_{01}, \ldots, f_{0k})| < 1 \quad \text{for} \quad 0 \leqslant k < s$$

and

$$|\phi_k(f_{00}, f_{01}, \ldots, f_{0s})| = 1,$$

are necessary and sufficient conditions for $F(z) = \sum_{0}^{\infty} f_{0k} z^{-k}$ to be a Schur-type function.

Spectral Analysis of Blast Vibrations from Large Explosions

C. Kurtuluş, M. Alpmen

Yildiz University, Kocaeli Engineering Faculty, Department of Applied Geophysics, Kocaeli, Izmit/Turkey

Abstract

Spectral analysis was performed on 3-component vibration records for 91 large blasts in four different lithologies (taconite, hematite, hard limestone, soft limestone) to characterize the properties of radiating elastic waves. Maximum particle velocity data show considerable scatter but the spectral maxima are consistently below 10.0 Hz over the distance range 300-2000 m. Signal durations are longer for hematite, and soft limestone than for taconite, and hard limestone with the longer signals being characterized by !"narrower" frequency spectra. The corner frequency of corrected amplitude spectra varies inversely with source duration, that is, long duration blasts, in agreement with observations for farfield earthquaqe waves. Powder factors for hard rock are higher than for soft rock because of the large taconite and hard limestone are higher than hematite and soft limestone. It was not possible to obtain a definitive relation-ship between maximum particle velocity and the powder factor because of the larger number of variables in these blasts.

Measurements contained in this study are significant because of the scarcity of published data pertaining to large blasts.

INTRODUCTION

Explosives are widely used to obtain fragments of rocks in quarrying and mining, to excavate tunnels and in construction projects. However, the usage of explosives can cause structural/

or other damage due to blast induced vibrations. This study is concerned with the vibrations generated by multi-hole surface (open-pit) explosions in four different lithologies, namely taconite, hematite, hard limestone, and soft limestone.

An explosive is a compound that reacts to heat or shock by decomposing rapidly into other compounds, mostly gases. The expanding gases develop great pressures which exceed the strength of rock immediately around the explosive charge. The resulting pressures crush a small volume of rock in the immediate vicinity with the remaining energy radiating into the surrounding rock as a pressure or a shock front at a speed which is generally between 10,000 ft/sec (3048 m/sec) and 20,000 ft/sec (6096 m/sec) (9) When such a compressional pressure front reaches a free surface (rock/air interface), it is reflected back into the rock as a tensional wave.

If the combination of these impinging and reflected waves gives a net stress great enough, the rock (whose tensile strength is less then its compressional strength) fra ctures and spalls from the free surface. The remaining impinging pressure wave then finds a new free surface to reflect from as a tensional wave and to create additional spalls. The process may repeat itself serveral times. At the same time, expanding gases from the explosion work their way through fractures, churn the pieces, and increase fragmentation by collisional processes.

Only a small volume of rock around a blast is bounded by a free surface close enough to absorb energy by rock breakage processes. The rest of the pressure front radiates outward amni- directionally in the from of elastic waves (4,7,8,9). Some of these waves move into and through the rock, others travel along the surface. This creates the vibrations that shake man-made structures.

A great deal of work has been carried out to establish damage criteria for small controlled blasts, (6,10,12,14,15) but this can not be said for the case of large uncontrolled blasts. In this study, an attempt is made to provide observational data in the form of maximum displacement, particle velocity, acceleration, frequency, and duration of ground motion, so that analyses can be carried out to establish the damage potential from large blast and to establish damage thresholds. The objective of this study, thus, is to characterize the vibrations generated by large commercial blasts in four different lithologies; taconite, hematite, hard and soft limestone.

ANALYSIS OF BLAST VIBRATIONS

2.1 Ground Vibrations

Ground vibration is usually described as a time varying displacement, velocity or acceleration of a particle in motion. If the variations are assumed to be simple harmonic, they can be represented as follows:

$$x = ACoswt \tag{2.1}$$

where x is displacement, t is time, A is amplitude of ground motion, and w is angular frequency. Velocity of ground motion is obtained from the first time derivative of x,

$$V = \frac{dx}{dt} = -Awsinwt \tag{2.2}$$

The second time derivative of x gives the acceleration of ground motion as

$$a = \frac{d^2x}{dt^2} = -Aw^2coswt \tag{2.3}$$

Peak values of velocity and acceleration correspond to the time when the trigonometric functions are equal to unity, and the relationships for peak values are then obtained as

$$|u| = A, \tag{2.4}$$
$$|V| = AW = A(2\pi f), \tag{2.5}$$
$$|a| = AW^2 = A(2\pi f)^2, \tag{2.6}$$

where f is the maximum peak frequency.

Vibrations consist of the complex interplay and superposition of body and surface waves convolved with random noise. However, effects of individual waves have largely been neglected in blast vibration studies because close-in recordings are usually made and, therefore, waves have not yet completely separated due to velocity differences.

2.2 Properties of Seismographs

In this study mechanical seismographs which have natural periods of 0.75 second and 0.55 critical damping have been used. Dynamic magnification of each earth frequency is found by multiplying static magnification by the magnification factor. The static magnification values of the seismographs are shown in Table 2.1

Table 2.1

Static Magnification Values of the Seismographs

Seismographs	L	Z	T
Seismograph No. 1	51.2	51.8	51.1
Seismograph No. 2	95	108	96

2.3. Blast Records

The blast records as mentioned earlier were obtained from 91 commercial blasts carried out in the four different lithologies; namely, 34 shots in taconite, 33 shots in hematite, 19 shots in hard limestone, and 5 shots in soft limestone. The range of total explosive charge, maximum explosive charge per delay, and distance for the four different litholigies are given in Table 2.2.

Table 2.2

Range of Total Explosive Charge, Maximum Explosive
Charge Per Delay, and Distance
for the Four Different Lithologies

Lithologies	Total Explosive Charge lbs.	Max. Explosive Charge Per Delay lbs.	Distance Feet
Taconite	4,644-628,985	315-7284	122-1097
Hematite	322-6,473	181-1,006	550-2,410
Hard Limestone	4,570-72,340	360-3,090	285-1,750
Soft Limestone	6,009-11,795	542-900	730-1,090

2.4. Power and Corrected Amplitude Spectra

The signals on the records were digitized at a 0.02-sec.
sampling interval to find their frequency spectra. The Nyquist
frequency (the observable highest frequency depending upon the
sampling interval in the spectrum) is found by using formula
2.7

$$f_N = \frac{1}{2 \cdot \Delta t} \qquad (2.7)$$

where Δt is the sampling interval.

$$f_N = \frac{1}{2 \times 0.02} = 25 \text{ Hz}. \qquad (2.8)$$

Therefore, frequency spectra were examined for frequencies up
to 25 Hz.

A fast Fourier Transformation (FFT) algorithm was used to
transform the sampled signals from the time domain to the
frequency domain. Transformation by FFT gives the real and
the imaginary parts of the sampled signals. If a_n is the
real and b_n is the imaginary part of the signal then, the
power spectrum of the signal is found by

$$P = \frac{a_n^2 + b_n^2}{N} \qquad (2.9)$$

where N is the number of samples. The amplitude spectrum of
the signal is found by

$$A = \sqrt{a_n^2 + b_n^2} \qquad (2.10)$$

To correct the magnitude values of the power and the amplitude
spectra, the dynamic magnification factors were determined for
each component, and actual ground motion was obtained by dividing the measured values by these factors.

2.5 Blast Hole Geometry

The desired result of blasting is to obtain small fragments
of rock in a compact post-blast pile while maintaining

radiated vibration levels as low as possible to minimize damage
to nearby structures.

Explosive charges, for this purpose are not detonated in
a single shot but are distributed among a number of holes
and detonated sequentially at short intervals of time (measured
in milliseconds). Various types of blasting patterns were
encountered in this study. Some of the common blast geometries
are shown in Figure 2.1-2.2, where solid circles represent the
blast holes and crosses represent the positions of millisecond
delays.

A charge of explosive is placed in a hole drilled into the rock.
Granular material, called stemming, is used to fill the top
section of the hole. The depth of stemming and depth of drill
holes varied for the four lithologies. The average depth of
stemming and the average vertical hole depth for the four
lithologies are given in Table 2.3.

Table 2.3.
Average Depth of Stemming and Avarage Vertical Hole
Depth for the Four Lithologies

Lithologies	Average Depth of Stemming (feet)	Average Vertical Hole Depth (feet)
Taconite	11.0	26.0
Hematite	17.0	27.0
Hard Limestone	13.0	43.0
Soft Limestone	40.0	63.0

ANALYSIS

In this section the time and the frequency characteristics
of the blast records, the powder factor values, and the
vibration comparisons for the four lithologies are discussed.

The vibration records obtained in this study have three
components of ground vibration, namely a Longitudinal (L)

288

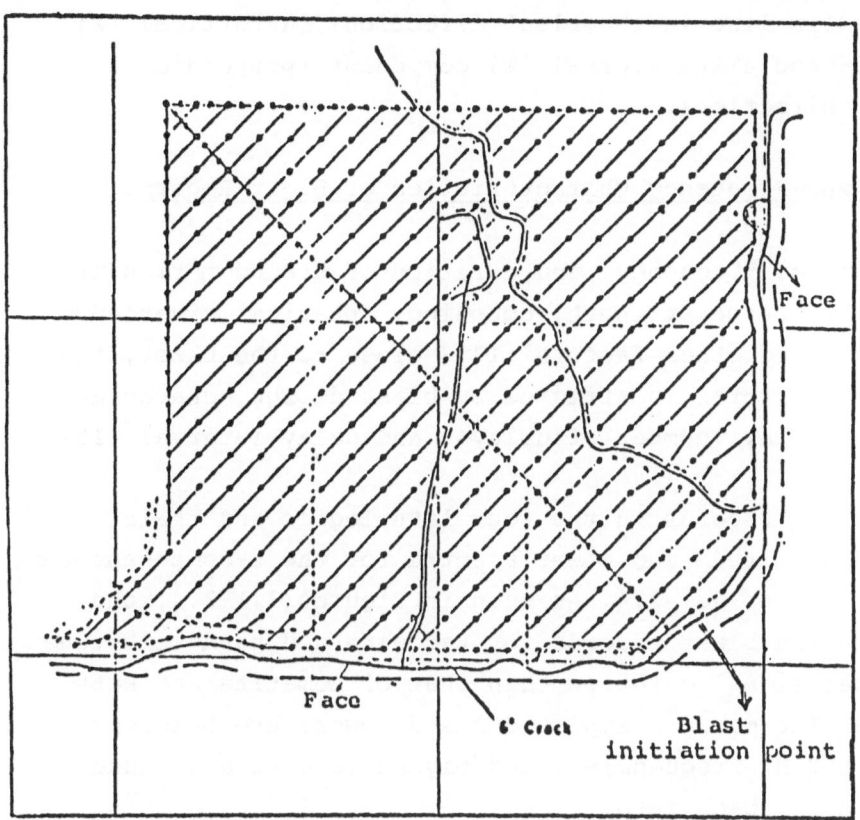

Figure 2.1 Plan Map Illustrating a Multiple Row-Blast
with Two Free Faces

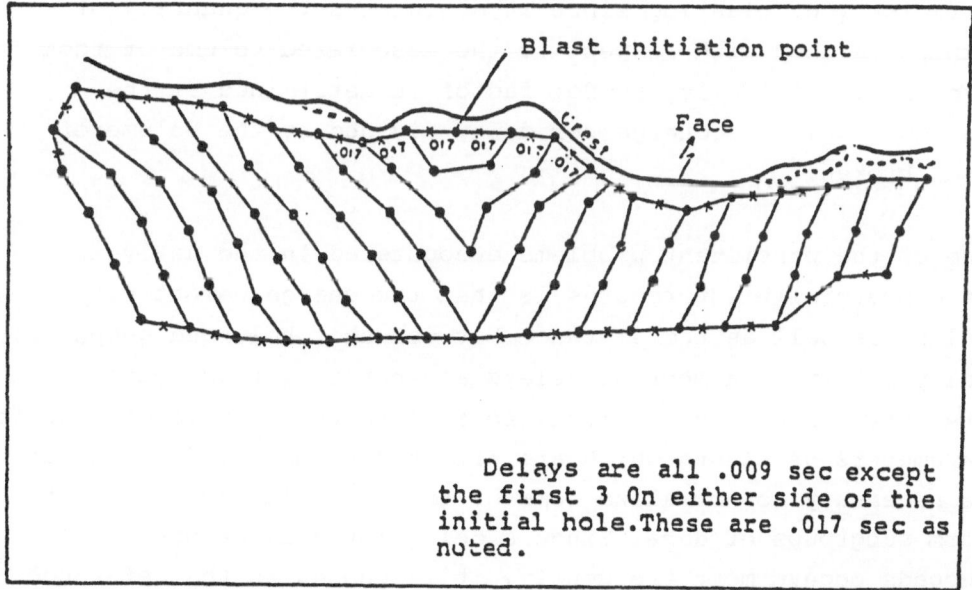

Delays are all .009 sec except
the first 3 On either side of the
initial hole. These are .017 sec as
noted.

Figure 2.2   Illustrative Plan Map of a Common Blast Geometry

component (in-line in the blast direction), a Vertical (Z)
component, and a Transversal (T) component (perpendicular to
the blast direction).

3.1 Time and Frequency Characteristics of Blast Records

The amplitude, frequency, and duration of ground motion are
affected by source strength, coupling, spherical divergence,
absorption, and other factors (13). close to the blast, the
vibration character is affected by blast design, charge and
weight per delay, number of delays, and delay interval (16).

Typical blast records in the four lithologies and their
corrected amplitude and power spectra for the z-component are
shown in Figures 3.1-3.4. As seen in Figures 3.1-3.4, the
signals on the blast records for taconite and hard limestone
have longer signal duration than that of hematite and soft
limestone. The maximum amplitudes and powers are observed to
be below 5.0 Hz frequencies, and beyond 10.0 Hz amplitudes
and powers approach zero.

3.2 Powder Factor

The term powder factor is widely used as a measure of the
efficiency of blasting since it designates the quantity of
explosive used with respect to the associated volume of rock
broken. Specifically, powder factor is determined as the
maximum explosive charge per delay divided by the volume of
rock broken.

One of the persistent problems encountered in the analysis
of commercial blast records is that the charge weight per
delay, as well as the distance between shot hole and geophone,
delay interval, number of delays and burden versus spacing
are different for each blast. To isolate the effects of specific
parameters of blasts which are affected by many variables, it
is necessary to normalize some of the variable factors and to
form subgroups of data. Since a delay interval of nine milliseconds occurs most frequently, it is chosen as the reference
value for the delay interval.

TACONITE

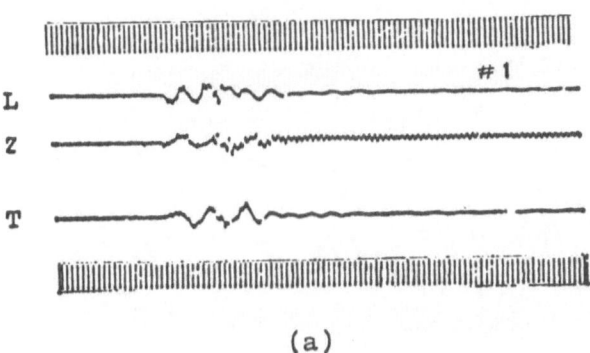

(a)

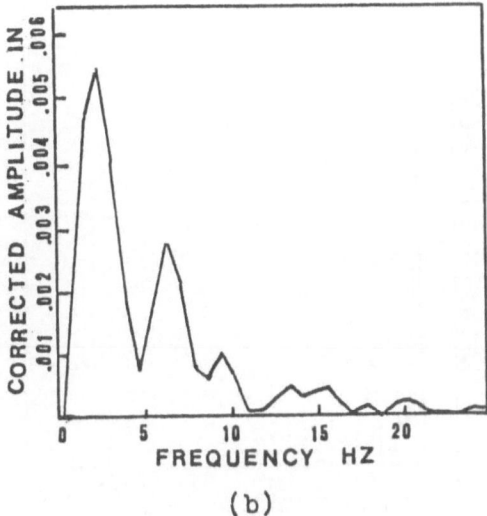

(b)

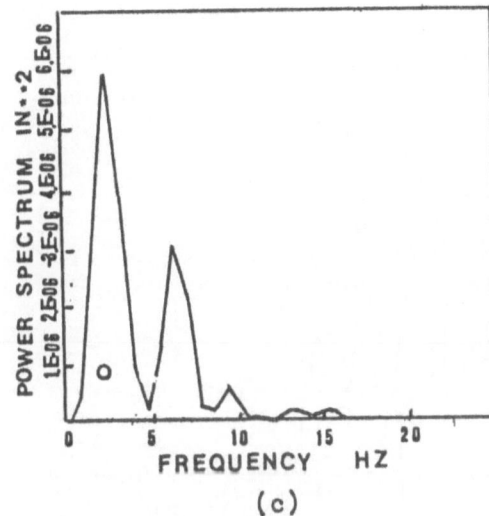

(c)

Figure 3.1   a- Record #1

b- Corrected Amplitude Spectrum of
   Record #1 for the Z-Component

c- Power Spectrum of Record #1 for the
   Z-Component

HEMATITE

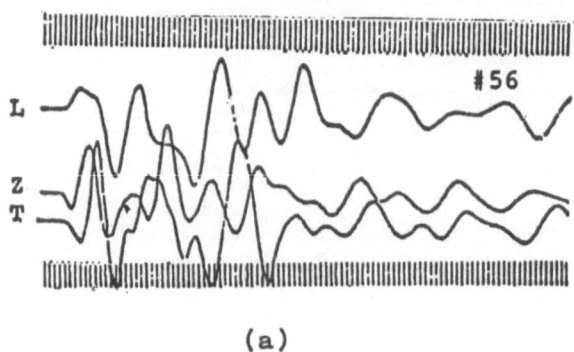

(a)

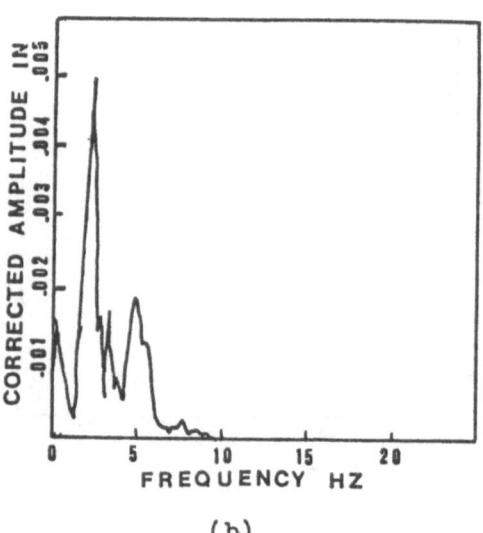

(b)

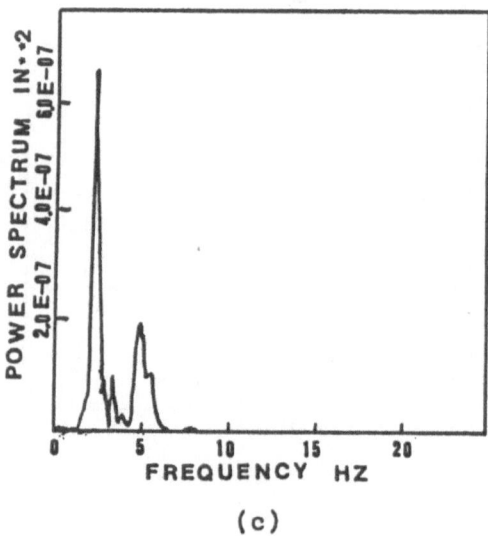

(c)

Figure 3.2 a- Record #56A

b- Corrected Amplitude Spectrum of
Record #56A for the Z-Component

c- Power Spectrum of Record #56 for
the Z-Component

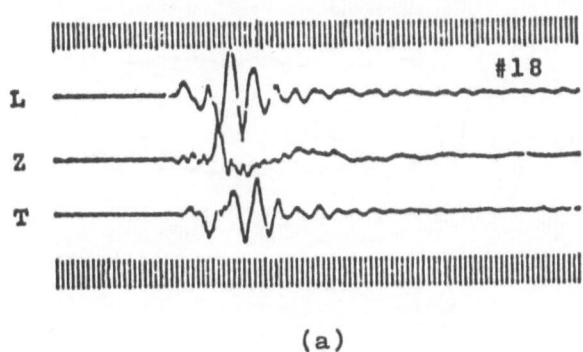

HARD LIMESTONE

#18

L

Z

T

(a)

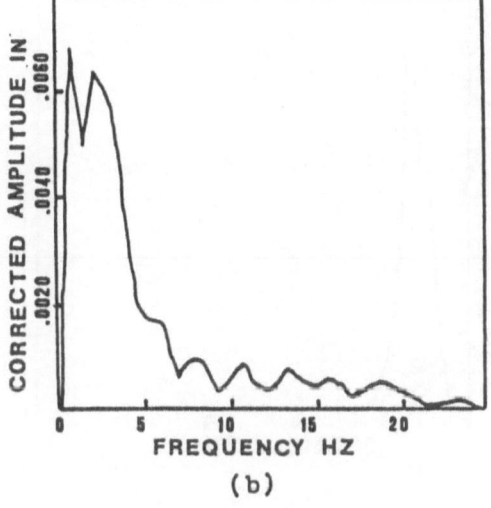

(b)

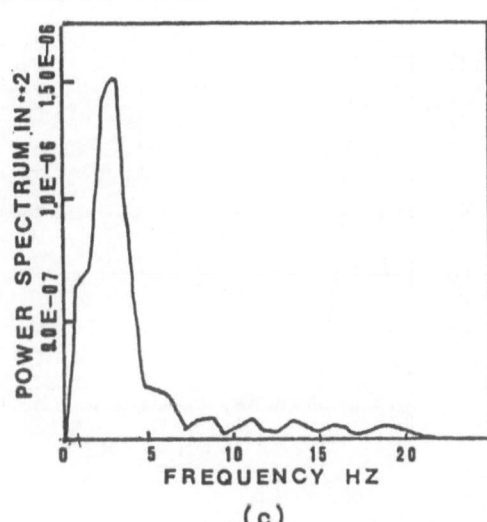

(c)

Figure 3.3 a- Record #18

b- Corrected Amplitude Spectrum
of Record #18 for the Z-Component

c- Power Spectrum of Record #18
for the Z-Component

SOFT LIMESTONE

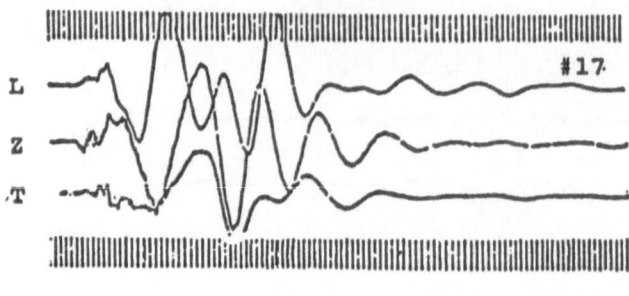

(a)

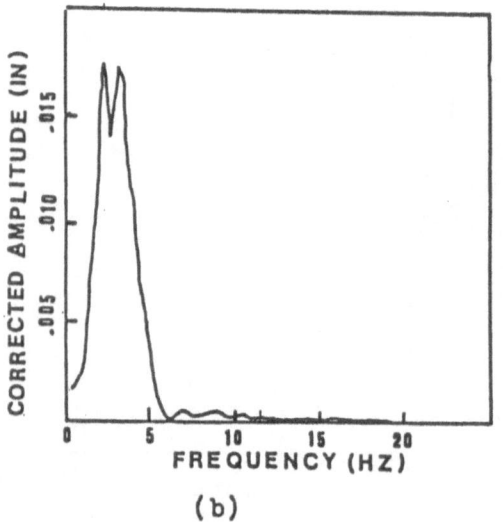

(b)

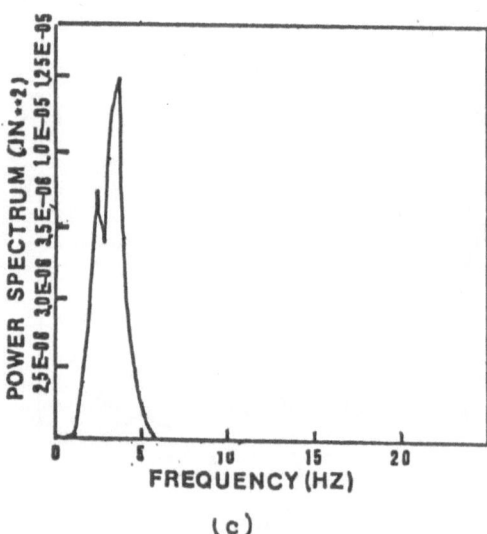

(c)

Figure 3.4 a- Record #17

b- Corrected Amplitude Spectrum
of Record #17 for the Z-Component

c- Power Spectrum of Record #17
for the Z-Component

To eliminate the effect of variations in maximum explosive per delay, maximum explosive per delay was normalized to 1000 lbs., and the same normalization factor was applied to the corresponding particle velocity level. No normalization was applied to other factors such as delay interval, number of delays, shot-geophone distance and burden vs. spacing.

The plot of the 2-component of particle velocity versus powder factor for the four lithologies are shown in Figure 3.5.

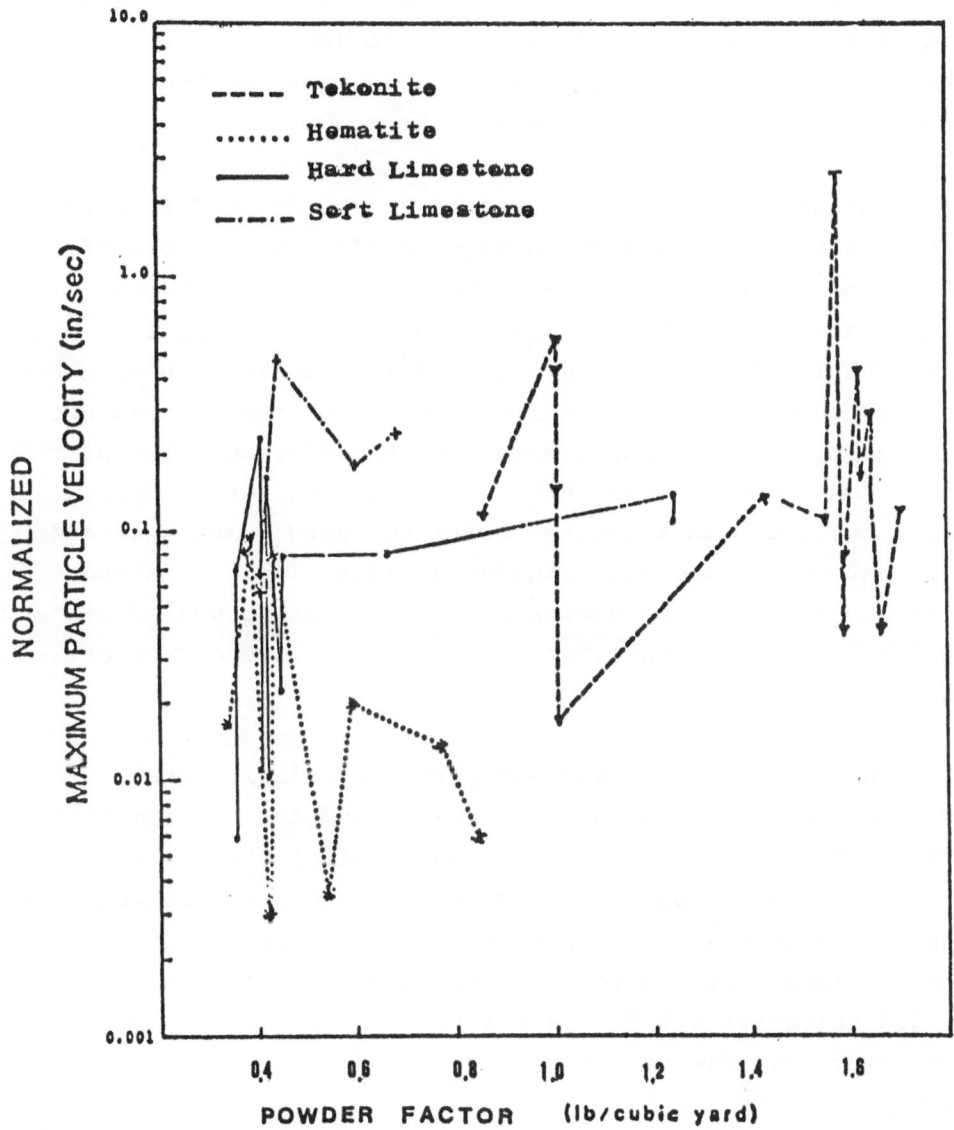

Figure 3.5 Plot of the Maximum Particle Velocity Versus Powder Factor for the Z-Component for all lithologies.

It is observed that taconite has the highest powder factor
and soft limestone has the lowest powder factor. The powder
factors for hematite are closer to that of soft limestone
whereas the powder factors for hard limestone are closer to
that of taconite.

3.3 Vibration Comparisons

Vibrations from the blast in the four different lithologies
were analyzed by using the square root scaled distance
criterion (which is the actual distance divided by the square
root of the maximum explosive charge per delay).

Square root scaled distance criterion is mostly used in blast
engineering to illustrate the damage levels. The purpose of
using scaled distance criteria is to eliminate the charge
effect on distances. But in case of the large commercial
blasts, the maximum explosive charge varies as well as distance.
Therefore, the square root scaled distance criterion does not
totally eliminate the charge effect on the distance. The plot
of the Z-component maximum particle velocities versus scaled
distance for shots in taconite, hematite, hard limestone, and
soft limestone are shown in Figures 3.6-3.9. It is observed
that particle velocities decrease with increasing scaled distance.
The particle velocity data for all rock types are characterized
by large scatter.

In Figures 3,10-3,13, the maximum particle velocity range
versus frequencies in shown. The maximum and the minimum lines
indicate the upper and the lower boundaries of the velocity
envelope and the median indicates the average particle velocity
between these boundaries. For example, in Figure 3.6 the
average maximum particle velocity is 0.062 in/sec at 1.0 Hz
and 0.014 in/sec at 0.5 Hz. In all Figures, it is observed
that maximum particle velocity increases with increasing
frequency.

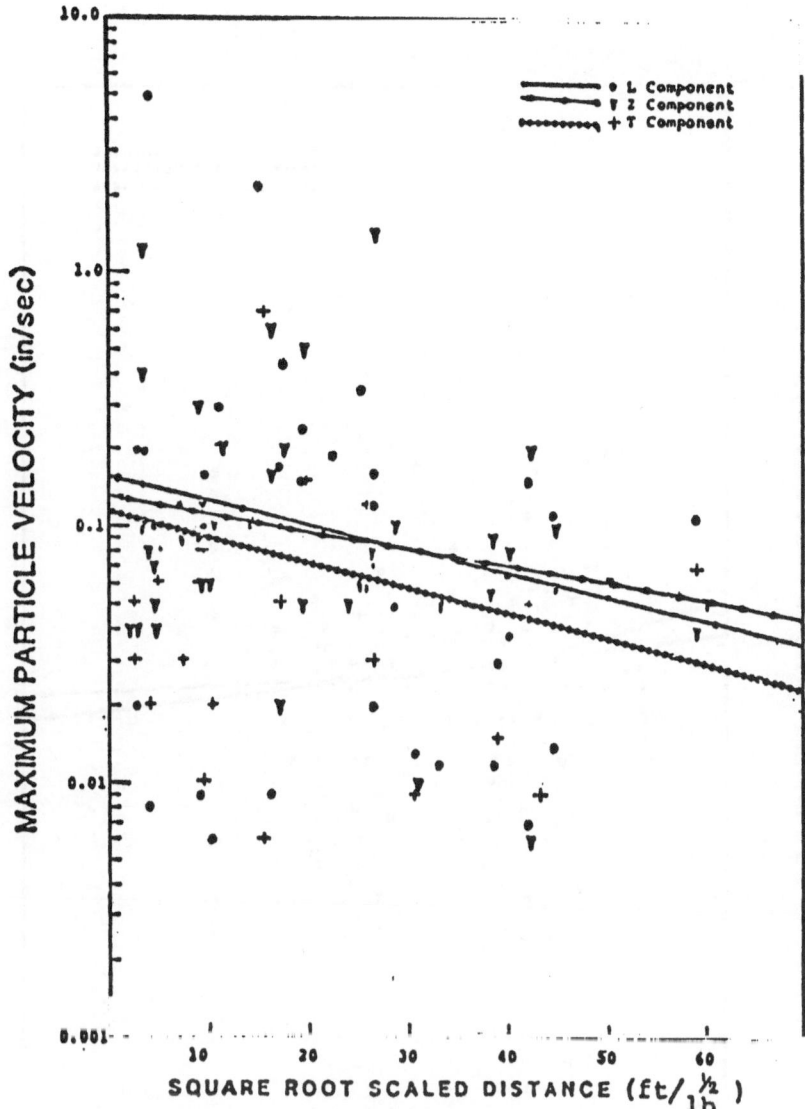

Figure 3.6   Plot of the Maximum Particle
Velocity Versus Scaled Distance
for Taconite

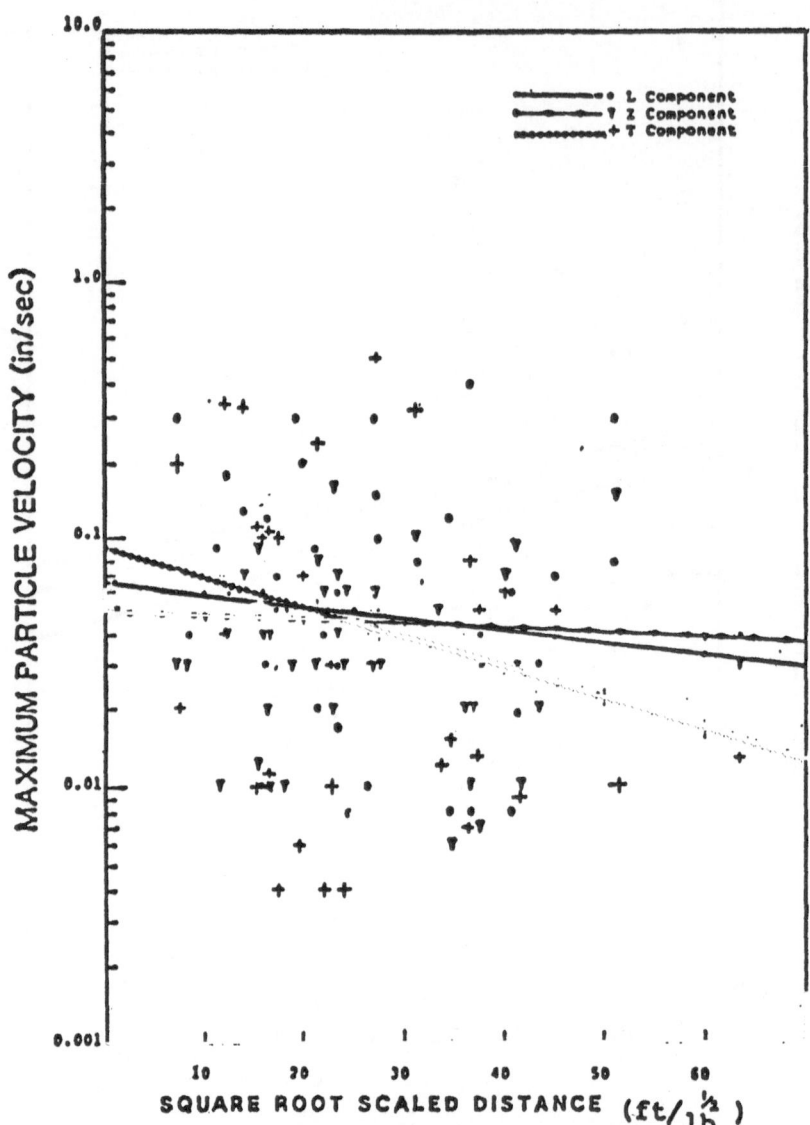

Figure 3.7   Plot of the Maximum Particle
             Velocity Versus Scaled Distance
             for Hematite

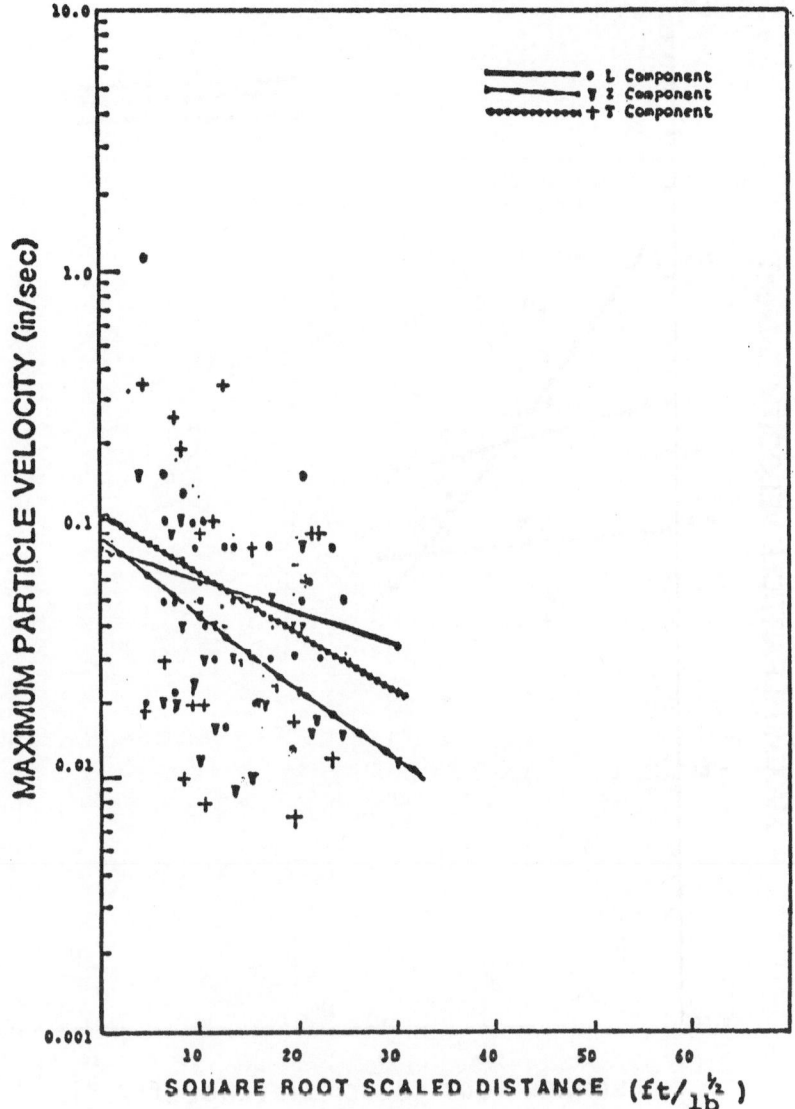

Figure 3.8 Plot of the Maximum Particle
Velocity Versus Scaled Distance
for Hard Limestone

299

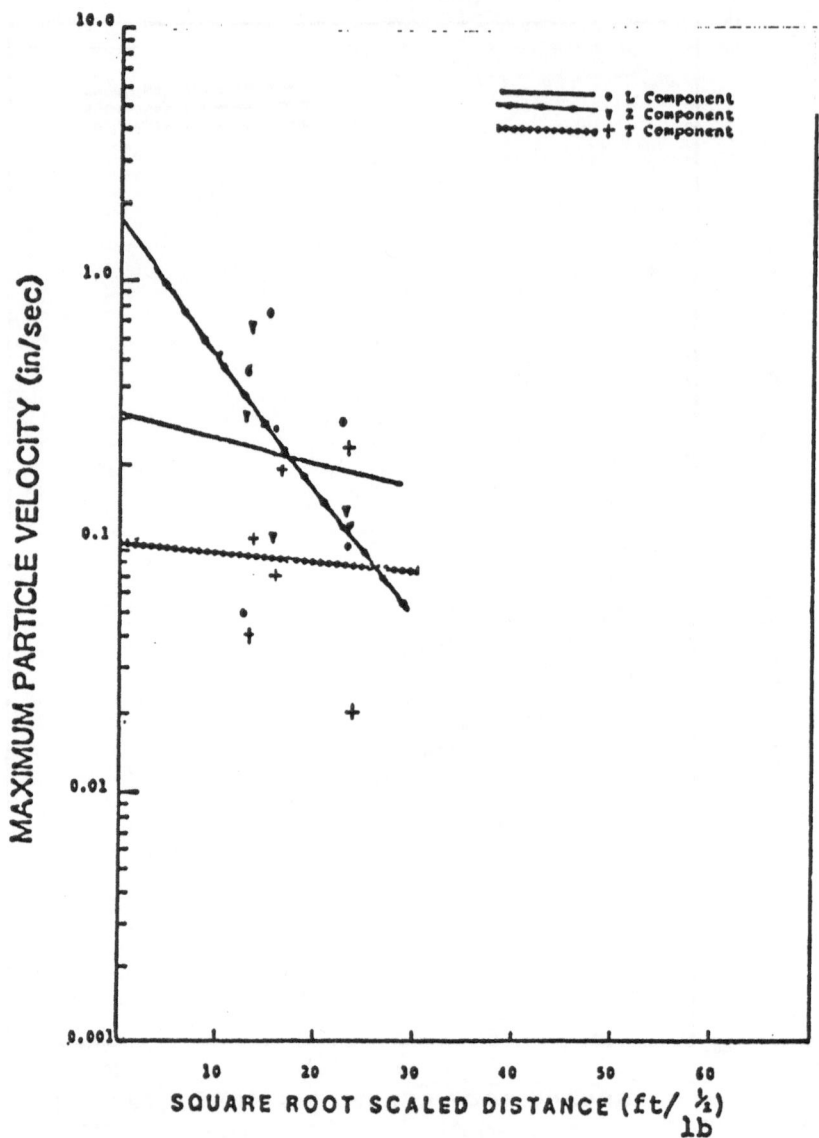

Figure 3.9   Plot of the Maximum Particle
             Velocity Versus Scaled Distance
             for Soft Limestone

300

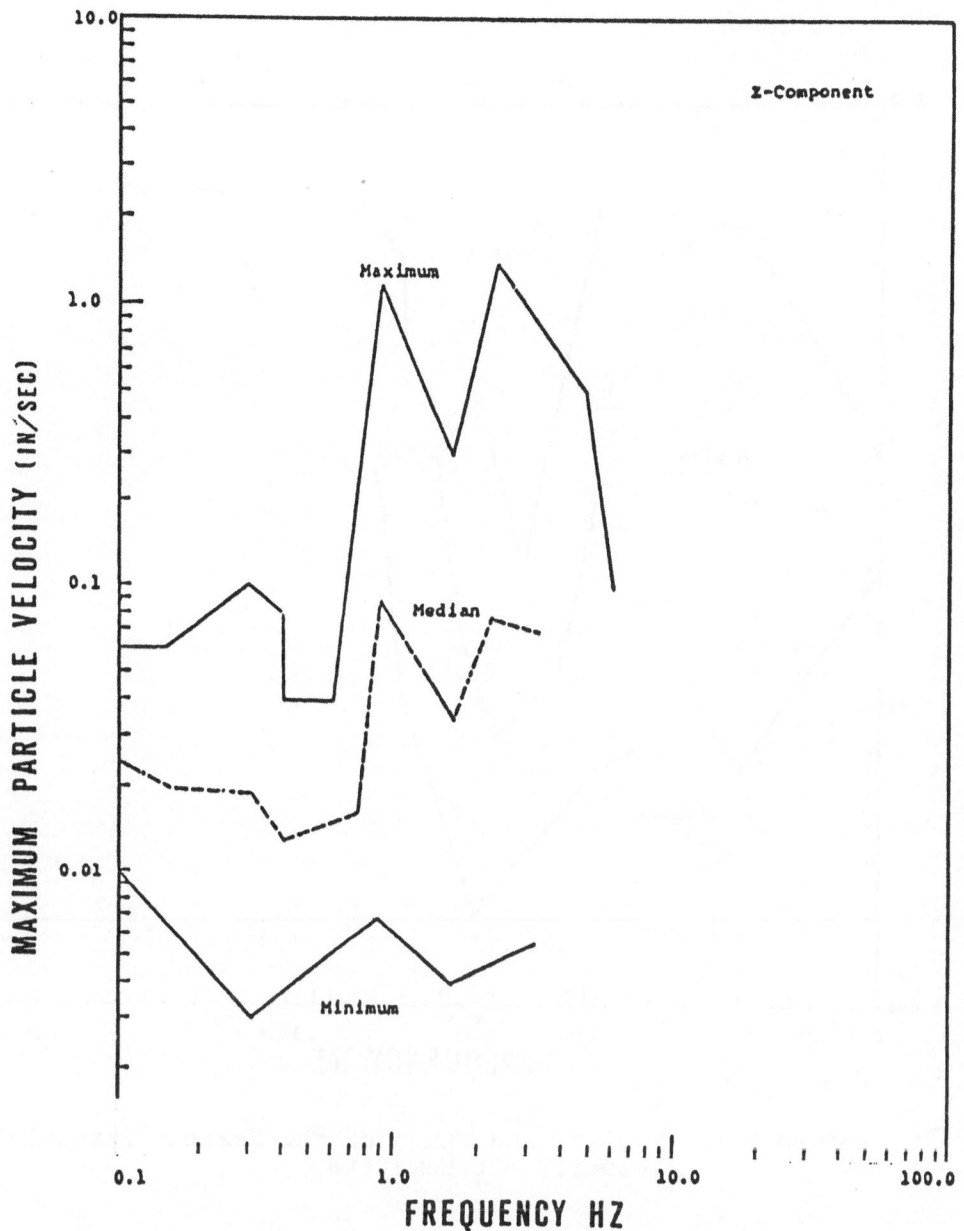

Figure 3.10 Median, and Range of the Maximum Particle
Velocity for Taconite

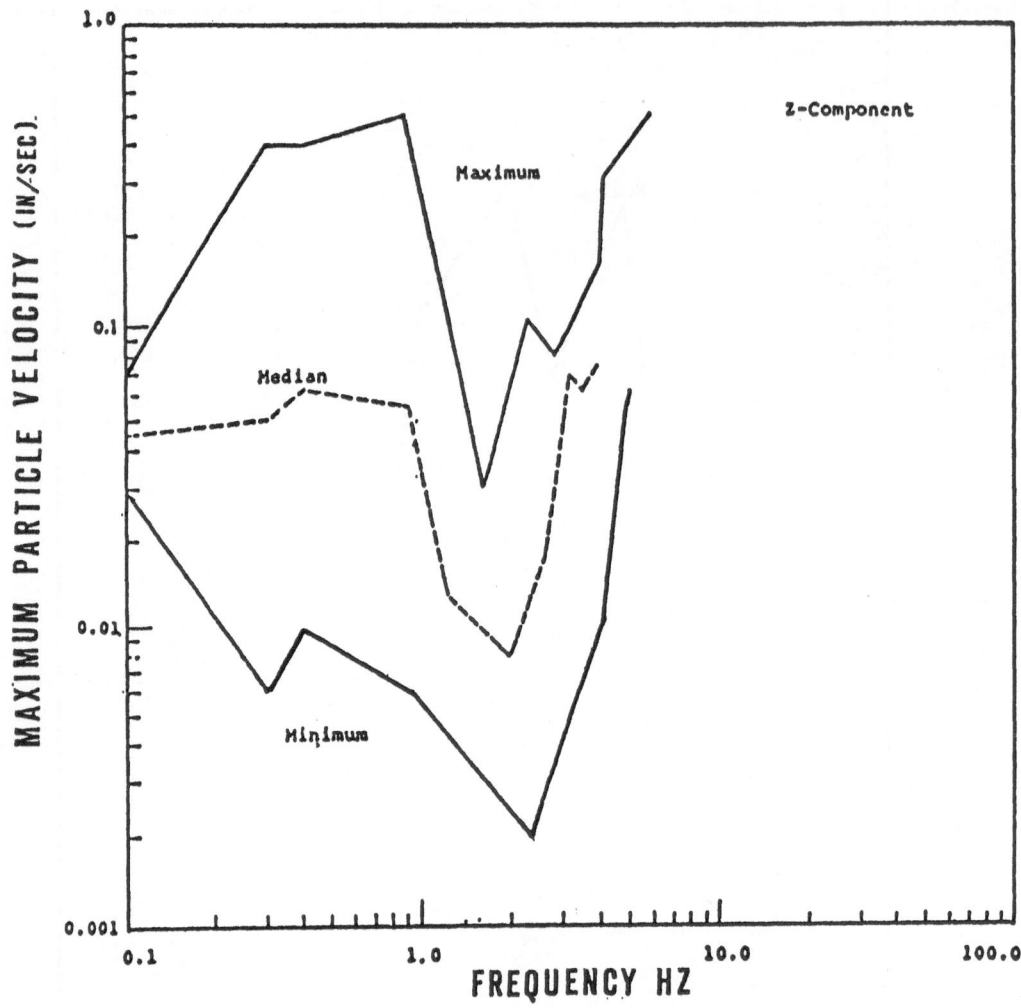

Figure 3.11   Median, and Range of the Maximum Particle
Velocity for Hematite

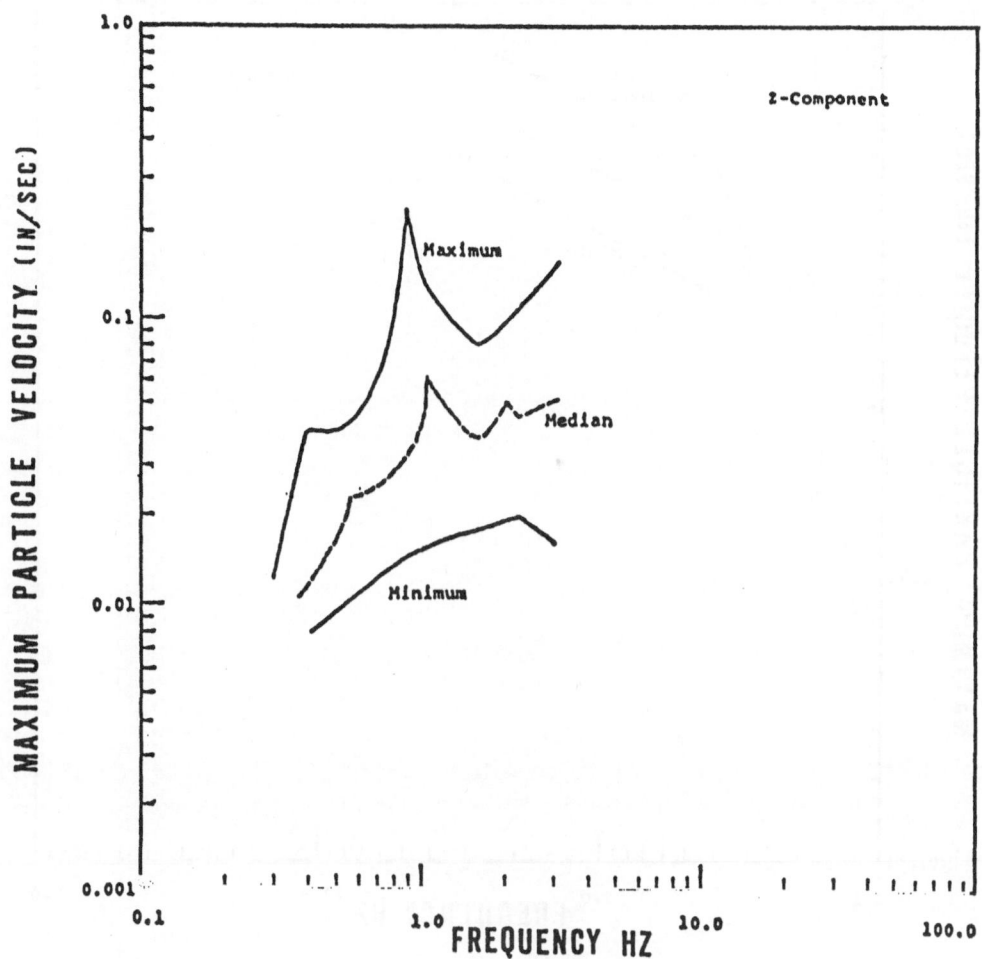

Figure 3.12   Median, and Range of the Maximum Particle
Velocity for Hard Limestone

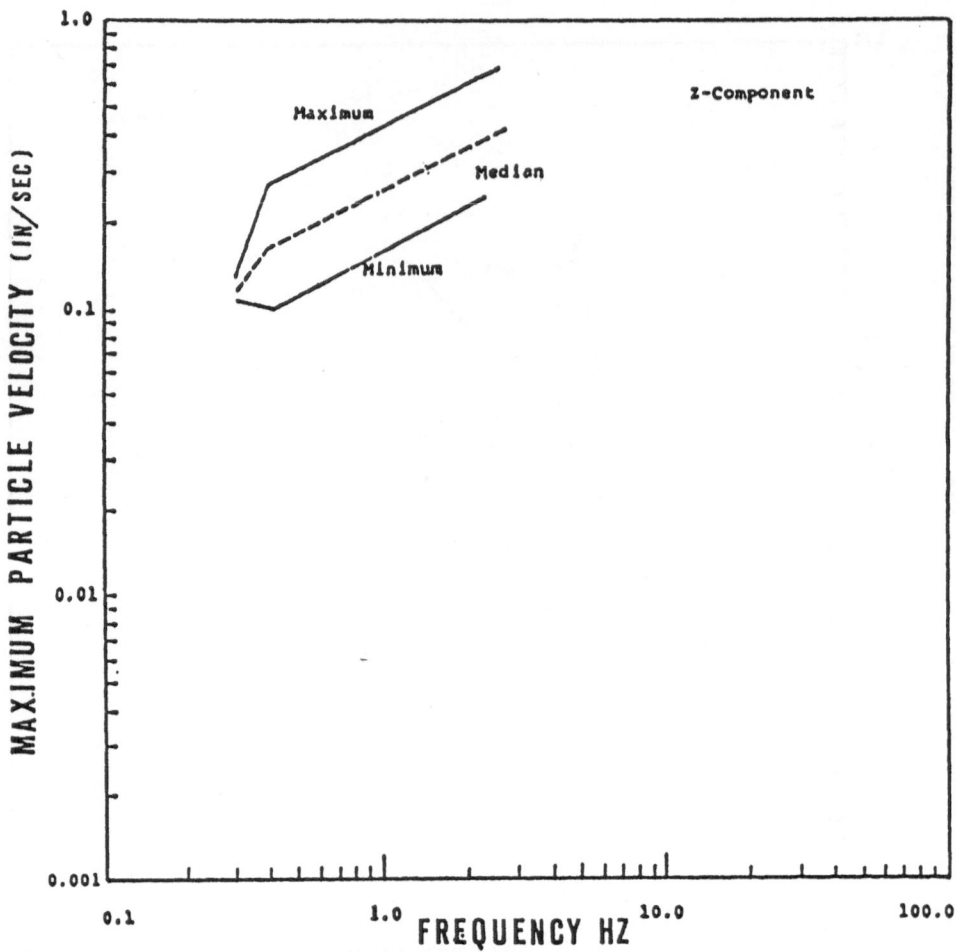

Figure 3.13   Median, and Range of the Maximum Particle
Velocity for Soft Limestone

3.5 Corner Frequency

The corner frequency concept is frequently used in seismology
to estimate the earthquake source configuration and the size
of the ruptured zone (1,2,3). The corner frequency is defined
as the frequency of a point where the low frequency and the
high frequency spectral trends of vibrational data intersect.
A hyphothetical plot of frequency versus amplitude illustrating
the corner frequency concept is shown in Figure 3.23. After an
earthquake, stress is relieved over a finite time or space, and
therefore, the frequencies of the radiated pulses are broadened
proportionally resulting in a corner in both the P-wave and the
s-wave spectra. An estimate of the ruptured zone can be obtained
from this corner frequency using the following equations:

$$r_0 = 2.34\, \alpha/W_\alpha \, , \tag{3.1}$$
$$r_0 = 2.34\, \beta/W_\beta \, , \tag{3.2}$$

where r_0 is the radius of ruptured zone, α and β denote the
p-wave and the s-wave velocities, and W_α and W_β denote the
p-wave and s-wave corner frequencies (11). Equation 3.1 and
3.2 show an inverse relationship between the radius of the
ruptured zone which is a measure of the source size and corner
frequency, i.e. the corner frequency decreases with source size.

The corner frequency concept was applied to blast vibrations in
this study to examine whether a corresponding relationship
between blast size and corner frequency did exist. Since the
spectral analysis of the blasts are really a combination of p,
s, and surface waves, corner frequencies of the p-and the s-wave
spectra could not be analyzed. Therefore, the corner frequencies
obtained are from the corrected amplitude spectra of the entire
blast record. It was difficult to establish the precise corner
frequency in the blast spectra because of their narrowness and
consequent diffuse low and high trends in many cases. The corner
frequency of some of the corrected amplitude spectra plotted
using arithmetic coordinates are shown in Figure 3.14 the result
of the analysis showed that the corner frequency appears to
decrease with increasing shot duration

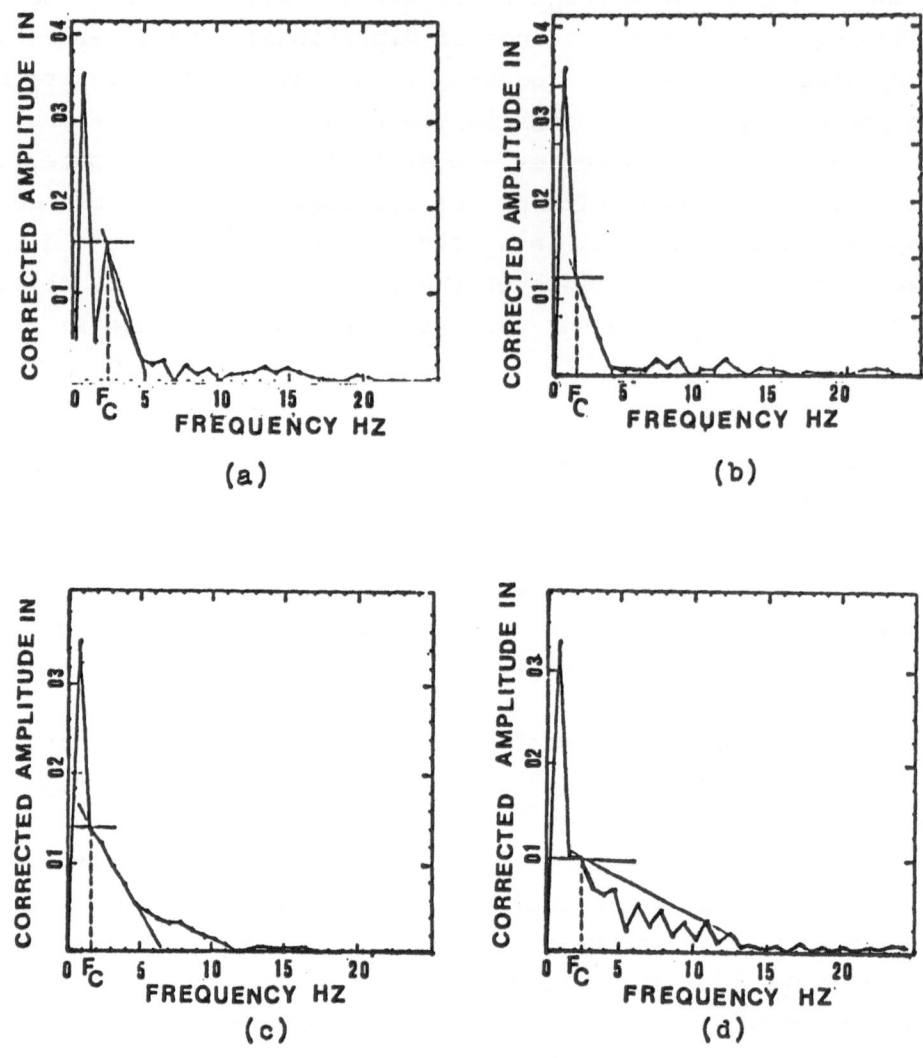

Figure 3.14   Corner Frequencies of Corrected Amplitude
Spectra for the Z-Component of   some records.

CONCLUSION

The seismic blast vibrations generated by multi-hole surface
(open-pit) commercial explosions in four different lithologies
(teconite, hematite, hard limestone, soft limestone) have
been analyzed and the following conclusions are drawn from the
data.

a- The maximum power and the maximum ground vibration amplitudes
   are below 10.0 Hz for all rock types.

b- The signal durations are longer in soft limestone and hematite,
   then in taconite and hard limestone.

c- The maximum particle velocity data for large commercial blasts
   are characterized by large scatter for all rock types examined.

d- Taconite has the highest powder factor, soft limestone the
   lowest powder factor, and hematite and hard limestone exhibit
   intermediate values of powder factor.

e- In general, large duration signals have "narrow" frequency
   spectra, and short duration signals have "broader" frequency
   spectra.

f- Corner frequency for most blast spectra varies inverseley
   with source duration, that is, long duration blasts exhibit
   a lower corner frequency than short duration blast.

g- Maximum particle velocity decreases with increasing distance
   and shows considerable scatter.

REFERENCES

1- Brune, J.N., 1970, Tectonic Stress and The Spectra of Seismic
   Shear Waves From Earthquakes, J.Geophys. Res. V. 75,
   P. 4997-5009.

2- Brune, J.N., Archuleta, R.J., and Hartzell, S., 1979, Far Field
   S-Wave Spectra, Corner Frequency, and Pulse Shapes. J.
   Geophys. Res., v. 84, p. 2262-2272.

3- Brune, J.N.. 1971, Correction, J. geophys. Res., v. 77, p. 5002

4- Duvall, I.W., 1953, Strain-Wave Shapes in Rock Near Explosions. Geophysics, v. 18, p. 310-323.

5- Duvall, I.W., and Petkof, B., 1959, Spherical Propagation of Explosion-Generated Strain Pulses in Rock, BuMines RI 5483.

6- Duvall, I.W., and Atchison, T.C., 1957, Rock Breakage by Explosives, BuMines RI 5356.

7- Fogelson, D.E., Duvall, I.W., and Atchison, T.C., 1959, Strain Energy in Explosion-Generated Strain Pulses. BuMines RI 5514.

8- Grant, C.H., 1980, An Emprical Method of Examing Energy Distribution in Blast Patterns. Soc. of Mining Engineers of AIME.

9- Hino, K., 1959, Theory and Practice of Blasting. Nippon Kayaku Co. Ltd. Japan.

10- Langefors, U., and Kihlstrom, B., 1973, The Modern Technique of Rock Blasting. John Willey end Sons.

11- Molnar, P., Tucker, B.E., and Brune, J.N., 1973, Corner Frequencies of P and S Waves and Models of Earthquake Sources. Bull. Seism. Soc. Am., v. 63, p. 2091-2104.

12- Olson, J.J., Fogelson, D.E., Dick, R.A., and Arlo D. Hendrickson, 1972, Vibrations from Tunnel Blasting in Granite. BuMines RI 7653.

13- Sheriff, R.E., 1980, Seismic stratigraphy. International Human Resources Development Corporation.

14- Siskind, D.E., Steckley, R.C., and Olson, J.J., 1973, Fracturing in the Zone Around a Blast Hole. White Pine, Mic. BuMines RI 7753.

15- Siskind, D.E., 1980, Damage to Residential Structures from Surface Mine Blasting. Soc. of Mining Engineers of AIME. Preprint No: 80-362.

16- Thonen, J.R., and Windes, S.L., 1938, Earth Vibrations Caused by Mine Blasting Progress Report 2. BuMines RI 3407.

Optimization Approach to the Earthquake Source Inverse Problem

A. S. Bykovtsev[1], V. A. Cheverda[2], V. G. Khaidukov[3]

[1] Institute of Seismology, Kursheed st. 3, Tashkent, 700128, USSR
[2] Computing Center, Ac. Lavrentjev st. 6, Novosibirsk, 630090, USSR
[3] Institute of Geology and Geophysics, University pr. 3, Novosibirsk, 630090, USSR

Abstract.

The paper deals with the model inverse problem of the determination of mechanical and geometrical parameters of the rupture propagating in an elastic medium. The optimization approach is used to solve this problem. The displacement field is calculated by means of the exact solution of the forward problem for the moving rupture of the complicated form in the infinite elastic medium. The character of wave field distribution in space and time, and main properties of the data misfit functional are investigated in detail for a simple example of the plane infinitely wide ground of the rupture. The possibility to solve such a kind of inverse problem in principle is established. Such kinds of problems could be of interest both in seismology to study earthquake mechanism and in practical applications to control the material destruction.

1. Introduction.

Determination of the geometrical and mechanical characteristics of the earthquake source is the basic problem not only for seismology, but for the Earth sciences in general. Its solution gives the possibility to reconstruct the main features of the seismic energy radiation process, which takes place because of the media destruction under a stress action due to the tectonical motion in the Earth's crust. So, the knowledge about an earthquake source mechanism allows us to make some principal conclusions about tectonical processes.

309

The most widely used approach here is the extraction, detailed analysis and interpretation of a separated component from the full wave field - arrival times, signs, amplitudes and polarizations for P- and S- waves, their spectra for some frequencies, amplitudes and phases for some modes of the surface waves and other characteristics (Aki and Richards, 1980). Such a separation of wave fields is possible only for the far zone. It is obvious, that at such distances a seismic wave field becomes very close to a point source field and it is absolutely impossible to determine the detailed structure of a real earthquake source from this kind of information. So, it is necessary to draw a wave field in the near zone (strong motion data) in order to study the detailed structure of the earthquake source mechanism. In such zones, however, it is impossible to separate a full wave field on the P-, S - and surface waves. Hence, it is necessary to develop a technique to do the full wave field inversion.

The first spatial inversion over the fault surface was done by Triffunac for the San Fernando, California, earthquake of February 9,1971. This work has demonstrated the wide possibilities of this approach and pointed out some problems in its applicability and usefulness. There were two main reasons for it. The first one is the overall complexity of the problem and its mathematical modeling involving the earthquake source and the wave propagation problem through the surrounding inhomogeneous medium. The second problem was associated with the lack of the near field data.

Full wave field inversion for strong motion data was performed by Olson and Apsel (1982), Hartzell and Heaton (1986), Archuleta (1984), Jordanovski (1986) and others. All these authors handled the problem of reconstruction of maximal values of shift components of the displacement vector on the known fault surface S. Based on a well known wave field representation by the equivalent body forces [1]

$$U_{\kappa}(\bar{x},t) = \int_{-\infty}^{\infty} d\tau \int_{S} B_{i}(\bar{\xi},\tau) \cdot C_{ijpq} \cdot n_{j} \cdot \partial \, G_{\kappa}^{\rho}(\bar{x},\bar{\xi},t-\tau) \Big/ \partial \bar{\xi}_{\kappa} \, ds$$

$$k = 1,2,3$$

310

where $G(\bar{x}, \bar{\xi}, t)$ is the Green function, $n(\bar{\xi})$ - is the normal to the surface S, $\bar{B}(\bar{\xi}, \tau)$ - is a displacement vector, c_{ijpq} - is the tensor of elastic parameters, the least square problem is formulated and reduced to a system of linear algebraic equations with respect to unknown slip components B_i .

2. Analysis of the displacement field.

We will consider a more complicated problem: to determine not only the displacement vector, but also the fault surface S and the velocity of a rupture propagation. An optimization technique is used to reconstruct these parameters.

Optimization of any kind calls for an effective numerical method to simulate a full wave field on the surface. Here we use an algorithm based on the analytical formulae, which describe the wave field for the rupture propagating in a homogeneous elastic space. This solution was first obtained by Madariaga (1978) for the Huskell's model (a rectangular fault with the shift component of the displacement vector in a homogeneous elastic space) and further developed by Bykovtsev (1986, 1987) for the piecewise - plane fault surface with

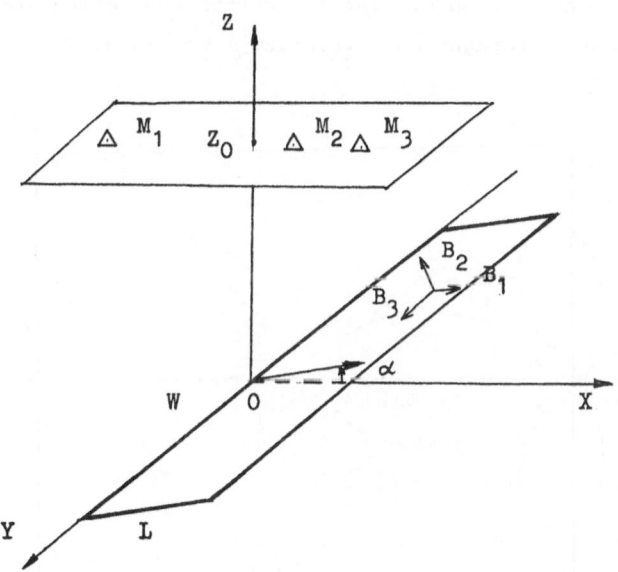

Fig.1. Geometrical and mechanical parameters of 2-D rupture ($W \gg L$): $Z0$ - hypocenter, α - dip angle, Vr - rupture propagation velocity, L - length, $B1$, $B2$, $B3$ - components of the displacement vector on the rupture.

branching.   The advantage of this method as compared with   the
Green   function   representation   is   that   instead   of   a
two-dimensional integration rather simple analytical formulae
can   be   used.   This   essentially   simplifies   wave   field
simulation, decreases computer time and other resources and
allows to organize an efficient iteration process of the data
misfit functional minimization.

Let us   start   with   a   2-D   model   situation: the fault
surface is   infinite   in   the   Y   direction   (for   practical
purposes width W of fault is rather more with respect to its
length L) and it is a segment at the X-Z plane (fig.1).   In
this   way   we will   attract   attention   to   some   principal
mathematical aspects of the minimization process.

At the   beginning   let us consider the main features of a
wave field generated by this fault.   The   method   allows   the
separation   of   P-   and S-   waves,   more   correctly   waves
propagating with the Vp and Vs velocities, respectively. These
waves   have a polarization only in their first arrivals, which
coincides with P- and S-waves.   Fig.2   demonstrates   radiation
diagrams   calculated from the first arrivals of P- and S-waves
for a rupture with pure   slip   displacement   and   pure   normal
displacement   vector   component.   Here   the antiplane shift is
taken equal to zero.   These diagrams are   calculated   for   the
rupture   propagation   velocity   Vr = 0.6Vs.   A   clear-cut

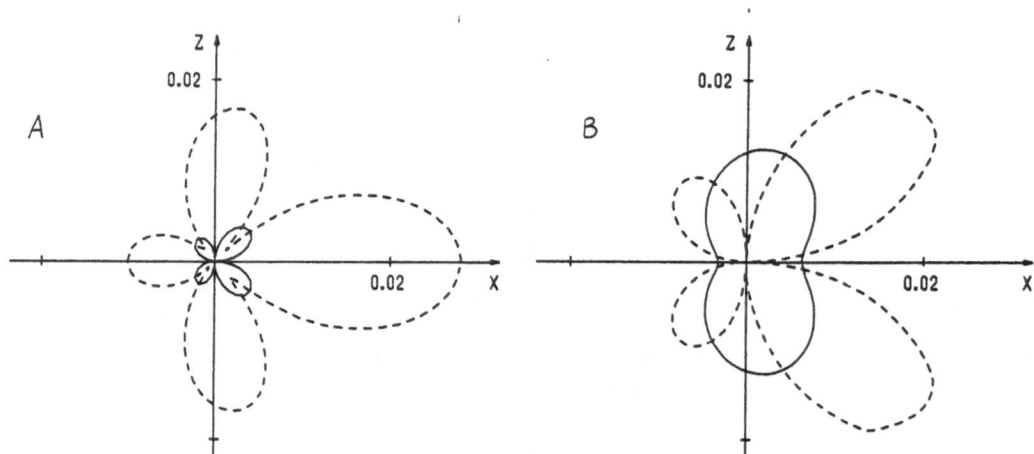

Fig.2. Radiation diagram for R = 46L. (A) - the rupture with
        a pure shift displacement; (B) - the rupture with a pure
        break displacement. Solid line - P-wave, dashed line -
        S-wave.

antisymmetrical effect exists everywhere , due to the rupture
propagation. The vector field of the elastic displacement
$\overline{U}(x,z,t)$ in the vertical plane orthogonal to the rupture
is propagation. The vector field of the elastic displacement
$\overline{U}(x,z,t)$ in the vertical plane orthogonal to the rupture
is demonstrated in fig.3 (a rupture with a pure shift
displacement) and in the fig.4 (a rupture with a pure normal
displacement). Diagrams are calculated for the time t equal to
a P-wave propagation time through the distance of about 46
earthquake source size and demonstrate complicated character
of polarization between P- and S-waves arrivals. Radiation
diagrams agree well with the ones adduced for similar models
in the book by Aki K. and Richards P.(1980). For models with
a more complicated rupture configuration and vector displacement, radiation energy diagrams have non-quadrant distribution

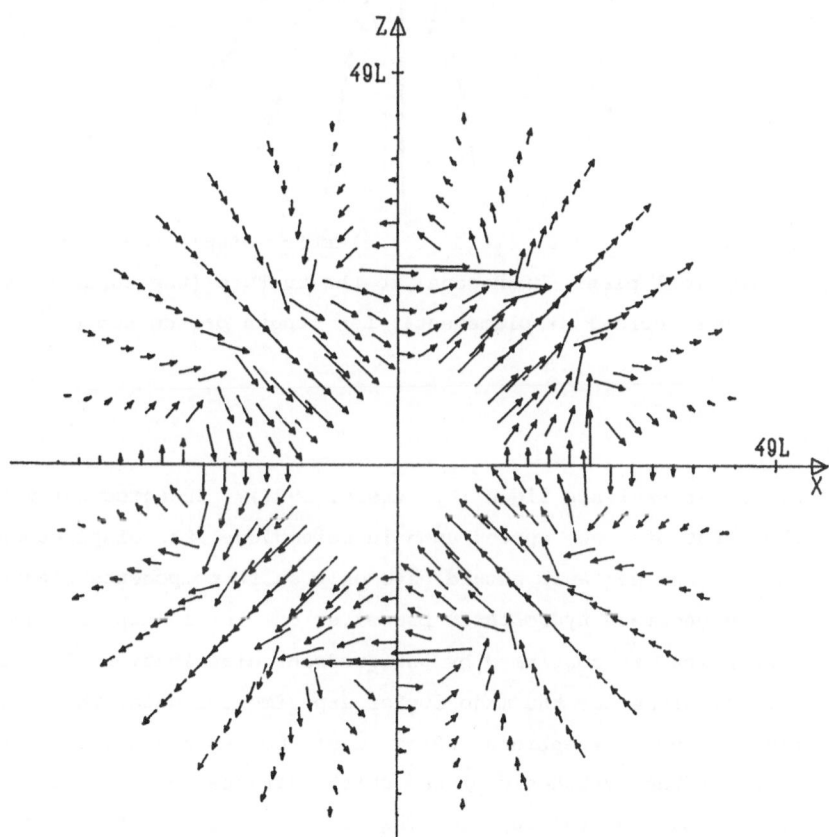

Fig.3. The vector field of elastic displacement in the
vertical plane, orthogonal to the rupture (the rupture with
a pure shift displacement). L - length of the rupture.

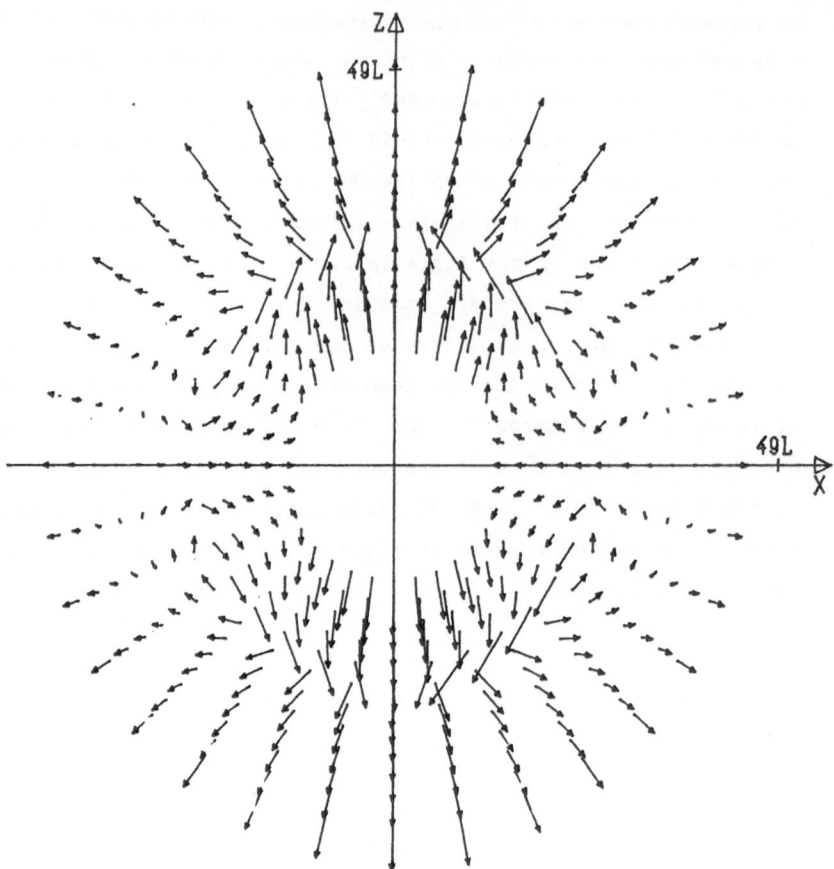

Fig.4. The vector field of elastic displacement in the
vertical plane, orthogonal to the rupture (the rupture with
a pure normal displacement). L - length of the rupture.

of first arrivals signs (Bykovtsev, 1987). In agreement with
the point of view agreed upon in seismology, the displacement
vector on the fault should have only shift components because
of the enormous hydrostatic pressure on these depths. This
assumption is confirmed by the quadrant distribution of first
arrival signs for the majority of experimental data. There are
though some exceptions which can't be explained within the
scope of the hypothesis of the shift character of the vector
displacement on the rupture. The situations are possible where
there arises vector displacement with normal component.

314

3. Least square inversion.

Let us consider now the following statement of the inverse problem: let a rupture start to propagate at the moment t=0 with the velocity Vr in an elastic space from the Y-axis in the direction given by the angle α with X-axis. The displacement along this rupture is given by the vector $\overline{B}$ with components B1,B2,B3 (see fig.1). This rupture is infinite along the Y-axis and has a finite length L in the (X,Z) plane. The problem is to determine rupture parameters and initial depth Z0 from the wave field $\overline{U}(M_j,t)$, recorded at N points of observations M_j on the Earth's surface. Let us denote the vector of parameters $\overline{\psi}$. In a given statement it has next the appearance $\overline{\psi}$ = (B1, B2, L, Vr, α ,Z0). As mentioned above we use an optimization approach. So, the solution of this inverse problem is the vector $\overline{\psi}^*$ providing with the minimum of the data misfit functional characterizing quadratic deviation between recorded and simulated seismograms for the current parameters:

$$\Phi(\overline{\psi}) = \sum_{j=1}^{N} \int_0^T |\overline{U}(M_j,t) - \overline{U}_s(M_j,t,\overline{\psi})|^2 dt$$

Here Us is the solution of the direct problem calculated by the method mentioned above. This problem is a standard non-linear optimization problem and the first question that arises here is if there are any local minima. Unfortunately, we couldn't investigate this problem theoretically for the general situation and so our efforts were concentrated on detailed numerical investigations of the data misfit functional.

4. Results and discussions.

The range of the parameter values was taken wide enough and normalized to unit magnitude.

As a model we took a rupture with the following unitless parameters from the pointed range:

Z0 = 0.5; α = 0; Vr = 0.625; L = 0.5; B1 = 0.25; B2 = 0.5. For this model, cross-sections of the data misfit functional were calculated for every pair of parameters (others were fixed) for different numbers of recording stations (N = 1,3,5). Here we shall demonstrate such cross-sections for a unique station, situated at the distance R = 50L from the rupture in the direction at 45 degrees with the fault surface.

Analysis of the wave field representation shows its linear dependence with respect to vector displacement components on the rupture. So, the data misfit functional must be quadratic with respect to B1, B2, B3. Our calculations demonstrated that the Hesse matrix of the data misfit functional is positive-definite and its cond number is small enough (fig.5). This situation may be useful to improve the convergence of the iterational process.

The most complicated is the functional dependence with respect to angle α. A stable and rather deep local minimum is present everywhere for value α distinguished from the true one for 180 degrees (0.5 for adjustment scale). There are also another local minima (not so deep as this one) which differ from the true values for the angles divisible by 90 degrees. Their structure roughly repeats the radiation diagram symmetry (figs. 6,7).

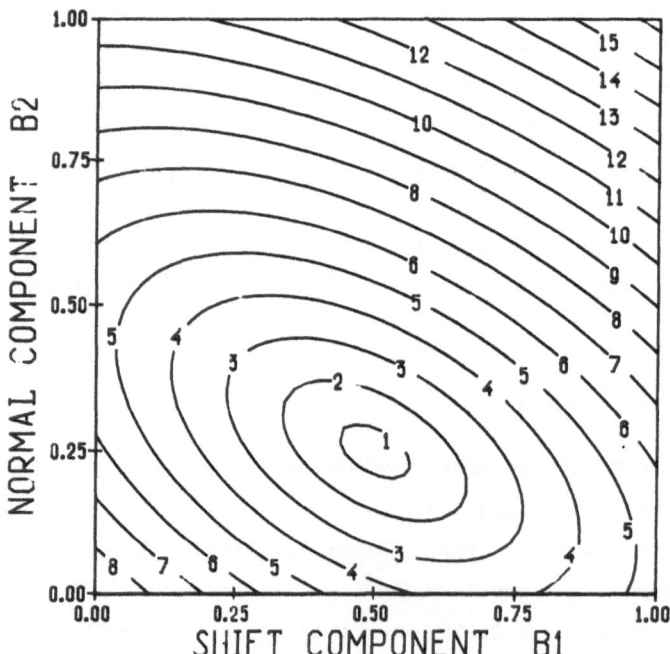

Fig.5. Cross-section of the data misfit functional, plane (B1, B2) in adjustment scale.

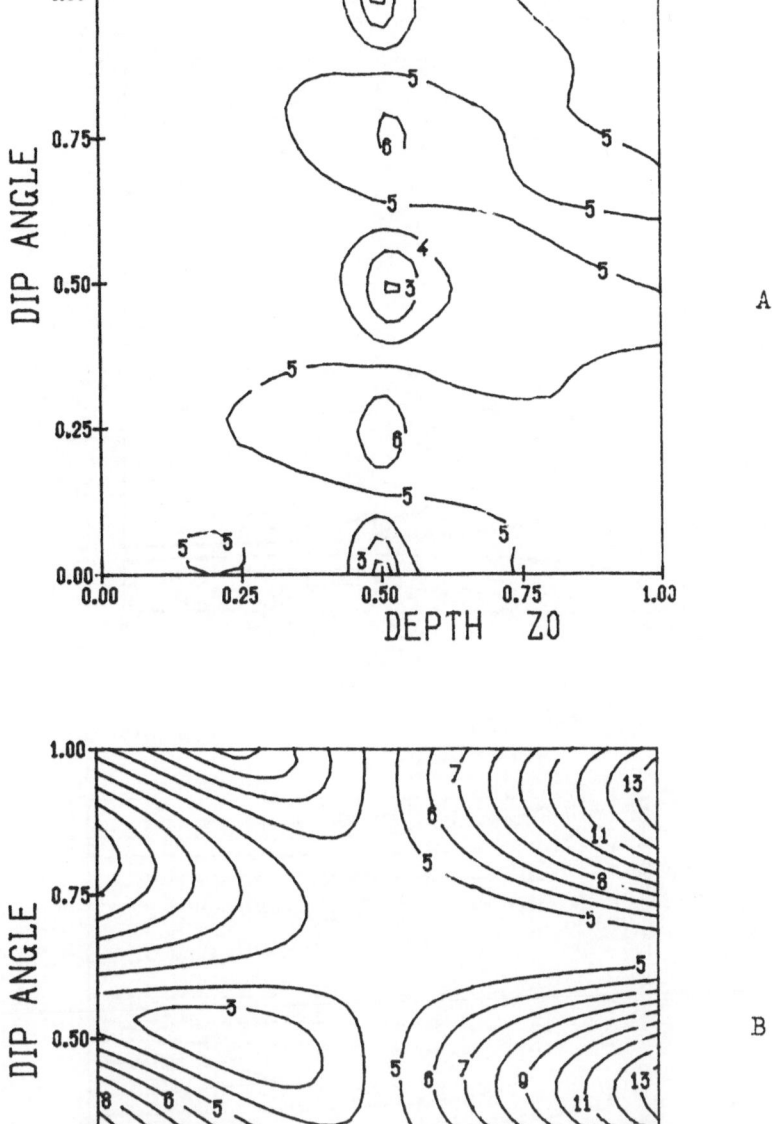

Fig.6. Cross-sections of the data misfit functional in adjustment scale. (A) - plane (Z,α); (B) - plane (B2,α).

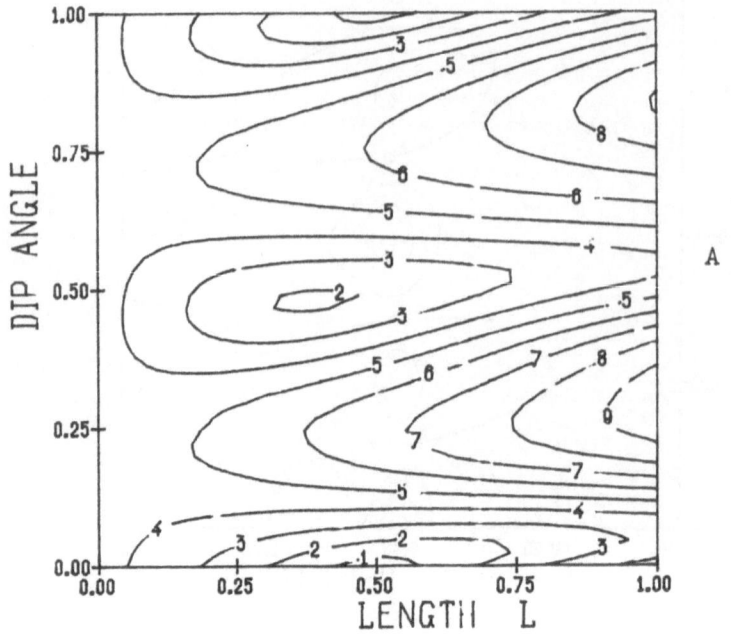

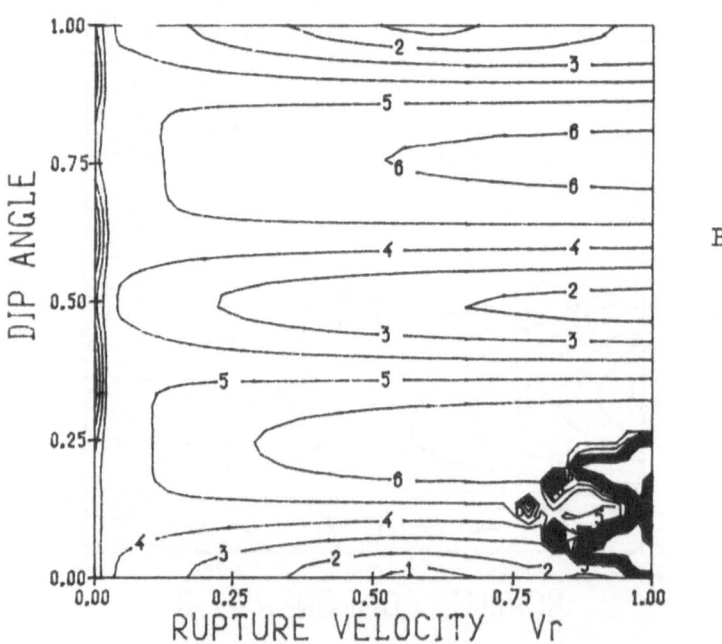

Fig.7. Cross-sections of the data misfit functional in adjustment scale. (A) - plane (L, α); (B) - plane (Vr, α).

There is a certain improvement with an increase of recording stations . Local minimum for $\alpha = \alpha$ true + 0.5 doesn't disappear but common smoothing takes place with such kind of data accumulation. Knowledge about local minimum distribution allows us to propose the algorithm, based on the exhaustion of local minima. Having compared the final results, we can choose the point of the global minimum.

A peculiarity of this data misfit functional is the presence of "ravins" and complicated structure of its level lines with respect to parameters Z0, L, Vr (figs.8,9). This fact, together with possible undifferentiability of this functional that is supposedly due to bends of level lines in some cross-sections, essentially hampers the choice of an effective iteration process.

Taking into account all these details, we have tried to use the symplex method (Nelder and Meed, 1964) to organize the minimization process. The main advantage of this method is the possibility to use no derivatives of the data misfit functional.

The next series of calculations were done to investigate the influence of the initial approximation on the stability of the iteration process and accuracy of the minimum point determination. We obtained the following results:

-if the deviation of initial approximation from true values doesn't exceed 20%, the iteration process converges very stably and rapidly enough to the true value;

-if these magnitudes are in the range 30 - 40%, there are some parameters which can't be determined with sufficient accuracy, though others are found with good precision. It's very astonishing that such parameters as depth Z0 and angle α are determined very stably and precisely. It may be explained by the fact that "ravines" of the data misfit functional with respect to these variables are rectilineal and parallel to others on the coordinate axis. Thus, if current parameters are hit in such a "ravine", Z0 and α are determined well enough;

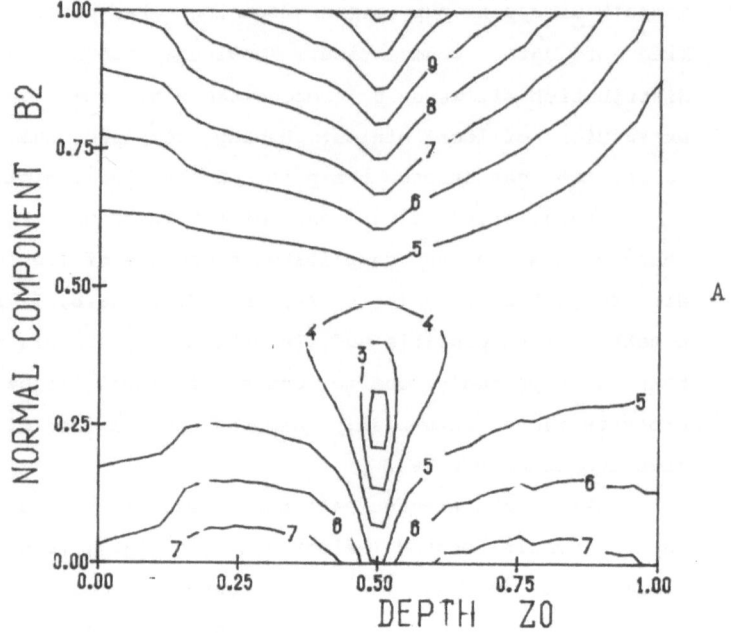

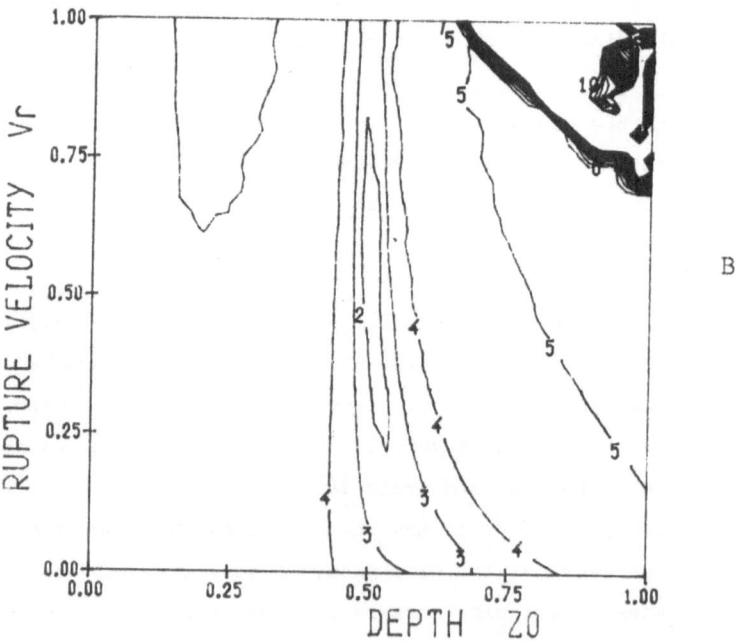

Fig.8.   Cross-section of the data   misfit functional in
adjustment scale. (A) - plane (Z,B2); (B) - plane (Z, Vr).

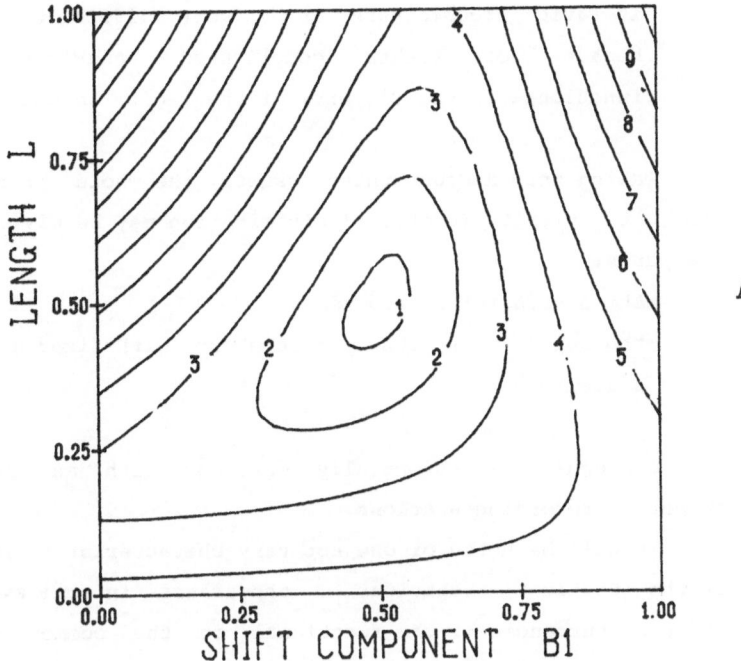

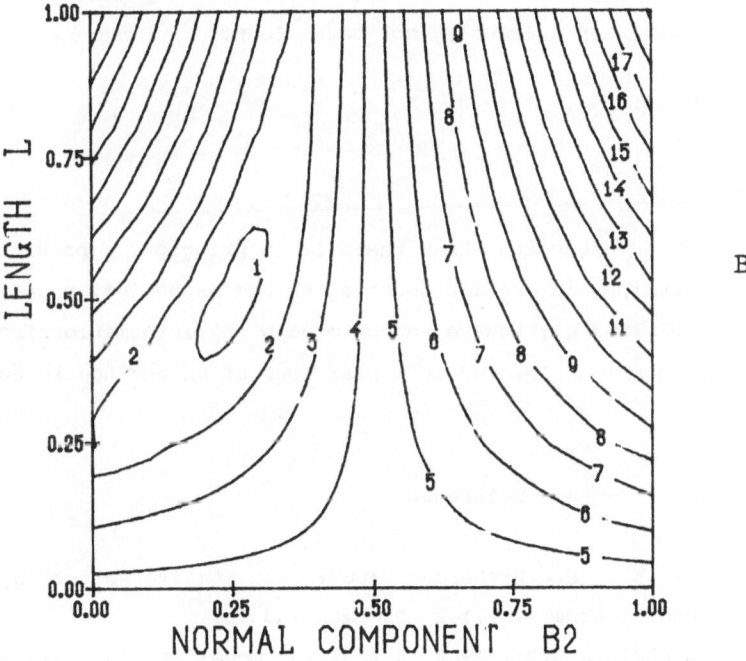

Fig.9. Cross-sections of the data misfit functional in
adjustment scale. (A) - plane (B1,L); (B) - plane (B2,L).

-for the initial approximation with a strong deviation
from the true parameter values, the convergence of the
iteration process here is a very difficult question
because of rather complicated behaviour of the
functional in the vicinity of the region boundaries.

Taking into account these results, the whole process of
the data misfit functional minimization may be divided into
two parts:
-the search for Z0 and α ;
-the search for other parameters with fixed depth and
angle.

Convergence is essentially improved with an increased
number of recording stations.
It will be noted of one not very characteristic property
of the functional behaviour - appearance in some events of
sharp disturbances of the functional on the common smooth,
stationary background. These "bifurcations" arise sometimes at
the edges of intervals for parameters Z0 and Vr , when
Z0 $\rightarrow$ 0 and Vr $\rightarrow$ Vs (fig.7B,8B). An explanation of this
effect has presently not been found.

5. Conclusion.

We establish that there is in principle a possibility to
determine in details geometrical and mechanical properties of
simplified earthquake source models by information from several stations located in a near zone of an earthquake source.

References

1.Aki K., P.G.Richards , 1980: Quantitative seismology.Theory
  and methods. vol.1,2, Freeman and Co.
2.Archuleta R.J., 1984: A faulting model for the 1979 Imperial
  Valley Earthquake, Jour. Geoph. Res., vol.89, no. B6,
  4559-4585.

3. Bykovtsev A.S., D.B.Kramarovsky ,1987: About propagation of the complicated default square. Exact 3-D solution. Prikladnaja matematica i mechanica, vol.51, 1, 117-129.

4. Bykovtsev A.S.,1986: Modelling of fracture processes occurring in the focal zone of tectonic earthquake. Proceedings of International Conference on Computational Mechanics, May 25-29, Tokyo, vol.1, 111-221-226.

5. Gill P.E., W.Murrey,M.H.Write , 1981: Practical optimization. Academic Press.

6. Hartzell S.H. and T.H.Heaton (1983). "Inversion of strong ground motion and teleseismic wave front data for the fault rupture history of the 1979 Imperial Valley, California,Earthquake", Bull. Seism. Soc. Amer., Vol.73, No. 6, 1553-1583.

7. Haskell N.A. (1969) "Elastic displacement in the near-field of a propagating fault", Bull. Seism. Soc. Amer., vol. 59, 2, 905-908.

8. Jordanovski L.R., M.D.Trifunac and V.W.Lee ,1986, "Investigation of numerical methods in inversion of earthquake source", Dept. of Civil Eng. Report No 86-01, Univ. Southern California.

9. Kasahara K., 1981,"Earthquake mechanics", Cambridge University Press.

10. Madariaga R., 1978, "The dynamic field of Haskell's rectangular dislocation fault model", Bull., Seism., Soc., Amer., v.68, 4, 869-887.

11. Nelder J.A., R.Meed, 1964, "Symplex method for functional minimization", The Computer Journal, 7, 308-313.

12. Olson V.H., R.J.Apsel , 1982, "Finite faults and inverse theory with applications to the 1979 Imperial Valley earthquake", Bull. Seism. Soc. Amer., vol.72, No.6, 1969-2001.

13. Trifunac M.D.,1974, "A three-dimensional dislocation model for the San Fernando, California, Earthquake of February 9, 1971", Bull. Seism. Soc. Amer., vol.64,1,511-533.

5. Geothermics

Reconstruction of the Surface Temperature History by the Least-Squares Inversion Theory

J. Šafanda

Geophysical Institute of Czechoslovak Academy of Sciences, 141 31 Prague, Czechoslovakia

Abstract

An algorithm for the extraction of past climate from the geothermal measurements, based on the general least-squares inversion theory, is dealt with. The procedure suggested enables us to reconstruct the main features of the surface temperature history for two simple models of the subsurface environment. The first supposes a transient heat conduction in a homogeneous halfspace, the other allows the transient heat conduction in a stratified medium composed of layers with different thermal conductivity and diffusivity.

Introduction

The problems of the past climate extraction from the geothermal measurements have been attracting ever increasing attention for more than one decade. In this connection such papers as Beck, 1977 and 1982; Shen & Beck, 1983; Vasseur et al., 1983; Lachenbruch & Marshal, 1986 and many others can be mentioned. In the department of geothermics of the Geophysical Institute in Prague this problem was encountered in connection with processing the temperature-depth logs obtained in Cuba during three joint Czechoslovak-Cuban expeditions in the eighties (Čermák et al., 1984; Čermák et al., 1990). A very conspicuous effect of a negative temperature gradient in the near-surface layer and the specific "U"-shape of the temperature-depth curve was recognized to be a general phenomenon in most of a few tens of boreholes measured. The complex interpretation of this anomalous behaviour of the temperature logs is addressed to in Čermák et al., 1990. The present study is intended to show in detail one of the methods used in this reconstruction of the surface temperature history.

327

The method of inversion for a homogeneous medium

Due to the lack of information on the actual depth dependence of thermal diffusivity within the boreholes investigated and due to the other vaguely known effects such as a multi-dimensional nature of the temperature field, its deformation by terrain unevennesses in some cases and so on, a homogeneous one-dimensional transient model was used as the first approximation of the reality. Despite its simplicity, this approach provides a fairly good insight into time scales and the general magnitude of temperature changes probably involved in generating the observed anomalies. Thus, the temperature must satisfy the equation

(1)　　$\partial T/\partial t = k \, \partial^2 T/\partial z^2$,

where T is temperature, t time, z depth and k is the thermal diffusivity. The initial condition is supposed in the form

(2)　　$T(z,0) = T_0 + G_0 z$,

with T_0 denoting the undisturbed surface temperature and G_0 the undisturbed temperature gradient. The boundary condition at the surface is allowed to vary with time according to a simple three-parameter function

(3)　　$T(0,t) = T_0 + \Delta T(t/t^*)^{n/2}$　for　$0 < t \leq t^*$　and $n = 0,1,..$

The solution of Eqs(1-3) evaluated at $t = t^*$ is (Carslaw & Jaeger, 1959)

(4)　　$T(z) = T_0 + G_0 z + \Delta T \, 2^n \Gamma(n/2+1) \, i^n erfc(z/2\sqrt{kt^*})$,

where $\Gamma(x)$ is the gamma function of argument x and $i^n erfc$ is the n-th integral of the error function. Equation (4) thus enables us to evaluate the ground temperature at the end of a surface temperature change ΔT of duration t^*. The form of the change (3) can be adjusted by the value of n ; $n = 0$ is a step change, $n = 2$ represents a linear one and so on.

To determine the surface temperature history, we employed the general least-squares inversion theory proposed by Tarantola & Valette (1982a,b). This method has been recently applied to various aspects of geothermics (Vasseur et al., 1985; Nielsen, 1986; Wang & Beck, 1987). In this approach, the set of temperature measurements $T(z_j)$, $j = 1,...,m$ is treated as m-dimensional vector T, which is related to a parameter vector p by the equation

(5) $T = g (p)$,

where g is an m-dimensional vectorial function whose components are given by (4). In our case, p contains four components ΔT, kt^*, T_o and G_o, which are to be estimated by the inversion of the non-linear system (5). The method regards p and T as random vectors, with the components having a priori joint probability density function of Gaussian form

(6)   const. $\exp\{-1/2[(T-T_o)^T C_{TT}^{-1}(T-T_o)+(p-p_o)^T C_{PP}^{-1}(p-p_o)]\}$,

where T_o and p_o are expectations of T and p, and C_{TT}, C_{PP} are the covariance matrices, respectively. In the problem considered, the a priori expectations of $T(z_j)$ are given by actual measurements. A priori expectations of G_o and T_o were obtained by the linear least-square fit to the lower undisturbed section of the temperature record. Our confidence in the a priori values enters the problem in the form of variances and covariances, i.e. elements of covariance matrices C_{TT} and C_{PP}, and serves to constrain the parameters during the inversion. For the parameters we are most confident with, such as $T(z_j)$, very small variances are imposed so that during the inversion they have the priority of remaining as close as possible to their a priori values. For the parameters we have least confidence in, such as ΔT, kt^*, very large variances are given leaving the parameter effectively unconstrained and easily adjusted by the inversion to satisfy those that are strongly constrained. For parameters of intermediate confidence, such as G_o and T_o, properly chosen variances are given so that they can be adjusted to a certain degree, but do not violate our knowledge.

Due to the non-linearity of Eq.(5), the a posteriori density function is no longer Gaussian. Nevertheless, it is possible to obtain numerically that vector (T_{opt}, p_{opt}) which corresponds to the maximum value of the conditional density function. The value of p_{opt} can be obtained iteratively by the following algorithm

(7) $p_{k+1} = p_o + C_{p_0 p_0} G_k^T (C_{T_0 T_0} + G_k C_{p_0 p_0} G_k^T)^{-1} [T_o - g(p_k) + G_k(p_k - p_o)]$

where G_k is the matrix of partial derivatives of Eq.(5) with respect to p at point p_k. A posteriori covariance matrices are also given as a first-order approximation after linearization of (5) and the relation for C_{PP} reads

$$(8) \quad C_{PP} = C_{P_0 P_0} - C_{P_0 P_0} G^T (C_{T_0 T_0} + G C_{P_0 P_0} G^T)^{-1} G C_{P_0 P_0}$$

where matrices G, G^T are computed at the point P_{opt}. Better a priori information helps reduce the a posteriori uncertainties in the optimum values. All given formulae are discussed in detail in Tarantola & Valette (1982a,b) and Wang & Beck (1987).

The values of a priori standard deviations (s.d.) of the estimated parameters were assessed as follows. The s.d. of the measured temperature at individual points – 0.5K – reflects not only the experimental errors given by the accuracy of the probe (< 0.05K), but also the other possible errors induced e.g. by measurements before the thermal equilibrium was attained, by a motion of water or air in the borehole and so on. The s.d. of the undisturbed gradient was taken as 10% of its a priori estimate and the s.d. of the original surface temperature equalled 2K. The s.d. of ΔT and t^* were chosen as 5K and 142 years, respectively. An a priori estimate of ΔT and t^* was obtained as a solution of the simplified problem, when the optimum values of ΔT and t^* were computed by the "classical" least-squares method for the same, but fixed G_0, T_0.

Strictly speaking, the shape of the surface temperature change was optimized with respect to the product kt^*. In the inversion the fixed value of k was kept. If the actual k is x-times greater, the corresponding t^* is x-times less.

The values of $n = 0,1,2,3$ and 4 were only considered in Eq.(3). The best match between observed and calculated temperature curves was achieved in most cases for $n = 0$, i.e. for a step rise of the surface temperature. The second most frequent value was $n = 4$. It was found that ΔT and t^* increased with increasing n. For $n = 4$, in most boreholes the rise amounts to less than 50% in ΔT and to about 300% in t^*, in comparison with values for $n = 0$. The differences of the residual sums of squares for the individual n's are small and, owing to the approximative character of the model, can be hardly used to the confident determination of the actual form of the surface temperature rise.

According to the location, the amplitude of the temperature increase ranges between 1K and 12K and the onset of the change varies from 15-20 years to 400 years. The warming was probably connected with the gradual deforestation of the island.

The method of inversion for a stratified medium

The results obtained for the homogeneous medium show that even the curves corresponding to the optimum shape of the surface temperature change do not fit the measured temperatures exactly. With a great probability, the observed differences are generated, besides the experimental errors during the temperature measurements, also by the thermal conductivity and diffusivity inhomogeneities. To handle the extraction of the past surface temperature changes in such a case, the above mentioned general least-squares inversion theory was extended to the transient heat conduction in a stratified medium. The extension is based on a numerical solution described in Nielsen & Balling (1985), which makes use of a concept of a medium stratification into unitary layers. A compact and efficient computational realization of the formalism has been compiled for the case of a step change of the surface temperature. Provided the thermal conductivity and diffusivity are known in the individual layers, the corresponding undisturbed gradients together with the surface temperature, the amplitude of the change and the time of its occurrence can be estimated from the measured temperature logs.

The application of the just mentioned extension is limited more or less to synthetic problems because in the practical cases our knowledge of thermal parameters is very poor. Nevertheless, in the future it may represent a promising tool for more detailed analyses.

References

Beck,A.E., 1977. Climatically perturbed temperature gradients and their effect on regional and continental heat flow means. Tectonophysics, 41, pp.17-39.

Beck,A.E., 1982. Precision logging of temperature gradients and the extraction of past climate. Tectonophysics, 83, pp.1-11.

Carslaw, H.S., Jaeger, J.C., 1959. Conduction of heat in solids. 2nd Ed., Oxford.

Čermák,V., Krešl,M., Šafanda,J., Nápoles-Pruna,M., Tenreyro-Perez, R., Torres-Paz,L.M., Valdes,J.J., 1984. First heat flow density assessments in Cuba. Tectonophysics, 103, pp.283-296.

Čermák,V., Krešl,M., Šafanda,J., Bodri,L., Nápoles-Pruna,M., Tenreyro-Perez,R., 1990. Terrestrial heat flow in Cuba. (submitted to the publisher).

Lachenbruch,A.H., Marshall,B.V., 1986. Changing climate: Geothermal Evidence from Permafrost in the Alaskan Arctic. Science, Vol.234, pp.689-696.

Nielsen,S.B., 1986. The continuous temperature log: method and applications. Ph.D. thesis, Univ. of Western Ont., London, Canada.

Nielsen,S.B., Balling,N., 1985. Transient heat flow in a stratified medium. Tectonophysics, 21, pp.1-10.

Shen,P.Y., Beck,A.E., 1983. Determination of surface temperature history from borehole temperature gradients. J.of Geophys.Res., Vol.88, No.B9, pp.7485-7493.

Tarantola,A., Valette,B., 1982a. Generalized non-linear inverse problems solved using least square criterion. Rev.Geophys.,20, pp.219-232.

Tarantola,A., Valette,B., 1982b. Inverse problem = quest for information. J.Geophys., 50, pp.159-170.

Vasseur,G., Bernard,Ph., Van de Meulebrouck,J., Kast,Y., Jolivet,J., 1983. Holocene paleotemperatures deduced from geothermal measurements.Palaeogeogr.,Palaeoclimatol.,Palaeoecol., 43, pp.237-259.

Vasseur,G., Lucazeau,F., Bayer,R., 1985. The problem of heat flow density determinations from inaccurate data. Tectonophysics, 121, pp.25-34.

Wang,K., Beck,A.E., 1987. Heat flow measurement in lacustrine or oceanic sediments without recording bottom temperature variations. J.of Geophys.Res.,Vol.92,No.B12, pp.12837-12845.

6. Joint Inversion of Geophysical Data

Empirical Covariance Functions between Seismic, Density and Gravity Data – an Important Constraint in 3D Gravimetric-Seismic Stochastic Inversion

G. Strykowski

Kort- og. Matrikelstyrelsen (National Survey and Cadastre) Gamlehave Allé 22, DK-2920 Charlottenlund, Denmark

Abstract

In gravity field approximation it is possible to describe the gravity field by a stationary stochastic process. This approach is used in gravity field modelling by Least Squares Collocation (LSC). The purpose of this paper is to include seismic and density information within the framework of the LSC approach. This is done by estimating (and later modelling) covariances between the involved quantities.

In this paper, 41 density and sonic logs were used together with gravity information, to obtain the empirical covariance functions between the involved residual quantities. The area of investigation lies in the North Sea, the horizontal dimensions are 3°x6° and the log data span from 200 m to 5000 m in depth. Regional reference models for density, reciprocal seismic velocity and gravity are modelled and subtracted from the data. A reference model relating the density and the reciprocal seismic velocity is also computed. The correlation coefficient between the density and the reciprocal seismic velocity is -0.66.

The results of this investigation show the characteristic correlation lenghts, the variances and the correlation coefficients for the involved residual quantities. The peaks in the variation of the characteristic parameters with depth are related to the geology in the area. The data show a strong correlation between the residual densities both horizontally and vertically. The correlation between the residual reciprocal seismic velocities is not as strong as for the residual density distribution, which indicates that this parameter varies much more than the density distribution. The correlation coefficients between the residual density and the residual reciprocal seismic velocity are large, both positive and negative. There is no correlation between the residual gravity and the residual density and between the residual gravity and the residual reciprocal seismic velocity. This may be due to the constructed reference model for gravity. The residual gravity may still contain strong signal generated from depths which are not covered by the log data.

The covariance functions presented in this paper are probably characteristic only for the area of investigation. They contain however 'large scale' information about the distribution of the involved physical parameters within the sedimentary bassin and thus perhaps about the bassin evolution (e.g. compaction, tectonics).

335

1. Introduction

The theory of stochastic processes was introduced in physical geodesy by Kaula (1959). He used the theory to interpolate between the measured values of gravity anomalies. In his approach, the gravity anomalies can be described as a second order homogenous random field on a sphere, i.e. as a family of random variables with finite second order moments, the mean value equal to zero and a covariance function which is invariant with respect to the group of rotations of three-dimensional space around the origin (cf. Lauritzen (1973)).

To estimate the covariance $C_{\Delta g \Delta g}$ between gravity anomalies for a spherical distance ψ, the following formula is used:

$$(1) \qquad \tilde{C}_{\Delta g \Delta g}(\psi) = \frac{\Sigma \; \Delta g(P) \cdot \Delta g(Q)}{N} \; , \;\; \psi(P,Q) = \psi \; ,$$

where $\psi(P,Q)$ is a spherical distance between points P and Q on a sphere and N is a number of such products. $\tilde{C}_{\Delta g \Delta g}(\psi)$ denotes an estimate of $C_{\Delta g \Delta g}(\psi)$.

The use of a stochastic model for the modelling the Earth's gravity field is not without problems (see Lauritzen (1973), Wei (1985), Sansò (1986)).

In practical applications, i.e. modelling the Earth's gravity field, the usual approach is to describe the covariance function for the Earth's gravity field by a simple mathematical model with few parameters. The parameters of the model are adjusted in such a way, that the model approximates empirical covariance estimates (computed by formula (1)) optimally in a least squares sense (cf. Knudsen (1987)). The covariance model function is known for all the spherical distances (and not only for some discrete values). This technique of determining the covariance model from measurements in a local area, is used in the modelling the gravity field by Least Squares Collocation (LSC). The justification of a covariance model lies in the ability of the method to give good predictions using the specific model (which can be veryfied by comparing the predictions with the measurements).

The general method of Least Squares Collocation in physical geodesy was developed by Krarup (1969). Most of the measurable quantities of the gravity field outside the masses (gravity anomalies, gravity disturbances, deflections of the vertical, geoidal hights,..etc.) are linear functionals of the anomalous gravity potential. The fundamental mathematical concept of the method is a Hilbert Space with a Reproducing Kernel (RKHS). The basic reproducing kernel is the kernel for the anomalous gravity potential.

The stochastic interpretation of the method comes in, when one interprets the reproducing kernel as beeing equivalent with the covariance function for the anomalous gravity potential. Once the covariance model in a local area is established, the covariances between all of the other quantities are easily obtainable by simply letting two linear functionals associated with the measurements act on the basic kernel (covariance propagation). This operation is equivalent to taking an inner product of the functionals in the dual space.

For a set of measurements y_i , $y_i = L_i(T) + e_i$, i=1,..,N, where $L_i(T)$ is a linear functional acting on the anomalous potential and e_i is a normally distributed error, the equation of Least Squares Collocation (LSC) is:

$$(2) \quad L(\tilde{T}) = \{C_{Li}\}^T \{C_{ij} + D_{ij}\}^{-1} \{y_j\}, \text{ where}$$

$L(\tilde{T})$ is an estimate of a quantity that one wants to predict $(L(T) = L(\tilde{T}))$.

$\{C_{Li}\}$ is a matrix of covariances between functionals L and L_i

$\{C_{ij}\}$ is a matrix of covariances between functionals L_i and L_j

$\{D_{ij}\}$ is a covariance matrix for normally distributed errors e_i and e_j

The solution minimizes simultaneously the L_2-norm of the vector of errors and the norm of T ($\tilde{T}$ is an approximation to T and an element of a RKHS). If both $\tilde{T}$ and T are elements of the same RKHS, the maximal error of prediction can be estimated. If the number of predicted model parameters is greater than the number of measurements (the underdetermined system), the LSC-solution is equivalent to the solution obtained by use of the generalized inverse, Sansò et al. (1986).

In formula (2), the predicted and the measured quantities do not have to be of the same type. In predictions by LSC one can use any of the above mentioned physical quantities and obtain consistent predictions. The method is thus suitable for an integrated approach.

The long-term purpose of this and the future investigations is to extend the scope of the LSC-method to also include density and seismic information. The method, if succesfully developed, can bridge the gap between two disciplines of Earth Science: Solid Earth Geophysics and Geodesy. In Solid Earth Geophysics (particulary in exploration) one is interested in structures within the Earth. The LSC-method could be used as an inversion tool to model e.g. the density distribution within the crust (in 3D!). In Geodesy, a new kind of data (density or/and seismic) could improve modelling of the Earth's gravity field in areas where gravity data are few or none.

To extend the scope of the LSC-method as mentioned above, some steps has already been taken. The first question is: how to construct a RKHS for density or reciprocal seismic velocity distributions within the Earth? A requirement to such a RKHS must be that if a covariance function for density distribution is propagated to the surface of the Earth by a Newton Functional (c.f. formula (3)), the obtained covariance function for gravity anomalies must agree with the estimated autocovariance function for gravity. Sansò et al. (1986) studied the problem and found, that the well known model for the subsurface, where the base functions are disjoint prisms with constant density and a minimalization of L_2-norm, gives wrong response in gravity covariances. The authors suggest another choice of base functions (the disjoint prisms with linearly varying density) and a minimalization criteria (a norm) which gives the right statistical properties for the covariance function for the anomalous gravity potential outside the masses. Another choices of base functions for the density distribution that satisfy the above critera are: quasi-harmonic functions (Tscherning and Strykowski, 1987) and overlapping prisms (Hein et al. (1988) and Tscherning (1989)).

In order to be of any use for the inversion purposes, the choice of base functions must also be realistic from the geological point of view (e.g. it must be possible using these functions to model geological structures that are known to occure in the subsurface). A numerical example from Rhine Graben in Germany showed that the quasi-harmonic functions have too strong cross-correlation between the residual surface gravity and the residual surface density (Hein et al. (1988)).

The present work is an attempt to gain expirience about the statistical properties of density and reciprocal seismic velocity distributions within the shallow crust of the Earth. It is an extension of an earlier investigation, Strykowski (1989). The empirical covariance functions shown here will give additional constrains on the choice of base functions for the residual density and the residual reciprocal seismic velocity distributions. If similar investigations were conducted in another areas, one could look for the similarities and the differences between the covariance functions and perhaps correlate the characteristic parameters of the covariance functions with the history of tectonics and the history of bassin development in the area of investigation.

In connection with this investigation, some questions arise concerning the choice of reference models for different types of data and concerning the validity and limits of the method in general. These questions will be shortly discussed in chapter 2. Chapter 3 contains formulas for computation of the covariance estimates, and creates background for understanding the numerical results of this investigation. The numerical experiment will be shortly described in chapter 4. Chapter 5 contains a discussion of the results. Chapter 6 contains the conclusions. Chapter 7 contains few remarks about how we intend to proceed with our investigations in the near future.

2. Some remarks about the background of this investigation

As it was mentioned in the Introduction, one of the constraints that one would impose on the construction of a RKHS for respectively density and reciprocal seismic velocity distributions is, that the covariance functions for these distributions inside the Earth, when propagated to the surface of the Earth by a linear functional, have statistical properties which are in agreement with what one actually observes. In case of the density distribution one could as a functional use the following linear relation between $\Delta\rho$ and ΔT:

(3) $$\Delta T = \int_\Omega G \cdot (\Delta\rho/r) d\Omega, \text{ where}$$

ΔT is the residual anomalous gravity potential, G is the gravitational constant, Δρ is the anomalous density distribution (a question of how to construct the reference models for different types of data is discussed below) and Ω denotes the support of the masses (it is in principle the volume of the whole Earth). The relation between ΔT and Δρ expresses simply the well known Newtons law for the gravity potential (equation (3) will throughout this paper be referred to as Newton functional).

The seismic (one way) transmission time τ to the reflector is a quantity that can be read from the processed seismograms. By analogy, the quantity τ will play the same role for the seismic signal as T plays for the gravity signal. The future investigations will involve studies of statistical properties of the parameter τ as a function of the spherical distance. To relate the residual quantities Δτ and the anomalous interval transmission time Δt (the physical quantity t is measured by sonic logs and can also be called the reciprocal seismic velocity or the slowness) a following linear relation (a linear functional) can be used:

$$(4) \qquad \Delta\tau = \int_0^{z_r} \Delta t \ dz, \text{ where}$$

z_r is a depth to the reflector, and 0 referres to the surface.

A very important issue, which has an influence on the covariance functions constructed from the residual data, is the issue of how to construct the reference models for the involved physical quantities. In the LSC-method one would like to subtract the low-frequency signal (compared to the size of the area) from the data. The prediction by LSC is thus reduced to compute the deviations from the subtracted model. After the subtraction of the regional reference model, the variance of the signal decreases and the constructed covariance functions are more isotropic (cf. Forsberg, 1986).

An additional requirement to the choice of a reference model is that the mean value of the residual field is zero or known. For any set of data a number of reference models that fulfill the above requirements can be constructed. To the authors opinion it is good, if possible, to relate the reference model to some corresponding physical cause which, at least in principle, can be modelled in another area. This is usefull if one wants to compare the covariance functions from different areas or if one wants to relate the covariance functions (constructed from the residual data) to the physical reality in the area. In the case of density distribution one could construct e.g. a linear model with two parameters expressing the increase of density with depth. This increase of density with depth can be attributed to the compaction of the sediments. Although one could model this increase of density in another way (e.g. an exponential model), the proposed linear model is mathematicly simple and clearly removes most of the low frequency signal in depth (as it can be seen from fig.5a).

The wisdom of using simple reference models is appreciated, when one in predictions intends to mix different types of data. The additional requirement to the choice of reference models must be the consistency. If a reference model is subtracted from the density distribution, one should remove the corresponding models from τ, T and t-distribution. To obtain strict consistency is in general difficult if not impossible (the true dependence between the seismic velocity and the density is complicated as one in general does not know how the elastic parameters of the media depend on density). A simple model for the t-distribution (that corresponds to the model proposed above for the ρ-distribution) is a depth dependent regional least squares model for the interval transmission time obtained from the log data. The proposed linear model for the t-distribution is also a model with two parameters (cf. fig.6a). The model expresses that on a large scale, the seismic velocity of the crust increases (the reciprocal velocity decreases) with depth. This increase in seismic velocity is mainly due to the compaction (dense materials have in general larger seismic velocity than the less dense materials).

In this investigation, the reference models for the ρ and the t-distribution are constructed in the way suggested in the above discussion. The remaining issue is how to construct a reference model for the ρ-t dependence. Such a model is particulary important if one wants to make predictions from mixed gravimetric-seismic data. One way of constructing such a model is to use the two depth dependent models described above. Another way is to compute a least squares reference model from the corresponding density and reciprocal seismic velocity data (cf. fig.7). Both models were computed. In chapter 4 it will be shown that there is almost no difference between the two models (this will be explained in chapter 5).

What are the reference models for T (or Δg, where Δg denotes bouguer anomalies) and τ-distribution? For both the gravity potential and the gravity anomalies a depth dependent density model corresponds to a constant gravity response. A subtraction of the normal gravity potential from the true gravity potential corresponds in a local area to a subtraction of a density model which depends only on depth. The subtracted density model (cf. fig.5a) is thus partly already included in the reduction of the gravity data.

An exact computation of the attraction of the subtracted density model requires that one limits the extention of the model in depth and that one can compute the gravitational attraction of the masses below this depth in the area of investigation. One has also to keep in mind that the reduction of free-air anomalies to bouguer anomalies corresponds to 'filling up the ocean with rocks' (which in turn corresponds to a density model depending on bathymetry).

From the above discussion it is clear, that a strict computation of consistent reference models for Δg and ρ-distribution is difficult. One way to make a practical shortcut through all these problems is to assume, that the reference model for gravity anomalies is the mean gravity value in the area of investigation. Although one could make a better gravity field reduction, by e.g. subtracting some higher order spherical harmonic expansion of the Earth's gravity field, the gravity field reduction requires a corresponding density model. Such a density model is not horizontally homogenous, and thus difficult to construct on the basis of the sparse density data.

One could of course assume that the low-frequency spherical harmonic correspond to a gravimetric response from depths which are larger than those covered by the density data. This assumption is however not generally true, as there exist shallow structures with low-frequency gravimetric response (topography, bathymetry, graben structures, fault zones, intrusions,...etc.). In this investigation the reference model for gravity is the mean value of bouguer anomalies. Although the gravity model is reasonable in terms of covariance function for gravity (cf. fig.4a), the reduction in the gravity data was not sufficient for the constructed cross-correlations between the residual gravity and the residual density data (as it will be discussed in chapter 5).

A subtraction of the reference model for the t-distribution (cf. fig.6a) corresponds to a contraction of the seismograms (τ becomes shorter). It is instructive to realize the meaning of the reference model for the t-distribution. The model gives a mean regional time-depth dependence.

In seismic processing, the time-depth conversion of the processed seismograms is often based on the log data rather than on the 'stacking' velocities. The smooth low-frequency variation in seismic velocity with depth is not seen as a reflection on the seismograms (only the sharp discontinuities i.e. the large vertical gradients are seen). The low-frequency variation of the velocity with depth will however influence the parameter τ and thus, if not taken into account, it will give the wrong time-depth conversion. One of the possible areas where the joint inversion of the gravity and the seismic data could be used is to model the velocity field based on the log data and the gravimetric data.

3. Empirical covariance functions for density, gravity and reciprocal seismic velocity

The anomalous gravity potential T is per definition a harmonic function outside the masses because it satisfies the Laplace equation. In gravity field approximation in a local area it is often possible (after some reductions) to regard the surface of the Earth as a sphere. One can approximate the anomalous gravity potential on that sphere by means of spherical harmonic functions. The approximation $\tilde{T}$ to T is then established in every point outside the reference sphere (outward continuation). In order to establish the auto-covariance function for gravity anomalies, it is enough just to establish the covariances on the sphere (cf. formula (1)). The covariances for points outside the sphere are given through the spherical harmonic expansion of the potential field (the covariance function C_{TT} can be expressed as an infinite sum of products of the fully normalized solid spherical harmonic functions (cf. Moritz (1980) or Tscherning (1984)).

On fig.1 a typical autocovariance function for gravity anomalies is shown (see Tscherning (1984)). The figure shows the covariance function with its characteristic parameters $C_{\Delta g \Delta g}(0)$, ψ_1 and ψ_0. $C_{\Delta g \Delta g}(0)$ is the variance of the residual gravity field, ψ_1 is the correlation lenght ($\psi_1 = 1/2 \cdot C_{\Delta g \Delta g}(0)$) and ψ_0 is the spherical distance where the covariance function becomes zero for the first time. In qualitative description of the covariance function one uses mostly the parameters $C_{\Delta g \Delta g}(0)$ and ψ_1. The pa-

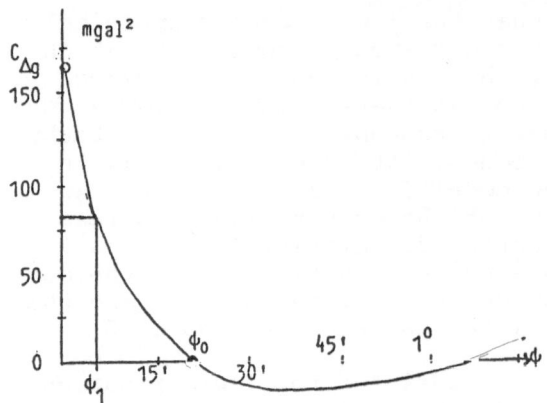

Fig.1 The characteristic parameters of $C_{\Delta g \Delta g}(\psi)$:
$C_{\Delta g \Delta g}(0)$ is the variance, ψ_1 is the correlation
distance and ψ_0 is the first zero point

rameter ψ_0 is more sensitive to the bias in the reference model
than the other two.

In the case of the anomalous density and the anomalous reciprocal seismic velocity distributions there is no natural choice
of the base functions corresponding to the solid spherical harmonic functions in the previous case. A natural generalization of
formula (1) will thus be:

$$(5) \qquad C_{\Delta \alpha \Delta \beta}(z_1, z_2, \psi) = \frac{\Sigma \, \Delta \alpha(P) \cdot \Delta \beta(Q)}{N} \; , \quad \psi(P,Q) = \psi \; ,$$

where α and β denotes what kind of residual data are used (e.g
$\alpha = \rho$ and $\beta = t$). z_1 and z_2 denote respectively the (positive)
depths to the points P and Q in the subsurface (at the surface of
the Earth the depth is 0). If $\alpha = \beta$, $C_{\Delta \alpha \Delta \beta}$ is called the autocovariance function for the α-type of observation. If $\alpha \neq \beta$,
$C_{\Delta \alpha \Delta \beta}$ is called the crosscovariance function between the α-type
and the β-type of observations. When z_1 and z_2 are fixed and
$z_1 = z_2$, the autocovariance functions constructed in this investigation resemble the one that is shown on fig. 1.

The characteristic parameters of the empirical autocovariance
functions computed by formula (5) will be as follows:
$C_{\Delta \alpha \Delta \alpha}(z, z, 0)$ is the variance of the residual α-type of observations at depth z, $\psi_1(z)_\alpha$ and $\psi_0(z)_\alpha$ correspond to ψ_1 and ψ_0 (cf.
fig.1) for the α-type of observations and for $z_1 = z_2 = z$.

Another characteristic parameter which will be used to show
both the autocovariance and the crosscovariance functions is the
correlation coefficient $\mathrm{corr}_{\alpha \beta}(z_1, z_2, \psi)$:

$$(6) \qquad \mathrm{corr}_{\alpha \beta}(z_1, z_2, \psi) = \frac{C_{\Delta \alpha \Delta \beta}(z_1, z_2, \psi)}{\sqrt{C_{\Delta \alpha \Delta \alpha}(z_1, z_1, 0) \cdot C_{\Delta \beta \Delta \beta}(z_2, z_2, 0)}}$$

where $\psi(P,Q) = \psi$, z_1 and z_2 are depths to respectively the α and
the β-type of data.

4. Computation of the Empirical Covariance Functions for Density, Gravity and Reciprocal Seismic Velocity, a numerical example from the North Sea.

This investigation was conducted in an area of 3°x6° in the North Sea and the western part of Jutland. The gravity data used were 365 bouguer anomaly values. The log data were: bulk density data (FDC-logs) and seismic interval transmission time data (SONIC-logs) from 41 logs from the area. Fig.2 shows the location of the boreholes.

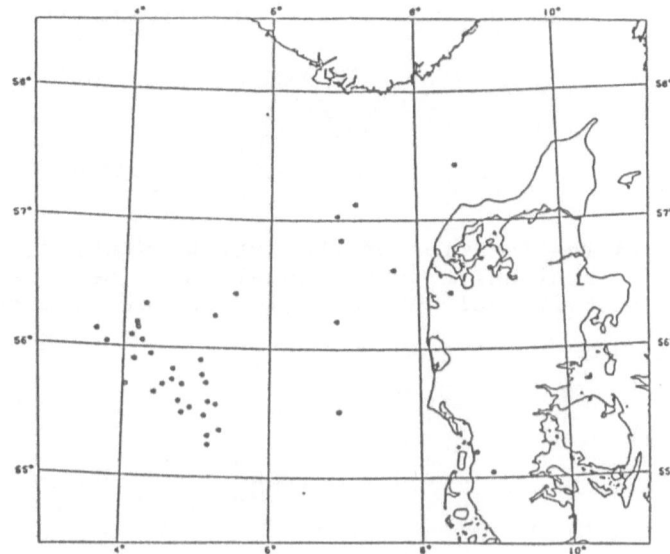

Fig.2

The location of
the boreholes

Fig.1 shows a typical covariance function for bouguer anomalies. Fig.3 shows the 365 gravity stations, which were chosen from a database containing the gravity data from the Scandinavia and the neighbouring waters. The criteria for choosing the gravity stations was their location. The gravity stations were chosen as close as possible to a regular grid of 0.15°x0.25° covering the area. The computation of bouguer anomalies was performed using the bathymetric information.

For the gravity data a reference model was computed. The model is the mean value of 6.22 mgal. Fig.4a shows the autocovariance function $C_{\Delta g \Delta g}(\psi)$ constructed from the residual bouguer anomalies. Fig.4b shows a histogram for the residual gravity values.

The log data were sampled with a vertical spacing of 0.5 feet (≈ 15 cm) and spanned vertically from some 200 m to some 5000 m below the Mean Sea Level (MSL). Generally the SONIC-log was measured all the way from the top to the bottom of the boreholes (exept for the short topmost part of the well). The density log (FDC-log) was only measured in the parts of the well of a comercial interest. Nevertheless, some of the density logs were almost 3000 m long.

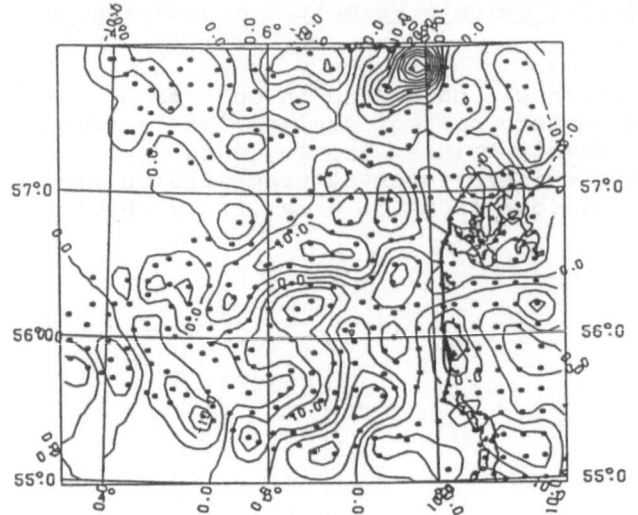

Fig.3 The location of the gravity stations and the
      residual bouguer anomalies (a mean value of
      6.22 mgal subtracted). Contour interval: 5 mgal

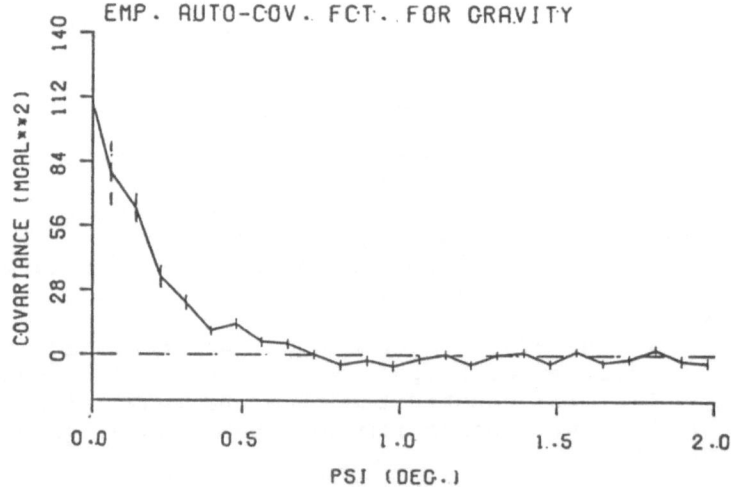

Fig.4a The Empirical Covariance Function for the Residual
       Bouguer Anomalies (the subtracted reference model
       is a mean value of 6.22 mgal).
       Variance: 110.7 mgal^2, ψ_1 ≈ 0.1 dgr., number of
       observations: 365

344

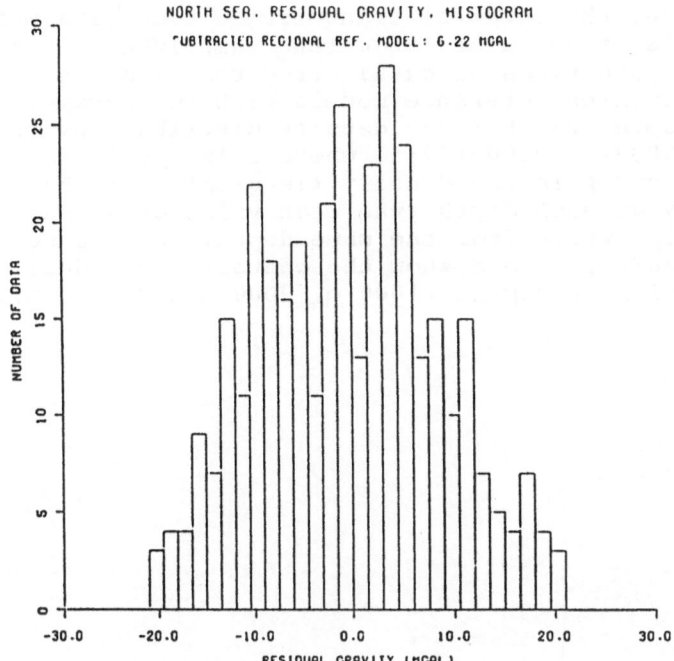

Fig.4b The Empirical Covariance Function for the Residual
Bouguer Anomalies. The histogram shows the residual
bouguer anomalies (the subtracted reference model
is a mean value of 6.22 mgal).

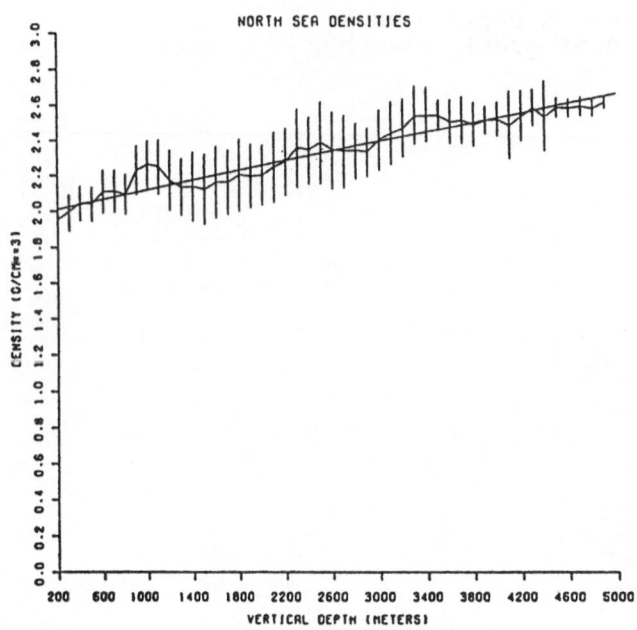

Fig.5a The mean regional model for density (cf. chapt.4):
$\rho(z) = 1.98331 + 0.000137 \cdot z$, where ρ is the
density (in g/cm^3) and z is the depth (in meters).

Both the density and the interval transmission time data were
averaged over intervals of depth of 100 m (the same intervals in
all the wells and for both types of data). From this averaged observations the least squares reference models with two parameters
were computed (cf. chapter 2). For the density distribution, the
model was: $\rho(z) = 1.98331 + 0.000137 \cdot z$, where z is the depth (in
meters) below the MSL and ρ is the density (in g/cm^3). Fig.5a
shows the mean density at each depth (the mean value of averaged
observation from all the wells from the same depth) and the constructed reference model. Fig.5b-d show the contoured residual
densities in three different depths (1000 m, 2000 m and 3000 m).

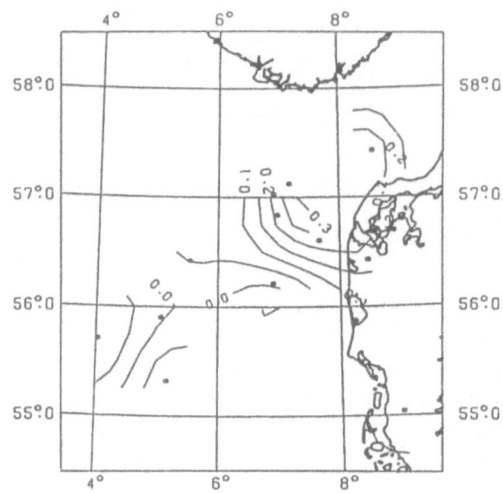

Fig.5b The residual density at depth 1000 m
     Contour interval: 0.05 g/cm^3

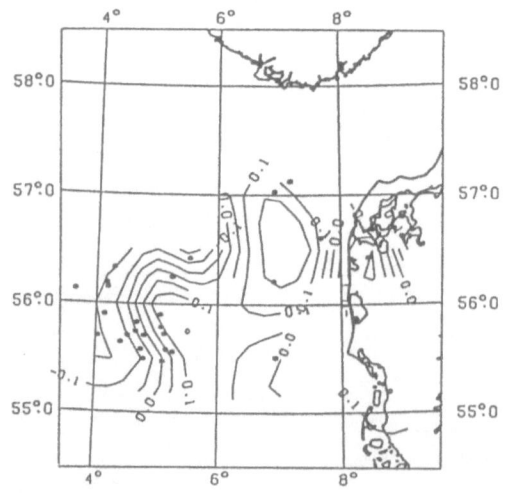

Fig.5c The residual density at depth 2000 m
     Contour interval: 0.05 g/cm^3

The reference model for the interval transmission time was :
t(z) = 160.494 - 0.01906 · z, where z is the depth below the MSL
(in meters) and t is the interval transmission time (in μs/ft).
Fig.6a-d for the interval transmission time corresponds to
fig.5a-d for the density distribution (the depths are the same).

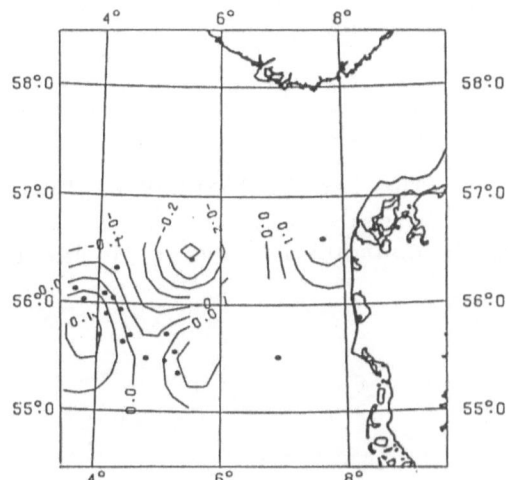

Fig.5d The residual density at depth 3000 m.
Contour interval: 0.05 g/cm³

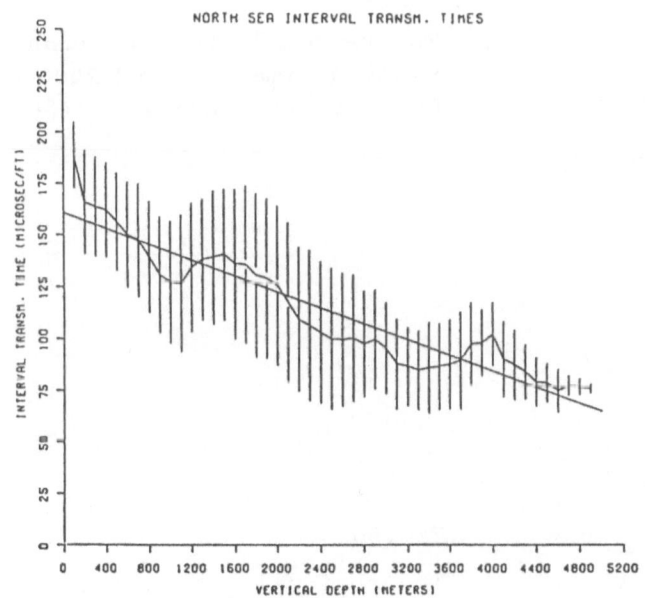

Fig.6a The mean regional model for the interval transmission time (cf. chapt.4):
t(z)=160.494-0.01906·z, where t is the interval
transmission time (in μs/ft) and z is the depth
(in meters).

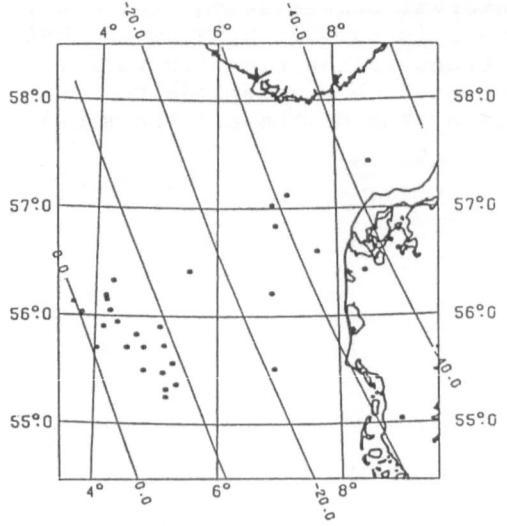

Fig.6b

The residual interval transmission time at depth 1000 m.
Contour interval: 10 µs/ft

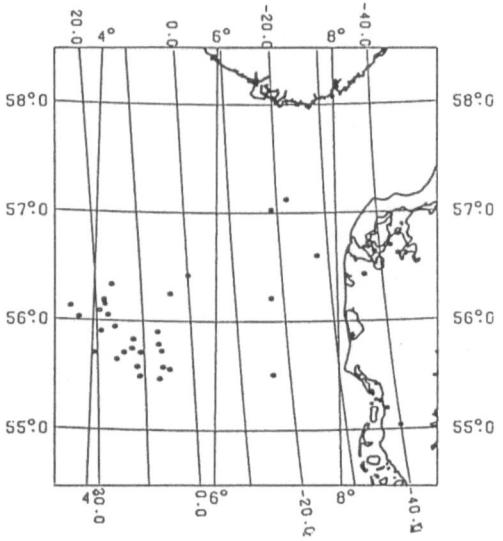

Fig.6c

The residual interval transmission time at depth 2000 m.
Contour interval: 10 µs/ft

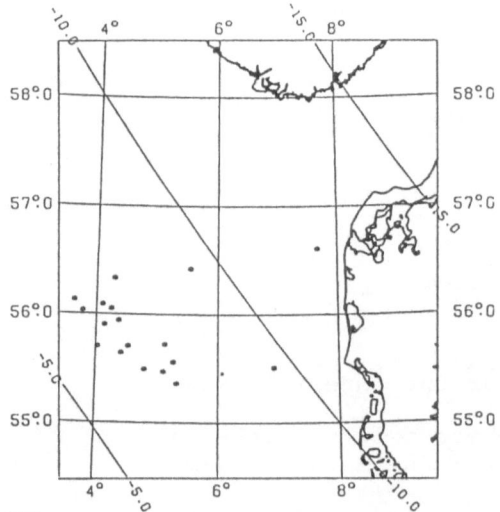

Fig.6d

The residual interval transmission time at depth 3000 m.
Contour interval: 5 µs/ft

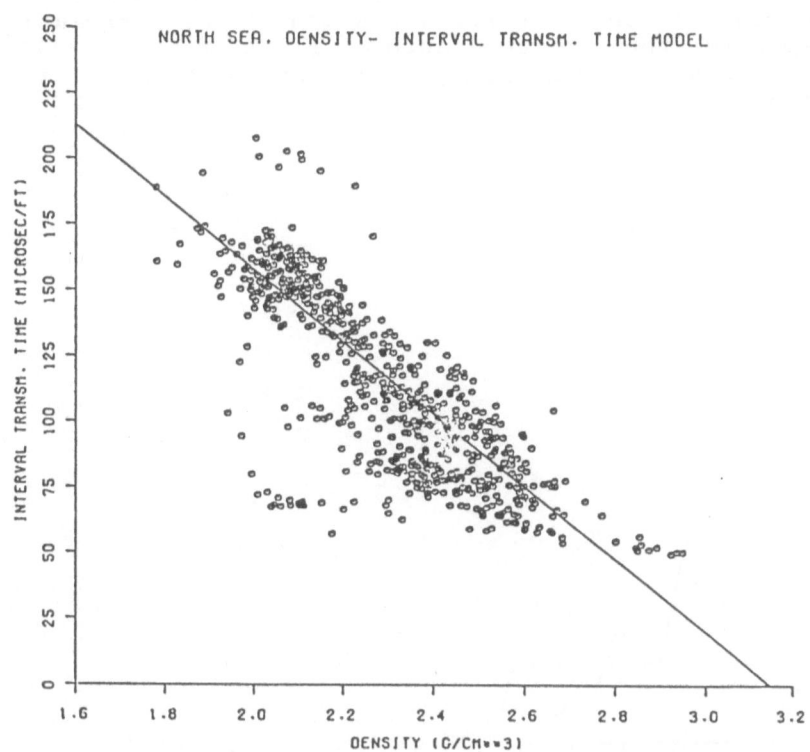

Fig.7 Density vs. interval transmission time.
    Least squares model: t(ρ)=432.572-137.439·ρ,
    ρ in g/cm³, t in μs/ft.

Fig.7 shows the reference model for the relation between the
density and the interval transmission time and the points from
which the model was constructed. The model was constructed in the
following way: the density data and the interval transmission
time data were first averaged over intervals of depth of 100 m.
The corresponding averaged values of density and interval transmission time from the same well and the same depth were used for
computation of the least squares model with two parameters. The
correlation coefficient between the density and the interval
transmission time distributions is -0.66. The least squares model
is: t(ρ) = 432.572 - 137.439·ρ, ρ in g/cm² and t in μs/ft. A corresponding model can be computed from the depth dependend regional reference models for density and reciprocal seismic velocity
(cf. above). The model is: t(ρ) = 436.018 - 138.921·ρ, ρ in g/cm³
and t in μs/ft.

    The residual density, seismic and gravity data were used for
estimation of the empirical covariance function (cf. formula
(5)). The gravity data were only given at the surface (at zero
depth). The vertical interval of depth for computation of the empirical covariance function was 200 m (which is equivalent to two
intervals of depth used for computation of the reference models).
The interval of the spherical distance was 25.0 km which is equivalent to 0.22483°.

Fig.8a shows the variation of the variance of the residual density with depth. Fig.8b shows the variation of the correlation lenght with depth for the empirical autocovariance function for density. Fig.8c shows a corresponding variation of the first zero point with depth.

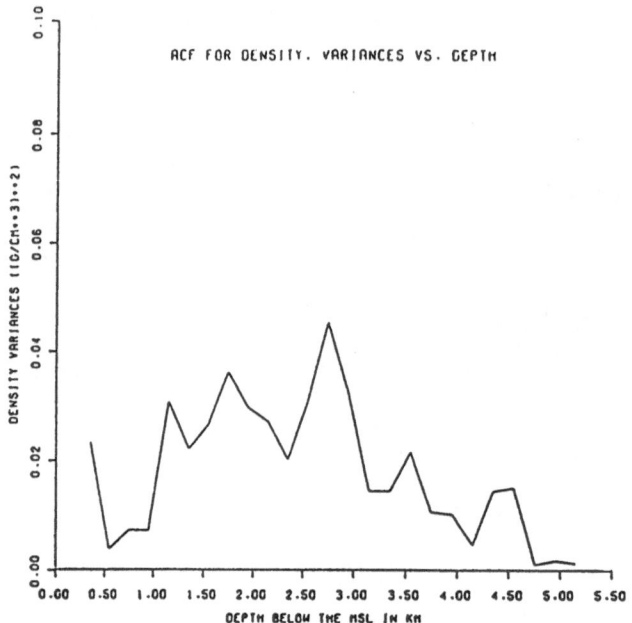

Fig.8a Variation of the variance of the residual density with depth ($C_{\Delta\rho\Delta\rho}(z,z,0)$, cf. chapt.3).

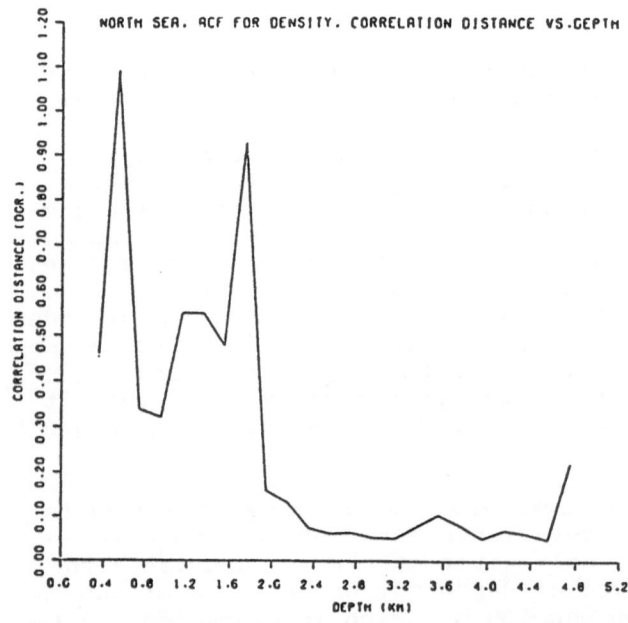

Fig.8b Variation of the correlation distance of the residual density with depth ($\psi_1(z)_{\Delta\rho}$, cf. chapt.3).

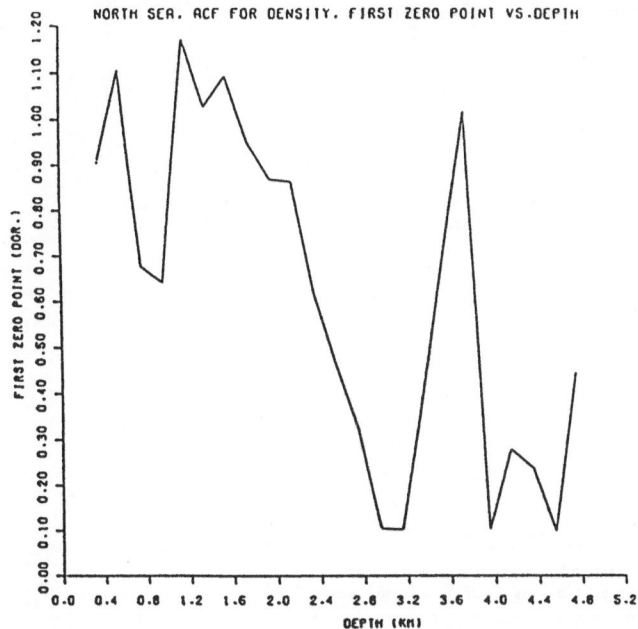

Fig.8c Variation of the zero distance of the residual
density with depth ($\psi_0(z)_{\Delta\rho}$, cf. chapt.3).

Fig.9a-c show variation of the same parameters as above for
the empirical autocovariance function for the residual interval
transmission time.

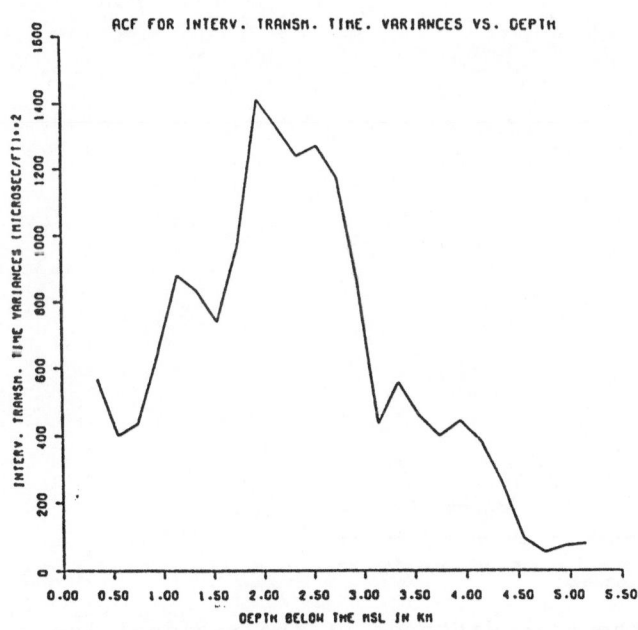

Fig.9a Variation of the variance of the residual interval
transmission time with depth ($C_{\Delta t \Delta t}(z,z,0)$, cf.
chapt.3).

351

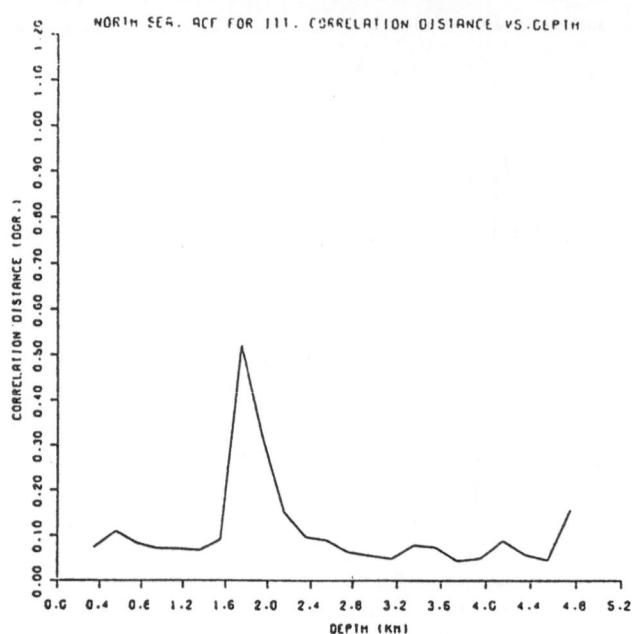

Fig.9b Variation of the correlation distance of the residual interval transmission time with depth
($\psi_1(z)_{\Delta t}$, cf. chapt.3).

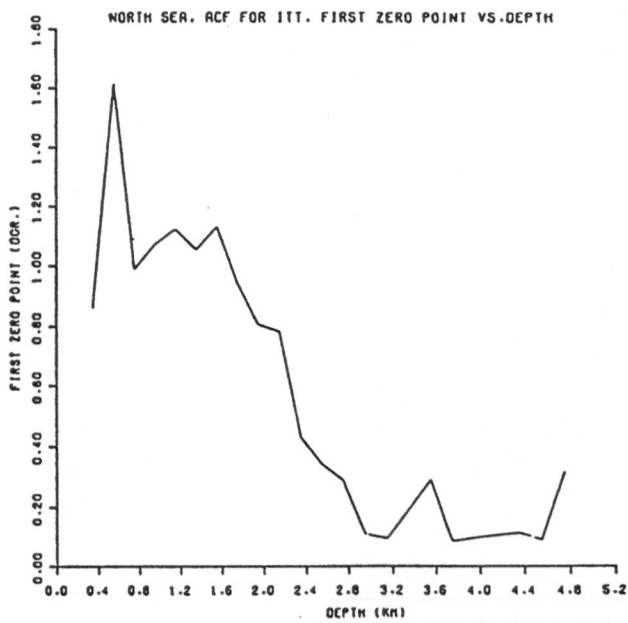

Fig.9c Variation of the zero distance of the residual interval transmission time with depth ($\psi_0(z)_{\Delta t}$, cf.
chapt.3).

Fig.10 shows the vertical correlation coefficient for the empirical crosscovariance function between the residual gravity and
the residual density data. Fig.11 corresponds to fig.10 for the
empirical crosscovariance function between the residual gravity
and the residual interval transmission time.

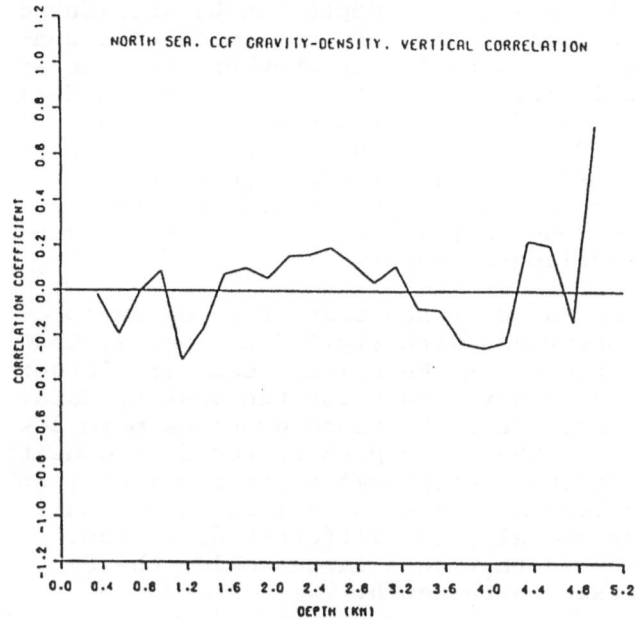

Fig.10   Variation of the vertical correlation coefficient
         between the residual gravity and the residual den-
         sity with depth ($C_{\Delta\rho\Delta g}(z,0,0)$, cf. chapt. 3).

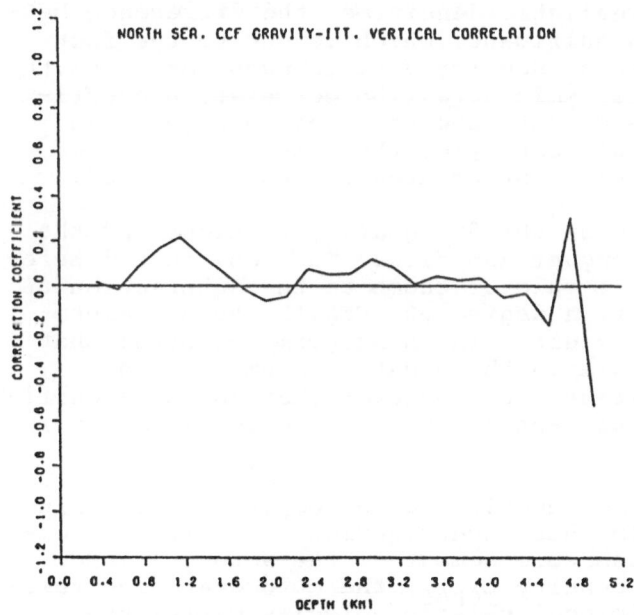

Fig.11   Variation of the vertical correlation coefficient
         between the residual gravity and the residual in-
         terval transmission time with depth ($C_{\Delta t\Delta g}(z,0,0)$,
         cf. chapt. 3).

5. Discussion

The reason for choosing the refernce models as it was done in this investigation was partly explained in chapter 2. The reference model for the density distribution (cf. fig.5a) removes most of the low-frequency variations of the density with depth. Physically, the increase of density with depth can be attributed to the compaction of the sediments. The reference model for density is a function of depth. The question is whether there is any significant horizontal low-frequency signal left in the residual data. Fig.5b-d show the contoured residual density in three depths: 1000 m, 2000 m and 3000 m. All figures contain both the negative and the positive residuals. The figures indicate that there is no significant regional tilt in the residual data. The conclusion must be that the constructed reference model for density is also a reasonable horizontal model.

Fig.6a show the constructed reference model for the reciprocal seismic velocity. By comparizon with fig.5a it is clear that the relative standard deviations for the seismic data are larger than the corresponding standard deviations for the density data. The residual reciprocal seismic velocity field contains more low-frequency signal in depth than the corresponding residual density field. Fig.6b-d show the residual reciprocal seismic velocity in depths: 1000 m, 2000 m and 3000 m. The data indicate a regional tilt of different steepness and slightly different direction. Physically, the model expresses that on a large scale, the P-velocity of the crust increases with depth.

Fig.7 shows the corresponding averaged values of density and reciprocal seismic velocity from all the wells. The computed least squares model is shown on the figure. As mentioned in the previous chapter a similar model was constructed from the depth dependend reference models for the reciprocal seismic velocity and the density. For the realistic densities, the difference between the two models is insignificant, which is due to the fact that the standard deviation of density data is much less than the standard deviation of the seismic data. The correlation coefficient of -0.66 between the density and the reciprocal seismic velocity expresses that the geometry of the seismic structures and the geometry of the density structures is 66% correlated.

Fig.3 shows the location of the 365 gravity stations and the contours of the residual bouguer anomalies. The constructed reference model corresponds by a rule of thumb to the subtraction of a global gravity model up to a degree 60. Fig.4b shows a histogram of the residual gravity data. The histogram indicates that the distribution of the positive the negative anomalies is slightly assymetric. A X^2-test shows however that on the significance level of 5%, one cannot reject that the residuals are normally distributed.

The empirical covariance functions were computed as explained in the previous chapter. In this investigation the following empirical covariance functions were computed: the autocovariance function for the residual density $C_{\Delta\rho\Delta\rho}$, the autocovariance function for the residual reciprocal seismic velocity $C_{\Delta t\Delta t}$, the autocovariance function for the residual bouguer anomalies $C_{\Delta g\Delta g}$, the crosscovariance function between the residual density and the residual reciprocal seismic velocity $C_{\Delta\rho\Delta t}$, the crosscovariance function between the residual bouguer anomalies and the residual density $C_{\Delta g\Delta\rho}$ and the crosscovariance function between the residual bouguer anomalies and the residual reciprocal seismic velocity $C_{\Delta g\Delta t}$.

$C_{\Delta g\Delta g}$:
The parameters of the autocovariance function for gravity (cf. fig.4a) are reasonable as compared to the standard parameters (cf. Tscherning (1984), fig.4-5, k=3).

$C_{\Delta\rho\Delta\rho}$:
Fig.8a shows the variation of the variance of the residual density with depth. The interval of depth between 1000 m and 3000 m had the best horizontal data coverage. In this interval, the density variance has three peaks. The peaks are at the following depths: 1.15 km, 1.75 km and 2.75 km. The variance of the residual density in this three depths is respectively: 0.0308 (g/cm³)², 0.0362 (g/cm³)² and 0.0456 (g/cm³)². Fig.8b shows the correlation lenght as a function of depth. In the interval of depth between 1.0 km and 3.0 km the figure shows the following: the correlation lenght has a constant value of some 0.5° between the depths of 1.00 km and 1.75 km, at a depth of 1.75 km the correlation lenght has it's peak value of 0.93°, below the depth of 1.75 km the correlation decreases from some 0.15° to 0.06°. Fig.8c shows the distance to the first zero point as a function of depth. As it was mentioned in chapter 3, this parameter is more sensitive to the bias in the data than the other two. For this reason no attempts of interpreting this figure will be made. The figure is shown here in order to make the description of $C_{\Delta\rho\Delta\rho}$ complete.

Before any interpretation of these peaks will be attempted, one should notice that the horizontal distribution of the wells cluster around the South-Western part of the area (cf. fig.2). This is the area of Central Graben. The geology of the area is complicated (graben structure, faulting). By looking at formula (5), it is clear, that the data from Central Graben will dominate the estimates of the covariances for small spherical distances. For the spherical distance zero, the data from Central Graben will outnumber the contributions from other parts of the area if the deviation between the residuals is sufficiently small. Thus the constructed covariance functions show for small spherical distances the situation in and around the Central Graben.

By going back to the log data one can draw a map showing the age of the sediments in these three depths. The age of sediments can be classified as follows: pre-Cretaceous, Cretaceous and post-Cretaceous. Also the top and the bottom of the Cretaceous layer can be marked seperately. The map at a depth of 1.15 km shows that the sediments in Central Graben are post-Cretaceous (Tertiary). The same map shows that in wells between 6° and 8° of longitude and in two northernmost wells on-shore, the sediments are Cretaceous. The sediments in the two southernmost wells on-shore contain the transition between the Cretaceous and the pre-Cretaceous sediments. The northernmost well (off-shore) contains the pre-Cretaceous sediments. By looking at fig. 6b one gets a clear impression that the tilt in the reciprocal seismic velocity is related to the age of sediments (the youngest sediments are in Central Graben). The map at depth of 1.75 km shows that the majority of the wells in Central Graben still contain some post-Cretaceous sediments. However the five southernmost wells in Central Graben contain the Cretaceous sediments. The map shows also that all the other wells contain the pre-Cretaceous sediments. In a depth of 2.75 km, the sediments from 4 wells in the north-western part of the danish part of the Central Graben are post-Cretaceous. There are not many Cretaceous sediments (only in 3 wells) and the transition between the post-Cretaceous and the Cretaceous sediments is seen in other three wells. All the other wells East of the 5° in longitude contain the pre-Cretaceous sediments.

From the log data one notices that in Central Graben in depths between 1000 m and 1200 m there are some discontinuities in density (usually a decrease in density with depth of an order of 0.2 g/cm³ inside the Tertiary layer). This post-Cretaceous density discontinuities are not that clearly seen on the corresponding sonic logs. The top of the Cretaceous layer is a strong seismic reflector (a typical jump in the P-velocity is from some 2000 m/s to some 3000 m/s, the corresponding jump in the interval transmission time is from 152 µs/ft to 102 µs/ft). The corresponding jump in density is not always seen. Usually the density increases with depth across the boundary between the post-Cretaceous and the Cretaceous layer (the discontinuity is mostly seen as an increase in the vertical gradient of density). In some few cases, the density decreases with depth across the boundary. The bottom of the Cretaceous layer is not particulary connected with any seismic velocity contrast. Often, the velocity decreases inside the Cretaceous layer (from the Upper to the Lower Cretaceous). For the density distribution the situation just below the Cretaceous layer is dramatic. Large density variations are seen (from 2.1 g/cm³ for the Permian salt to 2.8-2.9 g/cm³ for the anhydrite).

The peak in the variance of $C_{\Delta\rho\Delta\rho}$ at a depth of 1.15 km is most probably due to the density anomalies within the post-Cretaceous (Tertiary) layer in the area of the Central Graben. The peak in the variance of $C_{\Delta\rho\Delta\rho}$ at a depth of 1.75 km is a combination of density anomalies within the post-Cretaceous layer close to the transition to the Cretaceous layer (volcanics), and the topography of the top of the Cretaceous layer. The large correlation distance at this depth indicates that the geometry of the anomalous structures is much smoother that what can be seen on fig.5b-c. The peak in the variance of $C_{\Delta\rho\Delta\rho}$ at a depth of 2.75 km is due to the varying geology at this depth. All three sediment ages are represented. The short correlation lenght indicates that the geometry of the anomalous structures is more high-frequent than in the other two depths.

$C_{\Delta t\Delta t}$:
The variation in variance with depth for the empirical auto-covariance function for the interval transmission time (fig.9a) shows a peak at a depth of 1.15 km of 882 (µs/ft)², at depths between 1.75 km to 3.15 km the variance shows a broad peak with the maximum value of 1414 (µs/ft)² at a depth of 1.95 km. Fig.6b shows the correlation distance as a function of depth for $C_{\Delta t\Delta t}$. The figure shows a low correlation distance of 0.1° in all depths exept a correlation distance of 0.5° at a depth of 1.75 km.

An interpretation of the above mentioned characteristic parameters of the covariance function has to take into account how the horizontal tilt in the residual sonic log data (cf. fig.6b-d) influences the covariance function. The short correlation lenght in almost all depths indicates that the influence of this tilt is negligible. One cannot however exclude that the tilt has some influence on the magnitude of the variance.

The peaks in depths of 1.15 km and 1.75 km correspond to the peaks at the same depths for $C_{\Delta\rho\Delta\rho}$. If the variance of $C_{\Delta t\Delta t}$ is influenced by the regional tilt (the variance is too high), it also influences the correlation lenght (cf. fig.1) which becomes shorter. A comparison of fig.8b and fig.9b shows a shorter correlation lenght at shallow depths for the reciprocal seismic velocity data. This can mean that the seismic velocity varies more locally than the density data, but also that the above mentioned regional tilt influences the variance. Probably both effects play some role. The reason for choosing the age classification for sediments related to the Cretaceous period is, as explained above, the fact that the top of this layer is a strong seismic reflector. One would thus look for this reflector in the covariance functions. The peaks in both the variance and the correlation distance for $C_{\Delta t\Delta t}$ at a depth of 1.75 km can be related to this reflector. The lack of a peak for $C_{\Delta t\Delta t}$ at a depth of 2.75 km is due to the fact that the bottom of the Cretaceous layer is not associated with any particular seismic discontinuity.

$C_{\Delta\rho\Delta t}$:
A tabel showing both the crosscovariances between the residual interval transmission time and the residual density and the crosscorrelations (cf. formula (6)) has been computed. It is of course difficult to show such a function graphically (it is a 4D plot!). The impression from the tables is however that the correlations are both strongly positive and strongly negative. To some degree, the correlations can be influenced by the tilt in interval transmission time.

$C_{\Delta g\Delta\rho}$:
Fig.10 shows the vertical correlation between the residual gravity data at the surface and the residual density data in different depths. The correlation is as low as +/- 30 % (this means that there is no correlation at all!). The reason for this low correlation lies in the simple reference model that was subtracted from the gravity data. The residual gravity field may still contain a strong signal of low harmonic degree. This signal can be generated from depths which are not covered by the log data, so that there is no correlation between the two types of data.

$C_{\Delta g\Delta t}$:
Fig.11 corresponds to fig.10 for the crosscovariance function between the residual gravity and the reciprocal interval transmission time. The figure shows that there is no correlation between these two data types for the same reason as described above. Notice how the negative correlation between $\Delta\rho$ and Δt is seen from fig.10 and fig.11 (the peaks and lows of this two figures have opposite sign).

6. Conclusions

In this investigation a set of gravity, density and reciprocal seismic velocity data was analysed statistically with a purpose of constructing the empirical covariance functions between
these different types of data.

The first step was to construct the mean regional reference
models for these data. For the log data (density and reciprocal
seismic velocity), the reference model was a depth dependend
least squares model with two parameters. To construct this model,
the original observations were averaged over 100 m of depth and
the model was constructed from these averaged observations. The
constructed models were:
1) $\rho(z) = 1.98331 + 0.000137 \cdot z$, where z is the depth (in meters)
    below the MSL and ρ is the density (in g/cm³)
2) $t(z) = 160.494 - 0.01906 \cdot z$, where z is the depth below the
    MSL (in meters) and t is the interval transmission time (in
    μs/ft).
The reference model for the surface gravity data (bouguer anomalies) was the mean value of 6.22 mgal. Also a reference model
between the density and the reciprocal seismic velocity was constructed. The model was: $t(\rho) = 432.572 - 137.439 \cdot \rho$, ρ in g/cm³
and t in μs/ft. The correlation coefficient between the density
and the reciprocal seismic velocity was -66%.

The residual data were used for the construction of the empirical covariance functions. The empirical autocovariance function for density shows strong correlation both vertically and horizontally. The peaks of the variation of the variance and the
correlation distance with depth were correlated with the geology
of the area. The peak at a depth of 1.15 km was associated with
density anomalies within the Tertiary layer; the peak at a depth
of 1.75 km was partly due to the density anomalies associated
with volcanics at the bottom of the Tertiary layer and partly due
to the topography of the top of the Cretaceous layer; the peak at
a depth of 2.75 km was associated with the large density anomalies just below the Cretaceous layer.

The empirical autocovariance function for the reciprocal
seismic velocity show much shorter correlation lenght than the
empirical autocovariance function for density. This is partly due
to the fact that the seismic velocity varies more rapidly with
distance than the density, but also due to the horizontal tilt in
the sonic log data. The tilt has the same regional SW-NE trend as
the age of the sediments.

The empirical crosscovariance function between the residual
density and the residual reciprocal seismic velocity shows high
correlation coefficients both positive and negative.

The crosscovariance function between the residual gravity and
both types of log data show no correlation. This is due to the
simple reference model subtracted from the gravity data. The residual gravity data may contain a strong signal from below the
depths covered by the log data.

The covariance functions constructed in this investigation
are probably characteristic only for the area of investigation.

However, they contain information about the 'large scale' variation in the distribution of the density and the reciprocal seismic velocity in the sedimentary bassin. If a similar study was conducted in another area, one could relate the similarities and the differences between the characteristic parameters of the empirical covariance functions to the history of bassin development and the tectonics.

7. Future investigations

The future investigations will involve more log data from the area. The Danish Geological Survey will supply 50 more logs (both FDC and sonic). One of the problems concerning the reliability of the estimated covariances was the poor horizontal coverage by the log data. The additional log data will provide much better horizontal coverage.

New numerical tests concerning the reference model for gravity will be carried out. This can be done in two ways: one can choose a smaller area (eg. the Central Graben area) and do the same kind of computations as it was performed in this investigation or one can use a more detailed gravity model (a spherical harmonic expansion of the Earth's gravity field to a degree corresponding to the size of the area and the corresponding isostatic compensation). In both cases the crosscorrelation between the gravity and the density signal should increase.

A prediction test using the existing data and the constructed covariance functions will be carried out. Before this can be done one has to construct a model for the empirical covariance functions. Further theoretical work must be done to construct the Hilbert Space with a Reproducing Kernel for the density and the reciprocal seismic velocity distributions.

Acknowledgements
This investigation is sponsored by Norsk Hydro Udforskning, Copenhagen. Special thanks to my collegues P.Knudsen, C.C.Tscherning, O.Remmer and S.A.Petersen from Norsk Hydro, Bergen for many fruitfull discussions and suggestions.

References:

Hein, G., Sansò, F., Strykowski, G. and Tscherning, C.C.: On the Choice of Norm and Base Functions for the Solution of the Inverse Gravimetric Problem. Presented at 17. Conference on Mathematical Geophysics, Blanes, Cataluna, Spain, June, 1988

Forsberg, R.: Spectral Properties of the Gravity Field in the Nordic Countries. Boll. di Geod. et Sc. Aff., No. 4, 1986, pp. 361-383

Kaula, W.M.: Statistical and Harmonic Analysis of Gravity. Techn. Rep. Army Map Service, Wash.,D.C., 1959

Knudsen, P.: Estimation and Modelling of the Local Empirical Covariance Function using Gravity and Satellite Altimeter Data. Bull. Géod. 61 (1987) pp. 145-160

Krarup, T.: A Contribution to the Mathematical Foundation of Physical Geodesy. Danish Geodetic Institute, Publication No. 44
Copenhagen, 1969

Lauritzen, S.: The Probabilistic Background of some Statistical
Methods in Physical Geodesy. Danish Geodetic Institute,
Publication No. 48, Copenhagen, 1973

Moritz, H.: Advanced Physical Geodesy. H.Wichman Verlag.
Karlsruhe, 1980

Sansò, F., Barzaghi, R. and Tscherning, C.C.: Choice of Norm for
the Density Distribution of the Earth, Geophys. J. R. astr.
Soc., 123-141, 1986

Sansò, F.: Statistical Methods in Physical Geodesy. Mathematical
and Numerical Techniques in Physical Geodesy. Lecture Notes
in Earth Sciences, Vol. 7, pp. 49-155, editor Sünkel H,
Springer Verlag, 1986

Strykowski, G.: A Study of Statistical Properties of the Earth's
Density Distribution based on Density Logs from the North
Sea. Presented at II Hotine-Marussi Symposium on Mathemati-
cal Geodesy, Pisa, June 5-8, 1989.

Tscherning, C.C.: Local Approximation of the Gravity Field by
Least Squares Collocation, Proceedings from Beijing International Summer School, Beijing, China, 1984, Edited by
K.P. Schwarz, Division of Surveying Engineering, The University of Calgary, Canada.

Tscherning, C.C. and Strykowski, G.: Quasi-harmonic inversion of
gravity field data. Proceedings of V Seminar on Model Optimization in Exploration Geophysics, Berlin, 1987

Tscherning, C.C.: Density-Gravity Covariance Functions Produced
by Overlapping Rectangular Blocks of Constant Density, Presented at II Hotine-Marussi Symposium on Mathematical Geodesy, Pisa, June 5-8, 1989. Submitted to Geoph. Journ. Intern.

Wei, M.: On the Ergodicity of the Covariance Function in the Case
of a Plane, Bull. Géod. 59 (1985) pp.332-341

Iterative Simultaneous Inversion of Gravity and Seismic Traveltime Data: I – Formulas

J. Švancara, J. Halíř

Geofyzika s. p. Brno, Ječná 29a, 61246 Brno 12, Czechoslovakia

Abstract

Formulas described in this paper were designed for the use in least-squares tomographic inversion methods with improved flexibility of parametrizing the density-velocity model. The gravity effect of the considered volume Ω is the sum of gravity effects of 2-D or 2 1/2-D bodies of polygonal cross-sections which form the disjunctive decomposition of Ω. For this geometrical arrangement analytical and numerical ray tracing algorithms were developed assuming either linear or parabolic dependence of quadratic slowness (square of reciprocal wave speed) in Cartesian coordinates. Partial derivatives of calculated gravity and traveltime with respect to geometrical parameters, which are necessary for the control of the first derivative inverse algorithms, are given in an explicit form. Two expressions for the traveltime derivative with respect to vertex coordinates were obtained : one for the straight rays based on Fermat's principle and the other based on the geometrical approximation not affected by the curvature of the ray. Results of the numerical experiment based on the formulas are presented.

1. Introduction

The goal of simultaneous inversion of gravity and seismic data is the estimation of a subsurface density-velocity model the response of which is consistent with these two independent sets of geophysical data Lines et al. (1988), Starostenko et al. (1988). The joint inversion includes the coupling between density and velocity in the form of linear relationship. For the solution of this problem it is necessary to define algorithms for gravity and seismic modelling using the same parametrization of the geologic model. Our paper was primarily designed for interpretation of regional refraction data, but we suppose that the obtained results can be used for solving of other problems encountered in exploration seismology.

Quantitative interpretation of gravity data is usually accomplished by 2 1/2-D bodies of polygonal cross-section, which makes it possible to retain the simplicity of the 2-D modelling (Talwani et al., 1959). The solution of the direct and inverse gravity problem for 2 1/2-D geometry was described in many papers, including those of Rasmussen, Pedersen (1979), Enmark (1981) and Švancara, Halíř (1987).

Initial value ray tracing in heterogeneous media is often based on analytically computed ray segments within suitable cells with a simple velocity law (see Berryman (1989), Bishop et al. (1985), Langan et al. (1985)). These elementary cells in 2-D case may be triangular or rectangular. In areas of complex geological structure this method needs many elementary cells.

We present explicit formulas for the tracing of rays in 2-D bodies of general polygonal cross-section. The velocity distribution within individual bodies is approximated by quadratic slowness linear or parabolic function of the Cartesian coordinates. Rays are bent according to reflection/transmission laws at the polygon boundaries and travel along a curved path within the polygons. Inverse modelling of seismic arrivals is often based on the first derivative least-squares algorithms which require the knowledge of partial derivatives of the model response with respect to model parameters. Two approximative formulas for the traveltime derivative with respect to vertex coordinates were obtained.

This paper presents formulas necessary for joint inversion of gravity and seismic data while our method permits construction of a complex model without subdividing them into a large number of cells.

2. Gravity modelling

Calculating the gravity effect of volume Ω, we sum the effects of its density homogeneous parts which form the disjunctive decomposition of Ω

$$\Delta g^t(\underline{r}) \;=\; \sum_m \Delta g^t_m(\underline{r}) \,. \tag{2.1}$$

In the 2 1/2-D approach (Fig. 1.) these elementary parts are homogeneous bodies with polygonal cross-section and finite strike-length. According to Rasmussen, Pedersen (1979) for a set of 2 1/2-D bodies it holds

$$\Delta g^t(\underline{r}) \;=\; -f \sum_m \sigma_m \sum_{i=1}^{N_m} \iint_{S_{i,m}} \frac{\underline{z} \cdot \underline{n}_{m,i}}{|\underline{r} - \underline{r}_{dS}|} \, dS \tag{2.2}$$

where m — is the body index

N_m — number of polygon vertices

σ_m — the body density

$S_{i,m}$ — the i-th face

$\underline{n}_{m,i}$ — the outward directed unit vector normal to $S_{i,m}$

$\underline{z}$ — the downward directed unit vector.

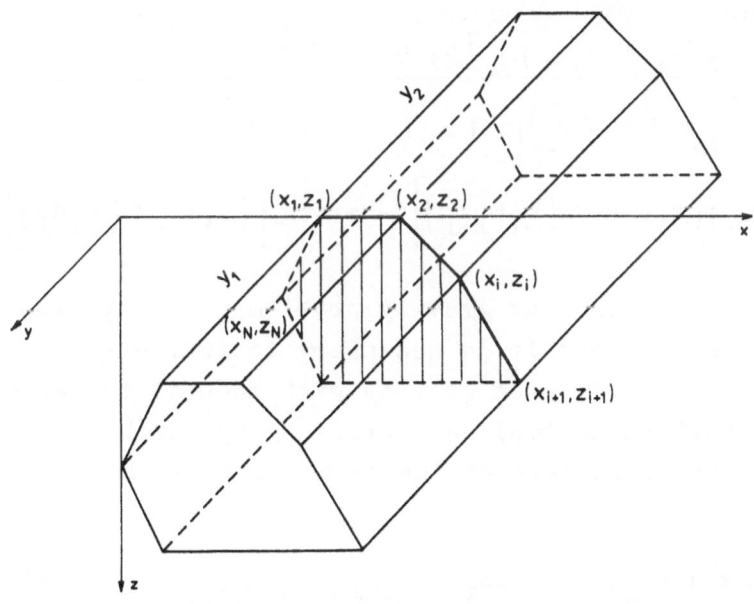

Fig. 1. Geometry of the 2 1/2-D body and description of the polygonal cross-section.

For simplicity of notation we drop the subscript on quantites associated with the m-th body and set $\underline{r} = (0,0,0)$. The analytical expression for the gravity effect is given in Rasmussen, Pedersen (1979) and Švancara, Halíř (1987). It holds

$$\Delta g^t = f6 \sum_{i=1}^{N} \left[\frac{\Delta x_i y \lambda_i}{d_i} + \frac{1}{D_i} \left(x_i z_{i+1} - z_i x_{i+1} \right) \cdot \right.$$

$$\left. \cdot \left(\Delta z_i \mu_i + \Delta x_i \theta_i \right) \right] \qquad (2.3)$$

if $i = N$ then $x_{N+1} = x_1$, $z_{N+1} = z_1$

$$\Delta x_i = x_{i+1} - x_i \qquad\qquad v_i = x_{i+1} \Delta x_i + z_{i+1} \Delta z_i$$

$$\Delta z_i = z_{i+1} - z_i \qquad\qquad u_i = x_i \Delta x_i + z_i \Delta z_i$$

$$D_i = \Delta x_i^2 + \Delta z_i^2 \qquad\qquad w_i = z_i \Delta x_i - x_i \Delta z_i$$

$$d_i = \sqrt{D_i}$$

$$R_i = \sqrt{x_i^2 + z_i^2 + y^2}$$

$$r_i = \sqrt{x_i^2 + z_i^2}$$

where x_i, z_i are the coordinates of the i-th vertex, y is the body length

and $\lambda_i = \ln \left[\left(v_i/d_i + R_{i+1} \right) / \left(u_i/d_i + R_i \right) \right]$

$$\mu_i = \ln \left[\left(r_{i+1} \left(y + R_i \right) \right) / \left(r_i \left(y + R_{i+1} \right) \right) \right]$$

$$\theta_i = \text{arctg} \left[\left(v_i y \right) / \left(w_i R_{i+1} \right) \right] - \text{arctg} \left[\left(u_i y \right) / \left(w_i R_i \right) \right].$$

For the solution of the inverse gravity problem we apply first derivative algorithms, hence it is necessary to derive relations for partial derivatives of gravity with respect to variable parameters. For the partial derivative of Δg^t with respect to the x-th and z-th coordinate of the i-th vertex it holds :

$$\frac{\partial \Delta g^t}{\partial x_i} = f6 \left(\Delta z_i \beta_i - \Delta z_{i-1} \gamma_{i-1} \right) \qquad (2.4)$$

$$\frac{\partial \Delta g^t}{\partial z_i} = -f6 \left(\Delta x_i \beta_i - \Delta x_{i-1} \gamma_{i-1} \right) \tag{2.5}$$

where

$$\beta_i = \left\{ \left(\Delta x_i v_i + \Delta z_i w_i \right) \theta_i - \left(\Delta x_i w_i - \Delta z_i v_i \right) \mu_i \right\} D_i^{-2} -$$
$$- \Delta z_i y \lambda_i d_i^{-1} D_i^{-1}$$

$$\gamma_i = \left\{ \left(\Delta x_i u_i + \Delta z_i w_i \right) \theta_i - \left(\Delta x_i w_i - \Delta z_i u_i \right) \mu_i \right\} D_i^{-2} -$$
$$- \Delta z_i y \lambda_i d_i^{-1} D_i^{-1}$$

and for $i = 1$ the functions $\lambda_{i-1}, \mu_{i-1}, \theta_{i-1}$ are taken with $x_o \equiv x_N$ and $z_o \equiv z_N$.

The partial derivative of gravity with respect to density is elementary because of their linear relationship.

The inverse gravity problem is defined by the task to find such geometrical and density parameters that the L_2 norm of difference between the measured and calculated gravity is mini-mized and the apriori information about the geological struc-ture is respected. We use alternativelly one of the three well-known gradient iterative algorithms : Levenberg - Marquardt - Meyer, quasi - Newton method with Davidon - Fletcher - Powell update and the conjugate gradient method.

3. Analytical ray tracing in polygonal bodies

In some media with simple velocity distribution the ray tracing may be performed analytically, which is the simplest and fastest solution. Particulary simple formulas are obtained (Červený, 1987) for the ray tracing in media with constant gra-dient of the quadratic slowness. Let us assume quadratic slow-ness linear in Cartesian coordinates x, z

$$1/v^2 = A + B_x x + B_z z \tag{3.1}$$

and define variable 6 monotonic along the ray by the formula

$$6(T) = \int_0^T v^2 dT \tag{3.2}$$

than the ray tracing system yields the following polynomial solution

$$x = x_0 + p_{x0}\, 6 + 1/4\ B_x\, 6^2 \tag{3.3}$$

$$z = z_0 + p_{z0}\, 6 + 1/4\ B_z\, 6^2 \tag{3.4}$$

with the slowness vector $\underset{\sim}{p}(p_x,\ p_z)$

$$p_x = p_{x0} + 1/2\ B_x 6 \tag{3.5}$$

$$p_z = p_{z0} + 1/2\ B_z 6 \tag{3.6}$$

and $p_{x0} = \sin\alpha/v_0$, $p_{z0} = \cos\alpha/v_0$, $1/v_0^2 = A + B_x x_0 + B_z z_0$
$$\tag{3.7}$$

where x_0, z_0 are Cartesian coordinates of the initial point ($6 = 0$, $T = 0$) and α is the angle between the coordinate axis z and the slowness vector at the initial point.

For the traveltime T it holds (Červený, 1987) :

$$T(6) = (A + B_x x_0 + B_z z_0)6 + 1/2(B_x p_{x0} + B_z p_{z0})\, 6^2 +$$

$$+ 1/12(B_x^2 + B_z^2)\, 6^3 \ . \tag{3.8}$$

Assume now a more general case of the quadratic slowness parabolic in Cartesian coordinates x, z

$$1/v^2 = A + B_x x + C_x x^2 + B_z z + C_z z^2 \tag{3.9}$$

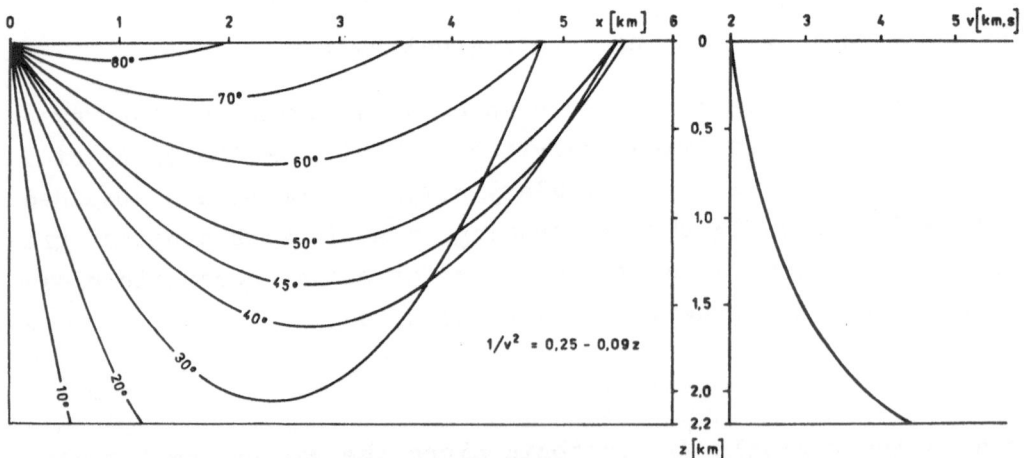

Fig. 2.　Analytical ray tracing in a medium with constant gradient of quadratic slowness.

than the ray tracing system reads

$$\frac{d^2x}{d\sigma^2} - C_x x = \frac{B_x}{2} \tag{3.10}$$

$$\frac{d^2z}{d\sigma^2} - C_z z = \frac{B_z}{2} \quad . \tag{3.11}$$

Solution of those nonhomogeneous second order linear diferential equations can be found in analytical form. (In the next we consider only eq. (3.11) only because (3.10) a (3.11) are analogical.) It holds :

$$z(\sigma) = -\frac{B_z}{2C_z} + K_{1z}e^{\sigma\sqrt{C_z}} + K_{2z}e^{-\sigma\sqrt{C_z}} \tag{3.12}$$

for $C_z > 0$ and

$$z(\sigma) = -\frac{B_z}{2C_z} + M_{1z}\cos\left(\sigma\sqrt{-C_z}\right) + M_{2z}\sin\left(\sigma\sqrt{-C_z}\right) \tag{3.13}$$

for $C_z < 0$.

From the initial conditions (3.7) we obtain formulas for constants K_{1z}, K_{2z} and M_{1z}, M_{2z}. It holds :

$$K_{1z} = \frac{1}{2}\left(z_0 + \frac{B_z}{2C_z} + \frac{p_{zo}}{\sqrt{C_z}}\right) \tag{3.14}$$

$$K_{2z} = \frac{1}{2}\left(z_0 + \frac{B_z}{2C_z} - \frac{p_{zo}}{\sqrt{C_z}}\right) \tag{3.15}$$

and

$$M_{1z} = z_0 + \frac{B_z}{2C_z} \tag{3.16}$$

$$M_{2z} = \frac{p_{zo}}{\sqrt{-C_z}} \quad . \tag{3.17}$$

The traveltime $T(\sigma)$ is given by the integral

$$T(\sigma) = \int_0^\sigma 1/v^2 d\sigma = \int_0^\sigma (A + B_x x(\sigma) + B_z z(\sigma) + C_x x^2(\sigma) +$$

$$+ C_z z^2(\sigma)) d\sigma \tag{3.18}$$

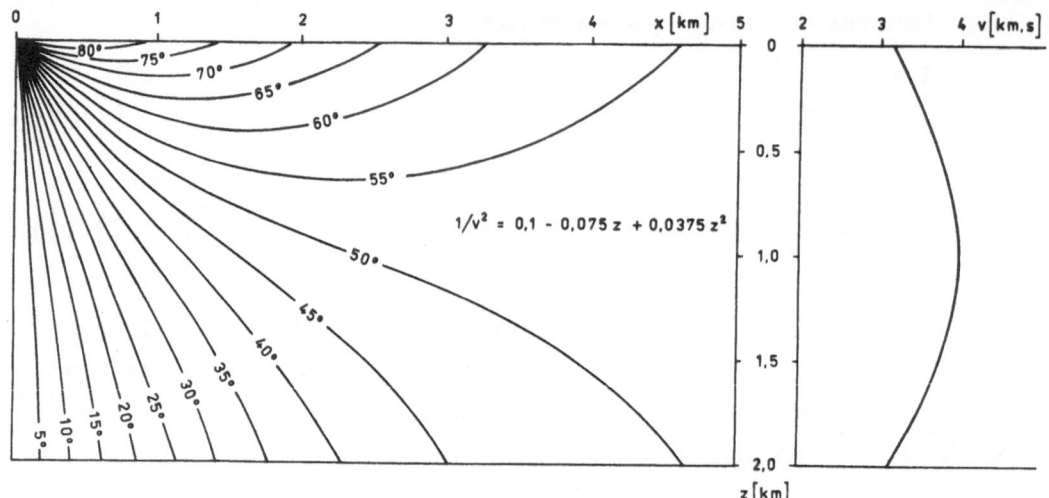

Fig. 3. Analytical ray tracing in a medium with quadratic
slowness parabolic in Cartesian coordinates given
by eq. (3.12) for $C_z > 0$.

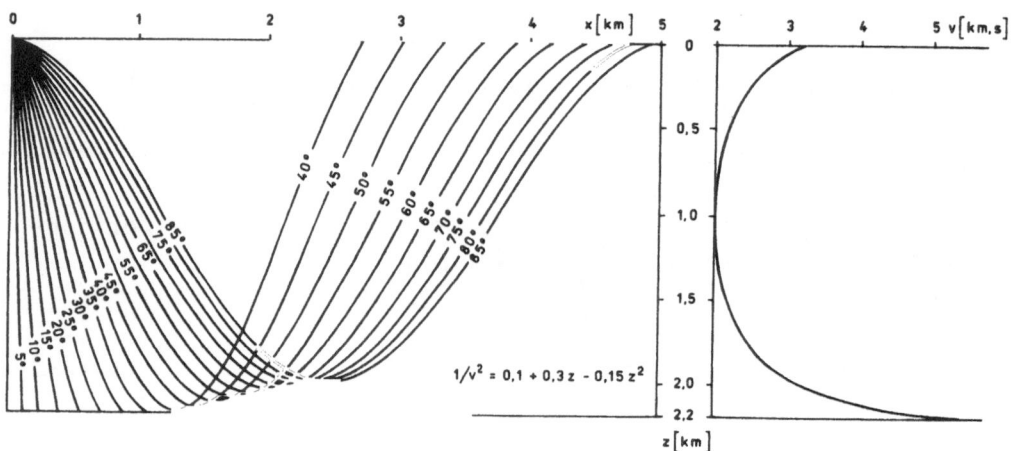

Fig. 4.   Analytical ray tracing in a medium with quadratic
slowness parabolic in Cartesian coordinates given
by eq. (3.13) for $C_z < 0$.

and can be evaluated analytically for both $C_x(C_z) > 0$ and
$C_x(C_z) < 0$.

The solutions (3.12) and (3.13) of the ray tracing system
yield suitable tools for examination of effects of both low-
velocity and high-velocity zones and even the waveguide phenomenon. Examples of analytical ray tracing in media with the
above described velocity distributions are given in Figures 2,
3 and 4.

Assume that the whole model is subdivided into a set of
2D bodies of arbitrary polygonal cross-sections. Analytical determination of the intersection point of the ray with the body
boundary is possible for the velocity law (3.1) i.e. constant
gradient of quadratic slowness. For the ray intersections with
the straight line z = ax + b forming the boundary (Fig. 5.) it
holds :

$$\mathcal{6}_{1,2} = \frac{-(ap_{xo}- p_{zo}) \pm \sqrt{(ap_{xo}- p_{zo})^2 - (B_x a - B_z)(ax_o+ b - z_o)}}{1/2(B_x a - B_z)} .$$

$$(3.19)$$

Because we do not know in advance which side of the body will
be intersected by the ray, it is necessary to find intersections with all sides. The actual intersection point of the ray
with the body boundary corresponds to the smallest positive
value of $\mathcal{6}$.

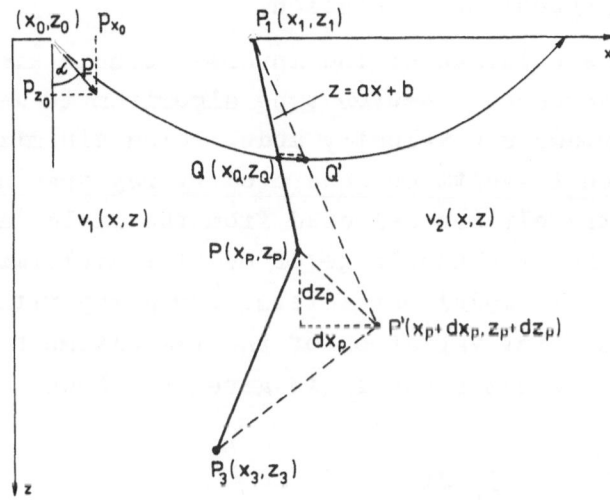

Fig. 5.   Geometry considered in the calculation of trav-
          eltime derivative with respect to vertex coor-
          dinates.

The behaviour of ray on a plane interface between two bodies is determined by the reflection/transmission laws. We denote $\underline{n}(n_x, n_z)$ the unit vector normal to the boundary, $\underline{p}_1(p_{1x}, p_{1z})$ the incident slowness vector and $\underline{p}_2(p_{2x}, p_{2z})$ the slowness vector of the reflected or transmited ray. According to Born, Wolf (1964) and Červený (1987) we can write

$$p_{2x} = p_{1x} - \left[\underline{p}_1\underline{n} \pm \text{sign}(\underline{p}_1\underline{n}) \sqrt{1/v_2^2 - 1/v_1^2 + (\underline{p}_1\underline{n})^2} \right] n_x$$

$$p_{2z} = p_{1z} - \left[\underline{p}_1\underline{n} \pm \text{sign}(\underline{p}_1\underline{n}) \sqrt{1/v_2^2 - 1/v_1^2 + (\underline{p}_1\underline{n})^2} \right] n_z$$

$$(3.20)$$

$$\text{where} \quad \underline{n} = \left(\frac{-a}{\sqrt{a^2 + 1}}, \frac{1}{\sqrt{a^2 + 1}} \right).$$

The upper sign in (3.20) holds for reflected waves, the lower sign for transmitted waves.

Formulas (3.1) to (3.20) permit initial value ray tracing and corresponding traveltime calculation for 2-B bodies of a general polygonal cross-section.

For the solution of the inverse seismic kinematic problem an iterative Gauss - Newton type algorithm is usually construc- ted that produces a velocity model which minimizes the differ- ence between traveltimes generated by ray tracing through the model and traveltimes selected from the field data. This algo- rithm requires the knowledge of model traveltime derivatives with respect to model parameters. For a ray with fixed end- points, the first variation of the traveltime for a perturba- tion $\delta\underline{p}$ of medium slowness is given by (Nowak, Lyslo, 1989)

$$\delta T \approx \int_0^6 \delta\underline{p} \, d6 \tag{3.21}$$

which is computed along the unperturbed raypath. Using this simplification we can for the velocity law (3.1) write

$$\frac{dT}{dA} \approx 6. \tag{3.22}$$

The traveltime derivative with respect to vertex coordinate x_p (Fig. 5.) is given by

$$\frac{dT}{dx_p} \approx \frac{\partial T}{\partial 6} \cdot \frac{\partial 6}{\partial a} \cdot \frac{\partial a}{\partial x_p} + \frac{\partial T}{\partial 6} \cdot \frac{\partial 6}{\partial b} \cdot \frac{\partial b}{\partial x_p} \qquad (3.23)$$

and analogical expression for z_p coordinate is valid. Expression (3.23) can be calculated analytically using equations (3.8) and (3.19), but the resulting formula is rather complicated. Therefore we present a simpler approximative formula based on geometrical considerations shown in Fig. 5. We assume that the traveltime perturbation can be approximated by

$$\delta T \approx \overline{QQ'} \left(\frac{1}{v_1} \pm \frac{1}{v_2} \right) \qquad (3.24)$$

where $\overline{QQ'}$ is distance between ray intersections with the velocity boundary in initial and perturbed position. The lower sign in (3.24) holds for the transmitted ray, the upper sign holds for the reflected unconverted ray where $v_2 = v_1$. Under this assumption we can easily derive :

$$\frac{\delta T}{\delta x_p} \approx \frac{(z_Q - z_1) \sqrt{1 + a_r^2}}{(z_p - z_1) - a_r(x_p - x_1)} \left(\frac{1}{v_1} \pm \frac{1}{v_2} \right) \qquad (3.25)$$

and

$$\frac{\delta T}{\delta z_p} \approx -\frac{1}{a} \cdot \frac{\delta T}{\delta x_p} \qquad (3.26)$$

where $Q(x_Q, z_Q)$ is the ray intersection point, $P_1(x_1, z_1)$ and $P(x_p, z_p)$ are the fix and variable boundary vertices and $a_r = p_z/p_x$ is tangent to the raypath at point Q. The sign convention is the same as for equation (3.24). On the synthetic example in Fig. 7. the traveltime perturbation caused by boundary displacement is compared with the value predicted on the basis of traveltime derivative approximation (3.25). Relative error in the estimation of traveltime perturbation is 8 %.

Under the assumption that the raypath can be approximated by straight line segments a more exact derivation of formulas

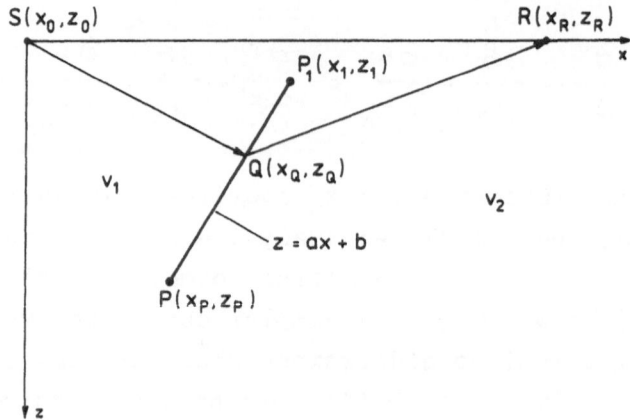

Fig. 6. Geometry used by the calculation of traveltime derivative with respect to vertex coordinates using Fermat's principle.

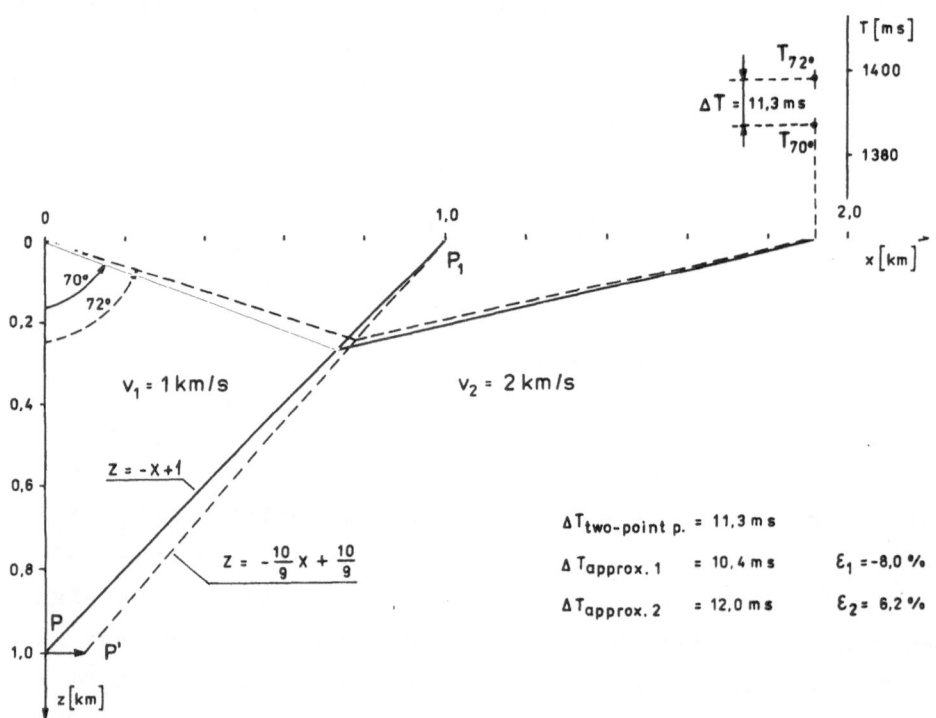

Fig. 7. Comparison of traveltime perturbation ($T_{two-point\ p.}$) caused by boundary displacement with the value predicted either by the formula 3.26.($T_{approx.\ 1}$) or by the formula 3.28.($T_{approx.\ 2}$).

for traveltime derivative with respect to vertex coordinates is possible.

Using the Fermat's principle, for the geometry shown in Fig. 6. we obtain

$$\frac{dT}{dx_p} = \left(\frac{z_Q - z_o}{v_1 \sqrt{(x_Q - x_o)^2 + (z_Q - z_o)^2}} + \frac{z_Q - z_R}{v_2 \sqrt{(x_Q - x_R)^2 + (z_Q - z_R)^2}} \right) \cdot$$

$$\cdot \frac{(z_1 - z_p)(x_Q - x_1)}{(x_p - x_1)^2} \tag{3.27}$$

and

$$\frac{dT}{dz_p} = - \frac{1}{a} \cdot \frac{dT}{dx_p} \tag{3.28}$$

where $S(x_o, z_o)$ and $R(x_R, z_R)$ are the source and reciever coordinates. The test on the synthetic example in Fig. 7. shows relative error 6 % in prediction of the traveltime perturbation caused by boundary displacement.

4. Conclusions

Presented formulas enable construction of an iterative algorithm for cooperative inversion of gravity and seismic traveltime data. Our approach was primarily designed for velocity inversion of refraction data, but the presented formulas describe also properties of the reflected rays because the reflectors can be associated with the body boundaries. For the cooperative inversion of seismic and gravity data we intend to adopt a special separable minimization algorithm, which respects the fact that the vector of unknown parameters consists of linear (density, velocity) and nonlinear (geometrical) parameters.

References

Berryman, J.G., 1989 : Weighted Least-Squares Criteria for Seismic Traveltime Tomography. IEEE Trans. Geosci. Rem. Sens. 27, 302-309.

Bishop, T.N., Bube, K.P., Cutler, R.T., Langan, R.T., Love, P.L., Resnick, J.R., Shuey, R.T., Spindler, D.A., Wyld, H.W., 1985 : Tomographic determination of velocity and depth in laterally varying media. Geophysics 50, 903-923.

Born, M., Wolf, E., 1964 : Principles of optics.
The MacMillan Company.

Červený, V., 1987 : Ray tracing algorithms in three-dimensional laterally varying layered structures, in Nolet, G., (ed.), Seismic Tomography. D. Reidel Publishing Company.

Enmark, T., 1981 : A versatile interactive computer program for computation and automatic optimization of gravity models. Geoexploration 19, 47-66.

Langan, R.T., Lerche, I., Cutler, R.T., 1985 : Tracing of rays through heterogeneous media : An accurate and efficient procedure. Geophysics 50, 1456-1465.

Lines, L.R., Schultz, A.K., Treitel, S., 1988 : Cooperative inversion of geophysical data. Geophysics 53, 8-20.

Nowack, R.L., Lyslo, J.A., 1989 : Fréchet derivatives for curved interfaces in the ray approximation. Geophysical Journal 97, 497-509.

Rasmussen, R., Pedersen, L.B., 1979 : End correction in potential field modeling. Geophys. Prosp. 27, 749-760.

Starostenko, V.I., Kostyukevich, A.S., Kozlenko, V.G., 1988 : Seismogravimetric method : principles, algorithms, results. Geophysical Journal 93, 295-309.

Švancara, J., Halíř, J., 1987 : Solution of the 2 1/2-D Inverse Gravity Problem Using Different Nonlinear Iterative Formulas, in Vogel, A., (ed.), Model Optimization in Exploration Geophysics 2. Friedr. Vieweg & Sohn Braunschweig/Wiesbaden.

Talwani, M., Worzel, J.L., Landisman, M., 1959 : Rapid gravity computations for two-dimensional bodies with application to the Mendocino submarine fracture zone.
J. Geophys. Res 64, 49-59.

Geophysical Data Inversion – Methods and Applications

edited by Andreas Vogel, Charles O. Ofoegbu, Bjorn Ursin
and Rudolf Gorenflo

*1990. VIII, 620 pages (Theory and Practice of Applied Geophysics,
Vol. 4; ed. by Andreas Vogel) Hardcover.
ISBN 3-528-06396-3*

This monograph contains a selection of papers presented at the Seventh International Seminar on Model Optimization on Exploration Geophysics held at the Free University of Berlin, February 8-11, 1989. The papers presented here cover the different methods of synthetic and real data sets. The volume covers a wide spectrum of the subject from basic mathematics, general theory and computer implementation of methods to their practical applications.

Vieweg Publishing · P.O. Box 58 29 · D-6200 Wiesbaden 1, Germany

Earthquake Prognostics

edited by Andreas Vogel, and Klaus Brandes

Hazard Assessment, Risk Evaluation and Damage Prevention. Proceedings of the 2nd International Seminar held in Berlin, June 24-27, 1986.

1988. X, 534 pages (Progress in Earthquake Prediction Research, Vol. 3, ed. by Andreas Vogel) Hardcover.
ISBN 3-528-06323-8

A prerequisite for the assessment of the earthquake hazard is a thorough understanding of earthquake generating processes obtained by intensive studies on seismotectonics, earthquake source physics, and earthquake-related ground deformation (Part I of the book).
Quantitative assessment of the earthquake hazard requires model simulation of earthquake generation, of expected focal mechanisms, of seismic wave radiation and propagation, and finally precalculation of the characteristics of groundmotion to be expected at a site designated for construction (Part II). Part III is dedicated to risk evaluation and damage prevention by earthquake engineering. Risk analysis includes seismic vulnerability and loss estimation.
Earthquake engineering is concerned with the design of structures in seismic regions. It deals with numerous facets of structural systems and details, numerical dynamic methods, soil dynamics, soil structure interaction, modeling of systems and materials behaviour.

Vieweg Publishing · P.O. Box 58 29 · D-6200 Wiesbaden 1, Germany

vieweg